Life, Earth, Colony

Life, Earth, Colony

Friedrich Ratzel's Necropolitical Geography

IAN KLINKE

UNIVERSITY OF MICHIGAN PRESS
Ann Arbor

Published in the United States of America by the
University of Michigan Press
Printed and bound by CPI Group (UK) Ltd, Croydon, CR0 4YY

First published August 2023

A CIP catalog record for this book is available from the British Library.

Library of Congress Cataloging-in-Publication data has been applied for.

ISBN 978-0-472-07617-8 (hardcover : alk. paper)
ISBN 978-0-472-05617-0 (paper : alk. paper)
ISBN 978-0-472-90351-1 (open access ebook)

DOI: https://doi.org/10.3998/mpub.11987710

Library of Congress Control Number: 2023937359

The University of Michigan Press's open access publishing program is made possible thanks to additional funding from the University of Michigan Office of the Provost and the generous support of contributing libraries.

Cover: Ant colony defending a plumeless saw-wort (Serratula lycopifolia) against a white-spotted rose chafer (Oxythyrea funesta) (Source: F. Ratzel, 1902, *Die Erde und das Leben.*)

Contents

Digital materials related to this title can be found on the Fulcrum platform via the following citable URL: https://doi.org/10.3998/mpub.11987710

Preface

Any attempt to come to terms with Friedrich Ratzel will need to begin with the controversy surrounding the work he bequeathed to the world. Too well documented is the impact that his *Lebensraum* (living space) concept had on the formation of national socialism in the 1920s. This notwithstanding, the genre of classical geopolitics, which his political geography helped to spawn, remains popular, not just in the Kremlin but in the Anglophone world too. Understandably, we often find in English-language geopolitical texts an explicit break with Ratzel. Geopolitics can only claim legitimacy if Ratzel and his followers are disavowed and some other "father" of geopolitics, most commonly the British geographer Halford Mackinder or the US naval strategist Alfred Thayer Mahan, are embraced instead.

In Germany, this split between a "dangerous" German school on the one hand and an "acceptable" Anglo-American tradition on the other is especially pronounced. I remember vividly an interview I conducted as a PhD student during the late 2000s in the villa of the former director of the German foreign intelligence service. When I asked him with simulated naivety whether he thought there was any purchase in geopolitics, he replied, without touching the tea his housekeeper had laid out, that the answer depended entirely on whether my question was about the German *Geopolitik* or the English *geopolitics*; the two were simply not the same. A few years later, the editor of the German weekly *Die Zeit* told me that German *Geopolitik* was emotionally charged nonsense and best avoided. And yet, in a move I have witnessed many times, he continued immediately to outline his own version of a geopolitics that was, or so he hoped, cleansed from any influence of this dangerous "Teutonic" tradition.

Those who have shown less hesitation with a Ratzelian brand of geopolitics have often faced resistance. The eminent historian Karl Schlögel, for instance, drew repeated criticism in 2003 for rehabilitating Ratzel. A decade later, an advisor to the German foreign ministry was condemned for floating the idea that Germany could find *Lebensraum* in Russia. In 2018, it was the far-right politician Björn Höcke who used Ratzel's term at a congress before asking his audience playfully, "Did I just say *Lebensraum*?" Although this taboo-breaking use of national socialist terminology provoked an outcry in the German and international media landscape, it did not harm his popularity. Over a century after his death, Friedrich Ratzel remains a locus of taboo and controversy.

I have to confess that I had to convince myself to write this book, not so much because I doubted its cause. Granted, there are problems with the biographical genre, such as a tendency for hagiography and hatchet jobs. But I hesitated primarily because I knew that Ratzel's oeuvre would be difficult to navigate due to its scope and disorderliness. Not only did the Leipzig geographer write incessantly. He also moved quickly between genres and writing projects, often submitting manuscripts for publication before he had found the time to carefully edit them. He also had a tendency to hide tensions and contradictions behind dense prose. Why, then, should I write a book about a thinker whose compulsion to put words on paper often resulted in a stream of consciousness rather than a carefully crafted body of work? Anyone who has even fleetingly read Ratzel, whether in German, English, or another language, will appreciate the need to bring order into the unruly ways in which his ideas were developed and presented. In fact, even his student, collaborator, and translator Ellen Churchill Semple recognized this. It is often difficult to distinguish between the moments in which Ratzel was simply reproducing, as he liked to do, an existing canon of work, and where he was genuinely original. Needless to say, I also worried about the politics of penning a book on the Leipzig geographer. Of course, writing about a politically contaminated body of thought is not to endorse it, but it is nonetheless to declare it worthy of attention, a move which is in itself political.

Ultimately, I decided to persevere with Ratzel, motivated in part by the observation that while there was a lot of talk about him, few seemed to have read his works. Ratzel had also made himself felt in my previous book, *Cryptic Concrete: A Subterranean Journey into Cold War Germany* (Wiley, 2018), which discussed the revival of geopolitical motifs in West German civil defense and military strategy during the 1950s and 1960s. I had been frustrated at the time by the absence of a detailed and modern monograph on the founder of political geography. As some readers will have noted, recent years have witnessed a burst of studies on key figures in geography's disciplinary canon. Book-length studies have appeared on Isaiah Bowman, Halford Mackinder, Ellen Semple, Karl Haushofer, Pyotr Kropotkin, Élisée Reclus, alongside readers dedicated to Carl Sauer, Doreen Massey, and David Harvey as well as examinations on thinkers who were not geographers but whose work centered on the question of space, such as Jakob von Uexküll, Carl Schmitt, and Henri Lefebvre. And yet, although Friedrich Ratzel remains a significant figure in the darker history of the geographical discipline, a critically minded English-language monograph on him has been curiously absent.

In what follows, I hope to convince my readers that it is imperative to continue to grapple with Ratzel. And while the book presents a whole range of reasons for doing so, there are three that stand out. First, Ratzel was no obscure scholar who rose to fame only in Weimar and Nazi Germany. He was a pivotal figure during his lifetime, a man who shaped debates both as an academic and a public intellectual in Imperial Germany. His writing thus allows an insight into German colonial fantasies and the impact they would have on the world. Second, Ratzel's materialism is not as antiquated as it may seem. Although his colonial and racialized writings are often rejected today, his lively theory of earth and life is invoked, often implicitly but increasingly explicitly, as a way of grounding conceptions of the state. This book shows, however, that the different strands of his work are inextricably intertwined, which raises crucial questions for work that operates within a neo-Ratzelian frame. Finally, Ratzel and his ideas have attracted significant interest from a range of political thinkers and movements in and far beyond Germany for more than a century. And while his reception has certainly been diverse, it has skewed heavily to the political right. Even today, and whether we like it or not, Ratzel seems to speak to a world that believes itself to be locked into a perpetual conflict for space.

This book is the product of four years of work on Ratzel and draws on a number of papers that have been published along the way. My readers will be pleased to know that none of the chapters in this book are a simple reproduction of my existing papers. In the earlier chapters, I have occasionally drawn on "*Lebensraum* and its Discontents," coauthored with Mark Bassin (*Journal of Historical Geography*, 2018). Chapter 4 feeds off sections of "Vitalist Temptations: Life, Earth and the Nature of War" (*Political Geography*, 2019) and passages from "On the History of a Subterranean Geopolitics" (*Geoforum*, 2021). Chapter 6 is an extension and deepening of "Friedrich Ratzel, Lebensraum and the Death Motif" (*Journal of Historical Geography*, 2018). Some of the discussion in chapter 3 also appears in a German-language chapter, coauthored with Mark Bassin, on Ratzel in North America (*Nomos*, 2021). A fuller examination of Ellen Churchill Semple's intellectual relationship with Ratzel has appeared in a further article (*Geographica Helvetica*, 2022).

This book is not the work of one person but the result of conversations with a vast array of colleagues, students, archivists, and friends. I am particularly grateful to Mark Bassin (Södertörn University) for conversations about the themes of this book. Mark and I co-organized a workshop in 2017 at Oxford, which led to a special issue on Ratzel, which was later published alongside Tul'si (Tuesday) Bhambry's translation of Ratzel's 1901 *Lebensraum* essay. Maan Barua (Cambridge University) has been very helpful in prompting me to pull out the semblances between Ratzel's work and various strands of contemporary new materialism. I would like to thank Ulrike Jureit (Hamburger Institut für Sozialforschung) and Patricia Chiantera-Stutte (Universita di Bari) for conversations on Ratzel and for inviting me to their marvelous Ratzel workshop series on Lake Como. My thanks go to Julia Verne (now at the Universität Mainz) for co-organizing a double session on Ratzel at the 2019 Royal Geographical Society annual conference in London. Encouragement to grapple with Ratzel has also come from Mike Heffernan (Nottingham University), Philip Conway (Durham University), Jason Dittmer, Alan Ingram, and my ever-skeptical former supervisor Felix Ciuta (all three at University College London). Klaus Dodds (Royal Holloway), Brad Garrett (University College Dublin), David Haney (Berlin), Benedikt Korf (Universität Zürich), as well as Ulrike Jureit and Maan Barua kindly read chapter drafts and helped me to clarify various argumentative strands. I have also benefited tremendously from email exchanges with Hans-Dietrich Schultz (Humboldt Universität zu Berlin). My thanks also go to Christopher Dreyer, Anna Pohlod, and Mary Hashman at the University of Michigan Press and to two sets of anonymous referees who did much to improve the quality of the manuscript. I am very grateful to Tim Ryder for supplying the index.

At Oxford's School of Geography and the Environment, I am indebted to Linda McDowell and Patricia Daley for conversations about the content of this book, Rich Grenyer for background on white-spotted rose chafers (among other things), Gillian Rose for a friendly dose of questions about the premises of my project, Tom Jellis (now at Bristol University) and Beth Greenhough for talking to me about vitalism and death, to Amber Murrey, Fiona McConnell, and Chih-Yuan Woon (National University of Singapore, then a visiting fellow at Oxford) for comments on a first draft of the book proposal, as well as to Colin Clarke for conversations about the history of the discipline. Britain Hopkins provided feedback on a number of chapters and intellectual inspiration in a wide variety of ways. At an earlier stage, it was Richard Powell (now at Cambridge University) who had encouraged me to read Ratzel in a more systematic way. Carole Angier gave a splendid talk in 2016 at SoGE on biographical research which helped me to think through my project. At St John's, I am grateful to Patrick Hayes, Barry Murnane, and Jack Langton for intellectual stimulation and feedback, to Jenny Richards and Britain for teaching our students while I was on sabbatical, as well as to the research committee for funding my fieldwork and the digitization of Ratzel's diaries and letters, which the outbreak of the Covid-19 pandemic necessitated.

Patient audiences in New Orleans, London, Oxford, Prague and Singapore played their part, as did the helpful archivists and librarians at the Institut für Länderkunde, the Bayerische Staatsbibliothek, the Bundesarchiv, the University of Kentucky Special Collections, the Royal Geographical Society, and the Max Planck Institut für Psychiatrie, as well as the many undergraduates whom I exposed to Ratzel, for which I can only apologize. David Hamann at Recherche-Dienste and Jost were kind enough to help me with the transcription of documents I struggled to decipher. Last but not least, there is Anna, who had to endure my dinnertime ramblings about Ratzel's deathly preoccupations. This book is dedicated to our daughters Valentina and Violet.

Friedrich Ratzel, 1887. (Source: Leibniz Institut für Länderkunde.)

Chapter I

Introduction

On 31 March 1897, Friedrich Ratzel was recovering from a cold in Prague. After eventually dragging himself out of bed, he was frustrated to find the museums closed and decided to visit the Jewish cemetery in Josefov instead. He was not disappointed, noting in his diary how he had been overwhelmed by the beauty of the site, which had served as a graveyard since the fifteenth century. Still under this impression, the German geographer decided to have a breakfast of goulash at a quarter past eleven. After a short rest, he sat down to write, putting the final touches on his "*Lebensraum* essay." He then ventured to the post office and sent his manuscript to the Frankfurt-based popular science journal *Die Umschau*, where it would appear two months later.[1] Nothing implies that he understood that *Lebensraum* would come to be seen as his most important conceptual contribution.

Given Ratzel's tendency for wordy publications, the essay was unusually short. In no more than five pages and in a sharp tone he expressed his long-held view that all phenomena of spatial expansion, from migration to colonization and war, were driven by an impulse inherent in the very nature of life, as it negotiated the earth's boundaries. The earth was subject to a general struggle for space, which affected plants, animals, and peoples alike and which emerged from a tension between the incessant movement of life and the finite resources of the soil. *Lebensraum* was a vital necessity for each individual or social organism. And while all species struggled for living space, the latter was always in flux, augmented and shrunk by geologic and climatic processes as well as the often-rapid expansion and retreat of other species and subspecies. There was from the start a deadly underside to his argument. In every period of the planet's history, there had been decay as well as emergence. "Continuously and everywhere," he noted, "the old lives side by side with the new." "Here it moves into a form of becoming, there it subsides and dies out," the man with a weakness for cemeteries added.[2]

Despite his tendency to coin a vast array of new concepts and compounds, Ratzel only rarely defined the terms he used. Much the same was true for *Lebensraum*, as the geographer Hans Schrepfer bemoaned in 1942, at a time when the term had become a *Schlagwort* (catchword) across wartime Nazi Germany.[3] What was striking about Ratzel's essay was not its conceptual precision, but rather the vibrant picture it painted of overlapping, interwoven, and three-dimensional habitats and the way in which it effortlessly moved from a discussion of flora and fauna to the realm of human societies and their polities. People and soil (*Volk und Boden*), he claimed, were organically united in the state. If a state were to transpose itself onto a different soil, it would necessarily become an entirely different polity. It was thus no surprise that the struggle between states was also a struggle for space, Ratzel noted, one in which there was growth and retrogression. He illustrated this argument by reference to the Polish partitions of the eighteenth century, which had left the Polish-Lithuanian Commonwealth divided between Prussia in the West, the Habsburg Empire in the South, and Russia in the East. "The Poles," he explained, "still exist, but the space they inhabit has ceased to be their space politically, and therefore their state has been exterminated [*vernichtet*]."[4]

More than a century after his death, there is still controversy surrounding Friedrich Ratzel (1844–1904) and his oeuvre. Ratzel was a tremendously prolific writer whose work spanned geology, history, biology, and ethnology, as well as journalistic and creative genres. He is best known today for developing a broadly Darwinian political theory that sought to explain the evolution of state formation and behavior by reference to geography. States, the trained zoologist argued, functioned like organisms that struggled for space under the influence of geomorphological and environmental factors. Migration, colonization, and war were ultimately epiphenomena of the same spatial logic that emerged from the tension between life as a creative force and the finitude of the earth's surface and resources. He termed his approach "political geography." It would later be shortened to "geopolitics" by his followers. Alongside the American naval strategist Alfred Thayer Mahan (1840–1914) and the British geographer Halford Mackinder (1861–1947), he is seen as the founder of this conservative tradition of political thought.

Although often dense in his writing, Ratzel was no ivory tower thinker. By the 1880s, he had become widely known in Germany as a public intellectual who vocally promoted settler colonialism and naval expansion. His work was read by large parts of the German bourgeoisie, attracting admirers such as the philosopher Friedrich Nietzsche, the socialist politician August Bebel, the chairman of the radical Pan-German League Ernst Hasse, and the founder of experimental psychology Wilhelm Wundt. His ideas quickly spread beyond Germany. He exchanged letters with Émile Durkheim and Franz Boas, the founders of sociology and anthropology, the main rivals of his *Anthropogeographie* (human geography). He was even read by the sociologist and civil rights activist W. E. B. Du Bois. And although Ratzel's ideas attracted criticism for their intellectual promiscuity and an environmentally determinist streak, they had many admirers among Europe's scientific community. One of Kilimanjaro's glaciers was named after him, as was an Arctic earthworm: *enchytraeus ratzeli*. And yet Ratzel is today known more for his influence on the formation of national socialism in the 1920s than for the mark he left on the late nineteenth century. Indeed, Werber argues, it was only the cataclysmic events of the First World War that made it possible for Ratzel's ideas to take off.[5] And while Ratzel may not be quite a "forgotten figure,"[6] it is certainly true that his reception in interwar and Nazi Germany has been better documented than either his ideas or his lifetime reception.

After his death in 1904, Ratzel's worldview was operationalized by the Swedish political scientist and glowing Ratzelian Rudolf Kjellén, who coined the term "geopolitics." In the late 1930s, it would shape the international thought of the political theorist and Nazi jurist Carl Schmitt. But it was the geography professor and retired major general Karl Haushofer who was most instrumental in establishing a school of German geopolitics (*Geopolitik*) in Ratzel's name. Frustrated by the terms of the Versailles Treaty, he believed that Germany could only restore its rightful position in world politics if it took seriously the lessons of Ratzel's political geography. Haushofer saw an opportunity to shape Germany's destiny when he was introduced to the as yet unknown watercolor painter and beer hall politician Adolf Hitler in 1924. Little is certain about the exact nature of the two men's encounters, but it is known that Haushofer left a copy of Ratzel's 1897 *Politische Geographie* in Landsberg prison, where the future Führer was writing *Mein Kampf*.[7] Both *Mein Kampf* and Hitler's unpublished *Zweites Buch* (Second book) exhibit traces of Ratzel's political geography and use the term *Lebensraum*.[8] It has thus been repeatedly argued that Ratzel's ideas had an important, albeit indirect, influence on the formulation of the Nazi war aims in Eastern Europe.[9] It is this trajectory that has ensured that his intellectual legacy is commonly seen as toxic, "even in the most forgiving interpretation."[10]

And yet there is more to Ratzel's geography than its relationship to Nazi geopolitics. Not only was his corpus of work immensely diverse, but his ideas also traveled on different trajectories. Ratzel inspired not just a German school of geopolitics, but he also shaped a generation of North American geographers, including most notably Ellen Churchill Semple and, though less directly, Isaiah Bowman, an advisor to Presidents Woodrow Wilson and Franklin D. Roosevelt. Indeed, Ratzel's political geography continued to find followers even after the fall of national socialism. In the 1950s and 1960s, West German nuclear strategists and civil defense planners used Ratzelian language to argue for the nuclearization and bunkerization of NATO societies.[11] Although the German weekly broadsheet *Die Zeit* noted in 1966 that Ratzel had become unfashionable in academic circles,[12] his spatial theory of survival and extermination lent itself to the thermonuclear age. Well into the later stages of the Cold War, US geopoliticians would cite him authoritatively.

In recent years, it has become a little quieter surrounding Ratzel. Anglophone military strategists have sought more consistently to cleanse geopolitics from Ratzel's name, a move that is commonly performed without necessarily disposing of his understanding of politics as a struggle for space.[13] When Ratzel's name is dropped in today's media landscape, he is most often invoked on different sides of the political spectrum as an advocate of aggressive national self-determination and *Lebensraum*.[14] Some observers are willing to see past the historical baggage. In these accounts, Ratzel is cast as an observer with foresight, be it about the rise of China, urban transportation systems or the nature of borders.[15] Perhaps most prominently, Ratzel has been excused by the German historian Karl Schlögel in his book *Im Raume lesen wir die Zeit* (In space we read time), a title he borrowed from the Leipzig geographer. There is, Schlögel claims, "no route from Ratzel to Hitler."[16] On the far right, Ratzel's ideas have survived comparatively unbroken. However antiquated Ratzel's ideas may appear, they seem to chime in a world of rising nationalism and reterritorialization.

This book follows three interconnected lines of inquiry into Friedrich Ratzel's life and work. The first is of a historical nature and seeks to assess his role as a public intellectual in Imperial Germany (1871–1918). I argue that although Ratzel was an elusive and shy individual who preferred to engage with the world through observation and writing, he was nonetheless an intellectual force, whose newspaper articles, popular science books, and activities in various colonial organizations shaped the German political landscape in profound ways. A second and more theoretical avenue sets out to examine and assess Ratzel's understanding of nature-society relations. I argue that Ratzel's concept of nature is complex and his attempts to fuse the natural sciences with the humanities often thought-provoking. Ultimately, however, it is impossible to disentangle Ratzel's geophilosophy from his geopolitics. A third line of inquiry seeks to reassess Ratzel's politics. I conclude that while Ratzel has variously been labeled a "national liberal" and "radical conservative,"[17] simply a "conservative" or a "conservative critic of modernity,"[18] it is difficult to place him ideologically with only the help of the German political spectrum. Rather, it is instructive to contextualize him also in his second home, post–Civil War America. I conclude that his work was consistently underwritten by a preoccupation with death, which gave expression to the powerful belief that "primitives" were self-annihilating even before white colonists had set foot on their soil. This necropolitical underside of his geography, I argue, is the as yet unrecognized driver of Ratzel's ideological project, and one that would unleash its full force only in the years after his death.

Life and Ideas

Friedrich Ratzel was born in 1844 in the Grand Duchy of Baden, a state that would be integrated into the German Empire in 1871. He left his parents' house at the tender age of fourteen to be trained as a pharmacist. Having always felt a curiosity for the natural world, Ratzel was unhappy with the prospect of life in the pharmacy and would, after some hesitation on his parents' side, convince them in 1866 to let him study zoology and geology instead. Spending time in the universities of Heidelberg, Jena, and Berlin, Ratzel came under the influence of the ideas of Charles Darwin, whose *On the Origin of Species* had recently been translated into German. He was also inspired by the biologist and philosopher Ernst Haeckel, the great popularizer of evolutionary theory in Germany, whose lectures he attended. After completing his thesis in 1868, Ratzel worked as a research assistant and journalist before volunteering to serve in the Franco-Prussian war of 1870–1871, during which he was wounded and forced to recover in a military hospital.

Upon his release from the armed forces, Ratzel embarked on a two-year expedition to the Americas, working as a travel writer for the *Kölnische Zeitung*. This journey to the United States, Mexico, and Cuba would shape his worldview profoundly. After returning to Germany, he took up a position at the University of Munich in 1875 and moved to Leipzig in 1886, where he worked until his death. A prolific writer, Ratzel moved through different intellectual phases in his career. Beginning with biogeography, he then turned to travel writing, ethnology, human and political geography, even autobiography, before resuming his interest in biogeography shortly before his death. His most notable works include the two-volume *Anthropogeographie* (Human geography, 1882 and 1891) and *Politische Geographie* (Political geography, 1897). The more encyclopedic three-volume *Völkerkunde* (literally "Ethnology" but translated into English as "The history of mankind," 1885, 1886, and 1888) is sometimes counted among his major works.[19]

Despite the plethora of genres and disciplines his work engaged with and in many cases shaped, it was not perhaps as diverse as it might appear. Rather, it coalesced around a number of recurring themes and arguments, including, most prominently, the influence of geology and climate on human history as well as the idea that decay and disappearance were always followed by new forms of becoming. Ratzel believed in the singleness of existence and thus tried to implode familiar conceptual binaries, such as space and time, nature and society, life and death. Like Alexander von Humboldt before him, he was convinced that science had its limitations and that geography needed to borrow from philosophy and art if it wanted to understand the world in a more comprehensive way. As we will see, he was not always consistent in this move to transcend binary oppositions and distinguished, for instance, quite rigidly between "natural peoples" (*Naturvölker*) and "civilized peoples" (*Kulturvölker*).

While Ratzel was driven by scientific curiosity, he was also a passionately political intellectual. Not only was his writing concerned with the state, read as an organism, but it was intended to guide a particular state, Imperial Germany, in its foreign and colonial policy. Having fought in the last of the German unification wars, Ratzel was a devoted nationalist. He was a member of the radical conservative Pan-German League (*Alldeutscher Verband*) and the equally influential German Colonial Society (*Deutsche Kolonialgesellschaft*), which relentlessly lobbied the German Empire to establish colonies. Regarding Europe as a politically saturated space, Ratzel advocated expansion to Africa as the only continent in which the German nation could still find new living space. Ratzel's model of German colonization was the United States. He idealized the white American frontiersman who journeyed westward into terra incognita. Germany too, he felt, needed a frontier. There was an important racial dimension to this settler colonial fantasy. Having adopted a broadly segregationist mindset in the 1870s, Ratzel believed that black Americans were in need of resettlement. Later, he would suggest that slave labor was a necessity in certain climatic regions.

Ratzel was convinced that war was a natural condition of world history. Indeed, he never spoke a critical word about the war that had left him an invalid, even remarking that it was "a shame that there are not more wars."[20] Fantasizing that the German navy would one day rival that of the British Empire, Ratzel was known as one of a number of "fleet professors" who enthusiastically promoted a larger German navy in the final decades of the nineteenth century.[21] In addition to his militarism, nationalism, and support for colonialism, Ratzel held many basic liberal beliefs. He spoke favorably of American democracy and supported the idea of women in the workplace. He used the terms individualism and cosmopolitanism in the affirmative and critiqued the racial theories of Houston Stewart Chamberlain and Arthur de Gobineau.[22] And yet his rejection of racial theory was only partial. Race was a fundamental vector in his work; it just had to be understood in environmental and cultural rather than narrowly biological ways.

In the later stages of his life, Ratzel became increasingly detached from public life. He was considered by many of his contemporaries a contentious figure because of his convoluted writing style, disregard for disciplinary boundaries and blurring of scholarly and political activity. Given his attempts to apply the lessons of evolutionary biology to world politics, he approached his scholarly and journalistic writing as intrinsically connected; nature and society were ultimately one. Ratzel's final project, a biogeography in which human society was but a chapter, remained unfinished, but it can be glimpsed from his final publications, particularly his 1901 *Lebensraum* essay and his 1902 two-volume *Die Erde und das Leben* (The earth and life).[23]

Convinced throughout his life that death was not simply the end but rather a new form of becoming, Ratzel was drawn, like many other evolutionary theorists, to the extinction of species and the disappearance of colonized peoples. Crucially, he operationalized this death motif in his political geography by claiming that states themselves emerged from the demise of other states, thus turning a sociobiological theory into a political one. There was nothing meaningless about death for Ratzel. Whether he was speaking of fallen leaves, dying soldiers, abandoned ruins, or indeed of declining empires, decay and death were to him no less than the highest and most beautiful expressions of life on earth. I use Achille Mbembe's term "necropolitics" to capture this deathly underside of Ratzel's *Lebensraum* idea in which biology, race, and sovereignty were fundamentally fused.[24] This morbid foundation would follow his *Lebensraum* concept as it gathered momentum in the decades after his death. The trajectory of Ratzel's necropolitical geography reveals like few other dynamics the continuities and shared political ground between the projects of settler colonialism, nineteenth-century liberalism, and early twentieth-century fascism.

Rereading Ratzel

Anglophone geographers and historians have over the past two decades produced a remarkable array of studies on controversial canonical thinkers who wrote in the broad tradition of geopolitics and political geography. This literature has covered figures such as Isaiah Bowman,[25] Halford Mackinder,[26] Ellen Semple,[27] Carl Schmitt,[28] and Karl Haushofer.[29] Scholars of international relations have on the whole been less inclined toward the biographical genre, but they too have published important books on the influential realists E. H. Carr and Hans Morgenthau.[30] While this flurry of studies counteracts a collective amnesia in which the ghostly figures from a discipline's past are sometimes "purposefully forgotten,"[31] it raises the broader question of the nature and politics of a disciplinary canon, particularly in relation to a disgraced thinker like Friedrich Ratzel.[32]

Many disciplines have a canon—most notably perhaps sociology, anthropology, and economics—but this does not mean that the moral orderings inherent in such crypto-ecclesiastic traditions are unproblematic.[33] Canonical history has repeatedly come under attack for being overinterested in male lineages of thought.[34] Driver and Baigent note moreover that there is a tendency to think of the geographical tradition in particular as made up of those who identified as geographers rather than the many surveyors, topographers, cartographers, geomorphologists, and geologists who long conducted geographical research inside and outside of geographical institutions.[35] Barry concurs and suggests looking beyond the discipline at the work of social and political theorists whose work pushed the geographical imagination into new directions.[36] Agnew is more skeptical about the need for a canon and argues that Ratzel's thought, for instance, has little application to today's world.[37] But even if we agree with this verdict, then we are still faced with the challenge of having to make sense of a world in which his ideas continue to echo, both on a theoretical and political level.

Probing the history of Ratzel's political ideas is a necessarily interdisciplinary endeavor. The field of international relations, for instance, has long had a difficult relationship with the tradition that was founded in Ratzel's name. It was the influential realist Hans Morgenthau who rejected geopolitics as a single-factor explanation in his 1948 *Politics among Nations*.[38] Outside the narrow field of strategic studies, there was little interest in geopolitics until the emergence of a critically minded geopolitics in the 1990s, one of whose goals was intellectual history.[39] And yet, as Ashworth has shown, Ratzel's ideas were never absent from the history of international relations but played a crucial role precisely as an "other" that made possible the advancement of a separate trajectory of political thought.[40] Exceptions notwithstanding,[41] international relations is today faced with a situation in which political realism remains a central object of scholarly concern but geopolitics is often ignored. This is a puzzling status quo, given the shared theoretical ground and frequent intersections between the two traditions.[42]

There are five existing academic monographs on Friedrich Ratzel, all of which, though to varying degrees, display a tendency for hagiography.[43] Steinmetzler[44] and Wanklyn[45] have both produced useful guides to Ratzel's work, but they largely ignore its political dimension. The same can be said of Buttmann,[46] although his study is exemplary in the meticulousness with which it addresses Ratzel's personal and professional life. Steinmetzler has produced the most thorough study of Ratzel's intellectual influences. Müller offers an insightful introduction to some of Ratzel's most crucial ideas,[47] even though large sections of the book are a collage of quotes rather than a critical discussion. Stogiannos sets out to salvage Ratzel and apply his oeuvre to contemporary world political events.[48] While the book features some careful conceptual analysis, its attempts to save Ratzel from the charge of environmental determinism fail to convince. None of these studies do justice to the afterlife of Ratzel's ideas.

The absence of a critically minded book on Ratzel has not, it should be emphasized, prevented a vibrant debate on the founder of political geography from emerging among geographers, colonial historians, and historians of ideas. Building on earlier investigations into Ratzel's geography in the context of nineteenth-century German science and its impact on the spread of geopolitical ideas in the interbellum,[49] more recent discussions have developed in a number of interesting directions. A first strand of work has probed deeper into the conceptual history and posthumous career of Ratzel's *Lebensraum* concept.[50] Colonial historians, moreover, have debated whether the origins of national socialist violence can be found in the atrocities committed by Imperial Germany in South West Africa during the early 1900s,[51] often providing iterations of what has become known as the "Windhoek to Auschwitz thesis" and which can already be found in an embryonic form in Hannah Arendt's 1951 *The Origins of Totalitarianism*.[52] They highlight that many leading national socialists had colonial biographies and modeled their plans for the colonization of Eastern Europe directly on those of the Wilhelmine empire. Although none of these observers have argued that Imperial Germany's atrocities were driven *primarily* by Ratzel's ideas, they have held that both his *Lebensraum* concept and his political activism empowered a radical colonial spirit in Germany, which in turn made this violence possible.

While critics have attacked the "Windhoek to Auschwitz thesis" for its thin evidence base,[53] they have not questioned the role that Ratzel played as an ideological intermediary between colonialism and national socialism. Jureit captures the consensus well when she reads Ratzel as "a transitional figure" who links "the imperial notion of Lebensraum with twentieth-century concepts of expansion and extermination."[54] And yet she leaves underexamined the question as to what role extermination (*Vernichtung* in German) played in Ratzel's oeuvre. Indeed, the genocidal campaign against the Herero in modern-day Namibia was fought by the colonial elite through a language of "race war" (*Rassenkampf*) and "a war of extermination" (*Vernichtungskrieg*), both phrases Ratzel had employed in his writing.

A separate strand of work has observed the interesting ways in which Ratzel's ideas chime with debates in contemporary social theory, in particular with regard to various new materialisms and Foucauldian biopolitics, the politics of life.[55] A dialogue between Ratzel and contemporary vitalism (or vital materialism), a theory that has found followers in both geography and international relations,[56] is particularly instructive. As part of a larger return to materialist concepts of political and more-than-human geographies,[57] these approaches seek to understand the social world not by reference to a fixed set of human agencies, but rather by examining the vibrancy and disruptive force of a whole range of organic and inorganic forces. This task necessitates, it is argued, both a bridging of the natural and social sciences, of physical and human geography, and an affirmative mode of critique that seeks to encounter the world through a politics of curiosity and wonder rather than explaining it, in the register of disenchantment, by reference to pregiven power structures.

Interestingly, Abrahamsson has spotted in these debates a reluctance to confront the more problematic figures who inhabit vitalism's own history. He points to lacunae relating to the ethical and political consequences that come with translating vitalist thought into political thought.[58] One of the thinkers Abrahamsson urges contemporary vitalists to re-examine is Friedrich Ratzel. As we will see, Ratzel did flirt with vitalist lines of argumentation, particularly toward the end of his career. He believed that life, in its insatiable creativity, erupted into social processes and that the earth system impacted slowly on the technological makeup of human societies. He too sought to bridge the divide between science and philosophy and framed the human as a mere effect of the natural world, consciousness as dispersed throughout the universe. It is with these debates in mind that I interrogate Ratzel's concept of space, which is significantly more complex than previously recognized, for Ratzel was interested in the earthly conditions of a wide variety of modes of spatial consciousness.

A further theoretical line of inquiry into Ratzel concerns his place in the history of biopolitics. Questions of biopolitics have preoccupied the humanities and social sciences for a number of years, particularly in relation to the work of the social theorists Giorgio Agamben and Michel Foucault.[59] Both Foucault and Agamben argue that we can find within the seemingly caring practice of biopolitics, a governmental concern for the health of populations, a second and significantly less benign form of power, which seeks to exclude forms of life that are ruled dangerous to the body politic.[60] Foucault holds that in a biopolitical age, which he sees as emerging first in the eighteenth century, even war and genocide are no longer legitimated by reference to a sovereign but executed in the name of a population's preservation. Agamben focuses his attention on the structure of sovereignty and the simultaneously legal and extralegal state of exception as its most crucial expression, a logic he traces from antiquity to the present. Both thinkers locate in the national socialist policy of encampment and extermination the most extreme form of biopolitics. Thanato- or necropolitics (the politics of death) emerges in this reading not as some kind of flaw in the modern state's care for its population, but as its constitutive underside.

Achille Mbembe pushes these arguments further by inserting colonial violence more firmly into this debate. Taking his cue both from Frantz Fanon's insistence that the very notion that "European progress has been built up with the sweat and the dead bodies" of colonized populations and from Hannah Arendt's suggestion that there was a link between Nazi extermination and earlier projects of European colonialism,[61] Mbembe points in particular to slavery as "one of the first instances of biopolitical experimentation."[62] Slave life, to Mbembe, is a form of "death-in-life" in which the political subject is dead and only the working body, reduced to a commodity, exists in a shadow-like state. In developing this line of thought, he poses the question about the place of the human body, "in particular the wounded or slain body" in colonial forms of domination.[63] Ratzel was himself a war invalid who was mesmerized by the figure of the slave. And while Mbembe may not have in mind the wounded body of a colonial enthusiast like Ratzel, the latter can nonetheless serve as a missing link in his broader history of necropolitics. Unlike Hannah Arendt, who was famously unwilling to extend her colonial boomerang to North America,[64] Ratzel never doubted that the United States of America was a colonial project at heart.

Although geopolitics and bio-/necropolitics are frequently said to be intertwined forms of power,[65] they are in fact often approached as analytically dissimilar phenomena. Geopolitics is thus understood as a discourse on international statecraft, while biopolitics is read as a practice of domestic politics. And yet there are perhaps few places in which the entanglement of geo- and biopolitics is as clear as in Friedrich Ratzel's work.[66] Indeed, even the most cursory reader will note the degree to which biology, geology, and politics are inextricably interlinked. It is thus no surprise that both terms—geopolitics and biopolitics—were coined by one of Ratzel's keenest followers. And while there is an emerging literature that attends to the role of death in the geographical tradition and political spatializations more broadly,[67] Ratzel's role as a necropolitical thinker is yet to be fully recognized.

With these debates in mind, this book seeks to make a number of contributions to the field. First, it inserts Ratzel's geography into the lineages of vitalist and biopolitical thought and proposes death as a fundamental key to his theoretical and political project. This allows for Ratzel's politics to be situated not just across a traditional political spectrum but within necropolitics. Second, it puts spotlights on his often-overlooked travel writing and journalism in order to reveal a more transnational Ratzel whose political and racial views were formed both by his American journey and in relation to the science and politics of the British Empire. Third, I seek to unpack the way in which personal experience shaped the confluence of colonial theory and practice in his work and how he was able to promote his ideas through his membership in a colonial elite. Fourth and finally, the book seeks to trace Ratzel's reception beyond the well-known episode of 1919–1945. This reveals a thinker who, while read in many places and in quite different corners of the political spectrum, always found his most receptive audiences and devoted disciples on the political right.

Approaching Ratzel

The chapters that follow set out to examine Friedrich Ratzel's ideas and their reception. My analytical focus is thus on key concepts and returning thought structures in his body of work, rather than on the individual. Biography is nonetheless a useful key to unlock the idiosyncrasies in a system of thought and a writer's persistence with particular themes and arguments. I thus pay attention not just to Ratzel's personal experiences and encounters but also to the places, institutional settings, and wider political context in which his ideas came to fruition and were subsequently taken up by others. Although this book is thus not designed as a classic biography, it faces many of the challenges that haunt this genre.

The first of these is the trap of presenting someone's thought as more coherent than it is. Any study of Ratzel has to confront the sheer volume of his writings, which prompted Harriet Wanklyn in 1961 to conclude that a full-scale biography was "perhaps out of the question."[68] At its core, his oeuvre encompasses around thirty books,[69] many of them multivolume, each volume often well over five hundred pages long. In addition to this, he published more than five hundred articles, obituaries, and shorter interventions as well as more than six hundred book reviews, some of which are contained in the two-volume *Kleine Schriften*, edited by Helmolt.[70] The challenge, however, is not merely one of volume but also of diversity. Ratzel wrote most of his texts for scholarly audiences, but many were targeted at a general readership, particularly the articles he penned for the daily national newspapers *Kölnische Zeitung* and *Allgemeine Zeitung* as well as the weekly national-liberal magazine *Die Grenzboten*. He published dense works of natural science and short, politically charged pamphlets, manuals for hikers, and snappy pieces on the quality of railway station pubs. His tone was variously didactic, dreamy, dry, or agitated. One of the challenges is thus to find a way to navigate what might best be described as a multiplicity of genres, held together only by a loose set of narrative frames and recurring arguments. I thus treat his thought not as a closed mathematical language but as an uneven, open, and imperfect structure, riddled with tensions and underwritten by moods and emotions that sometimes burst through his detached observations. Perhaps unsurprisingly, existing studies have focused their attention on Ratzel's major works, his human and political geography, thereby neglecting the lesser-known parts of his oeuvre. This study opts for a more comprehensive approach but recognizes the need for a certain selectivity. Ratzel, for instance, wrote at length about matters of astronomy and physical geography, but was seldom original when he did so.

The archive I draw on encompasses letters, diaries, and notebooks as well as a small number of unpublished manuscripts, all of which his family and colleagues preserved after his death and which now lie mainly in the Bavarian State Library in Munich and the Leibniz Institute for Regional Geography in Leipzig. An examination of these documents aids my discussion, even though the main focus is on his body of published work and that of his interlocutors. Ratzel managed to publish almost all of the article and book drafts that survive to this day. His lecture manuscripts broadly covered the topics he dealt with in his writing and are thus less noteworthy than his diaries and letters. Ratzel's diaries cover his time as a journalist and most of his academic career. Entries are rarely political in content and often brief, especially toward the end of his life. They address the banalities of everyday life: what he ate, whom he had coffee and cake with, letters he wrote and received on that day, physical exercise, and the weather. In the final years, he noted repeatedly his deteriorating health. Ratzel made a note of what he was writing and sometimes the content of conversations with others.

Like many of his contemporaries, Ratzel exchanged letters with a vast number of academic peers, former students, publishers, friends, and family members, including most importantly his wife Marie Elisabeth Ratzel (née Wingens) (1843–1926). Ratzel's letters, written in the German *Kurrent* script, paint the picture of a man who had a significant social circle but who avoided professional encounters, preferring in later years to hide away in the tranquility of his rural summer residence in Bavaria. Personal letters, particularly to his family, were often filled with platitudes. And yet, he sometimes used these letters to air new ideas and observations. And although Ratzel's letters and diaries provide an intimate insight into his character, they too are less revealing than his autobiographical account, sections of which semifictionalized. First serialized anonymously in *Die Grenzboten* shortly before his death, his memoirs were collected posthumously under the nebulous title *Glücksinseln und Träume* (Islands of happiness and dreams).[71]

A second challenge is posed by the issue of translation. Ratzel's *Völkerkunde* is to date the only of his more well-known books to have been translated into English, as *The History of Mankind*.[72] *Politische Geographie*, sometimes thought to be his magnum opus, is as yet unavailable to an Anglophone audience, even though the 1896 article "The territorial growth of states," translated by Ellen Semple for *Scottish Geographical Magazine*, addresses many of the book's central ideas.[73] Semple produced three further abridged translations of chapters 13–15 of *Politische Geographie* for the *American Journal of Sociology*,[74] early drafts of which Ratzel had discussed with her and given his blessing to.[75] In 1988, Ratzel's travel journals were translated by Stewart Stehlin and appeared under the title *Sketches of Urban and cultural Life in North America*.[76] Tuesday (Tul'si) Bhambry's translation of his *Lebensraum* essay followed three decades later.[77] The majority of quotations I employ in what follows are thus my own translations. I have tried to be as true to the original as possible, letting Ratzel sound as variously archaic, flowery, pompous, convoluted, judgmental, or vulgar as he does in German.[78]

A third set of challenges concerns the question of context. Withers has recently re-emphasized the importance of "the material and epistemological conditions that lie behind the making, shaping and consumption of texts."[79] In this spirit, it is crucial not just to attend to the boundaries of scientific knowledge within which a thinker like Ratzel produced his oeuvre, but to the institutional and financial conditions of his own life and of those he engaged with.

Most of all, I consider the changing sociopolitical context in which he wrote, not just in Germany but in his second homeland, the United States. This is not an attempt to collapse the question of textual meaning back into context, but rather one of highlighting that, however original and eccentric his ideas were, they were always in a dialogue with recent events. Ratzel visited the United States less than a decade after the end of the Civil War, a period during which the South's slave economy was giving way to an emerging system of segregation. He arrived two years after the rise and fall of the first Ku Klux Klan (1865–1871), arguably the world's first proto-fascist organization,[80] and at a time when lynching of black Americans had become an endemic condition of American society. Germany too witnessed epochal changes during Ratzel's lifetime, such as rapid industrialization, urbanization, and the meteoric rise of nationalism. Under emperor Wilhelm I's reign (1861–1888) and his chancellor Otto von Bismarck, Prussia fought a string of wars against its European neighbors that resulted in territorial annexations and, ultimately, the unification of Germany at the Palace of Versailles in 1871. In the 1880s, Germany established a colonial empire in Africa and the Far East, something Ratzel enthusiastically supported. The grasp for European hegemony was pushed further under Wilhelm II's reign (1888–1918). Germany's last emperor believed, much like Ratzel, that Germany should challenge British naval hegemony. As we will see, Ratzel's ideas were popular among an elite of colonial intellectuals, explorers, and administrators, with many of whom he entertained personal connections.

A fourth challenge thus concerns reception. In grappling with Ratzel's concepts and ideas, it is imperative both to unearth the web of texts on which Ratzel drew, as well as the ideas he spawned, tracing how "[m]eaning bleeds, as it were, from one text to another."[81] Existing studies have tended to consider Ratzel through a national frame, concentrating on the ways in which his ideas found inspiration in an earlier tradition of German geography and were subsequently received and spread by other Germans. In an attempt to tease out Wardenga's "histories of transfer and connectedness,"[82] this book traces how he was taken up and resisted across Europe and North America. Livingstone reminds us moreover of the significance of place in this messy production of textual meaning.[83] Nineteenth-century scholars were read not only in different national contexts but in particular architectural settings, from laboratories to private estates. Although only little is known about the locations in which Ratzel was read, the question of place is nonetheless relevant in a different sense; for we do know which places shaped his thought, from the military hospital to the ghost towns along the American frontier, in which he mused about the beauty of ruins.

Fifth, we have to consider the politics of writing a book-length study about an individual thinker, even where this task is guided by questions of intellectual history rather than by the aim of writing a traditional biography. The gendered dynamic to canonical history is powerfully highlighted by Rose in 1995. There is, she argues, a tendency for two models of writing to dominate. The first is what she calls the dutiful son model, in which writing subjects affirm their own authority by references to a great man. Then there is the rebellious son model, which "construct[s] a paternal line only to revolt against it."[84] Rose is not writing specifically about biography, but the dangers of falling into one of these traps are perhaps particularly prevalent in this genre. A critical engagement with Ratzel, or so I hope, does not necessarily mean to stylize him as some kind of paternal figure of authority, be it benign or malign. In fact, I seek to point out on a number of occasions the points where Ratzel's authority cracked and failed, giving way to other political forces and voices. The issue of gender haunts this book in other ways too. We have already noted the invisibility of women in the geographical canon and the same is true for other traditions of thought.[85] Ratzel almost exclusively socialized and engaged intellectually with other men. And yet there were at least two important women in his life, his wife Marie, a financially independent woman who after his death would oversee the Munich parts of the Ratzel estate,[86] and Ellen Churchill Semple (1863–1932), Ratzel's popularizer in the Anglo-American world. They will feature in chapters 2 and 7 respectively.

Engaging in colonial history through a reading of an individual colonial intellectual, scientist, or explorer raises further challenges. Too often do the experiences of the colonized disappear behind the shadow of the towering white man whom the colonial archive sought to preserve. Driver suggests counterbalancing this problem by unearthing "hidden histories," which bring to the fore "the agency of indigenous peoples and intermediaries in the history of exploration."[87] This requires the bringing out of those who remain only partially visible in the archive or who were only accidentally recorded. This task of "looking at familiar material with fresh eyes" is a "notoriously tricky business."[88] Although Ratzel rarely described the colonial subjects he encountered on his journey to the Western Hemisphere in any detail, some of their lives can be glimpsed from offhand comments and the occasional vignette.

Finally then, a few words on the challenges inherent in the history of ideas. Despite the latter's continued popularity in the history of science, an exclusive focus on ideas has long been challenged by histories that emphasize embodied practices, such as drawing, fieldwork, and the use of technology. An interest in popular culture, particularly in relation to geopolitics, has moreover drawn attention away from intellectuals and their concepts.[89] There has been a growing concern that the field has been "obsessed with tracing the intellectual genealogies of its concepts (as singular words) that the practical lives to which the concepts can help provide tentative understanding have been increasingly forgotten."[90] And yet, as Elden highlights, the history of ideas is a necessary but hardly sufficient instrument within the larger toolbox of historical inquiry.[91] A student of German colonial militarism might thus want, for instance, to read this book alongside work on Prussian military culture,[92] colonial-themed war toys,[93] or the African soldiers who were recruited into the ranks of the German colonial army.[94] It would be a mistake to locate in Ratzel the answer to all questions surrounding German colonialism.

If the tendency for overstatement of the role of individual thinkers and their ideas is perhaps the most serious pitfall of intellectual history, it is not the only one. Livingstone flags up in particular the hazards of presentism and internalism, of viewing the past only through the prism of the present and of explaining change within an intellectual tradition through internal drivers and factors.[95] He is right about the perils of internalism, and in what follows I seek to present Ratzel's ideas not just in the light of the geographical canon on which he drew but in that of the broader milieu in which he wrote. On the question of presentism I am less inclined to agree. As Rose has argued, "the writing of certain kinds of pasts is legitimated by, and legitimates, only certain kinds of presents."[96] Aware of the danger of a crypto-positivist epistemology, in which the historian curiously stands in some kind of ahistorical and apolitical bubble, Livingstone concedes that it is impossible to write the present entirely out of the story.[97] And yet he seems to want to insist that a history that searches for intellectual origins and precursors that "foreshadow" contemporary events is itself a cardinal sin.

There are a number of questions and concerns in this book that might be deemed presentist under this definition. Some of the prisms I use, particularly perhaps that of white supremacy, may smack of a present-day interest in the unresolved problems of settler colonial societies. And yet whiteness and racial hierarchy were clearly Ratzel's obsessions too. It would, moreover, be a mistake to treat the epochal and cataclysmic events of the First World War as an insurmountable barrier to an inquiry into the history of geopolitics. For the conceptual history of *Lebensraum* demonstrates important continuities, however much Ratzel's concept, which he himself had only borrowed, was ultimately inflected and transformed. And yet, in other ways, Livingstone is right to be cautious about prolepsis. American geographers sought in the 1940s to salvage a "legitimate" form of geopolitics by disassociating their own thought from its Ratzelian origins and exaggerating Ratzel's influence on Hitler.[98] Ratzel, in this view, might emerge as a proto-fascist and a proto-fascist *only*. If done in this spirit, a biography of Ratzel would risk adding to a reductive reading of the late nineteenth century as an age of reactionary darkness in ways that might overlook alternative, and perhaps more progressive, traditions of geographical and political thought.[99] And yet, as this book shows, Ratzel's narrative was often coated in a language of cosmopolitanism and civilizational progress. It is this fundamental compatibility of liberal progressivism and exterminism that has survived into the present and that makes an examination of Ratzel all the more imperative. As Barry notes, the most crucial function of a canon is precisely "to open up new possibilities for thinking politically in the present."[100]

In carrying these reflections further, the book proceeds as follows. The second chapter discusses Friedrich Ratzel's life within the sociopolitical and intellectual context of Imperial Germany, the state he served for as a soldier and scientist. It draws on his letters, diaries, and autobiographical writings as well as eulogies and other forms of engagement by his contemporaries. I also reflect on his scholarly network and politics. Chapter 3 takes a close look at Ratzel's framing of slavery and race in settler colonial contexts. I argue against the view that race was an afterthought in his work and show that his detailed descriptions of white settlers and nonwhite populations reveal an infatuation with the American frontier and the absorption of a typically North American form of racism. Chapter 4 addresses the foundations of Ratzel's geophilosophy. Particular attention is paid to his sources of inspiration, especially the work of Carl Ritter, Ernst Haeckel, Charles Darwin, Herbert Spencer, and Moritz Wagner. I conclude that while he ultimately struggled to produce a *theory of everything*, his monist conceptions of life and earth, time and space, were complex and foreshadowed intellectual currents that would take root in the humanities and social sciences only significantly later.

The subsequent chapter turns to the most widely known part of Ratzel's oeuvre, his political geography. I begin by unpacking Ratzel both as a theorist and practitioner of colonization before discussing his geopolitical writings and concept of war, the latter of which he read as a force of nature. In this, Ratzel sought to transcend one of his intellectual heroes, the Prussian military strategist Carl von Clausewitz. Chapter 6 explores death as the most crucial interpretive key to Ratzel's geography. Drawing on the work of the panpsychist philosopher Gustav Fechner, Ratzel questioned the distinction between organic and inorganic forces and developed an aesthetic theory of death. I examine Ratzel's fascination with cannibalism, the ruins of exterminated civilizations, and the supposedly self-annihilating "primitive," out of which the central arguments of his political geography arose. Chapter 7 explores Ratzel's influence beyond his grave. I begin with his lifetime reception. I then trace the way in which his ideas traveled through Rudolf Kjellén and Karl Haushofer into Nazi ideology, before falling out of fashion by the 1940s. I examine Ratzel's reception in the Anglo-American world in the first half of the twentieth century before addressing Ratzel's stealthy rebirth in Cold War nuclear strategy and civil defense planning. The chapter ends with a discussion of Ratzel's contemporary reception. Attention will be paid to more than a century of critiques of Ratzel's work. The conclusion ties together the book's three—historical, theoretical, and political—lines of inquiry. It poses the question of what it means to think with and against Ratzel in the present.

CHAPTER 2

Soldier and Scientist

In July 1870, Friedrich Ratzel was electrified by the rumors of an impending war. Too old to be drafted, he would later describe the mood as a sweltering thunderstorm that loomed over the lecture theaters, driving ecstatic students into battle. He decided to enlist and was within days stationed on the river Rhine alongside his fellow volunteers, eagerly awaiting the first lightning bolt to cut through the sticky air. As a comrade sang a morbid tune over a round of cherry brandy, Ratzel noted that it was beautiful to have been brought so close to the edge of death.[1]

The Franco-Prussian War of 1870–1871 was a short but truly modern war. Even more than the American Civil War, it featured the large-scale use of railways to move soldiers, equipment, and injured bodies to and from the frontlines. This ultimately benefited the victorious German armies, as did the use of breech-loading firearms. After the French army had been defeated, a sustained guerrilla campaign emerged, the most formidable Europe had seen to date.[2] This was a war in which the distinction between army and nation had become truly blurred, a war that was fought on both sides with an unprecedented nationalist fervor. During the Siege of Strasbourg, in which Ratzel was deployed, the city center was shelled and turned into the kind of urban ruin that would come to haunt Europe in the twentieth century.

For decades, Ratzel barely reflected on his experience of active service.[3] And yet the "looming storm" of 1870–1871 had left a lasting impression. When delivering a lecture series on political geography in the 1890s, he stated that war between two states was but a "historical thunderstorm," brought about by the "gathering clouds" of geographical forces.[4] When he finally did decide to pen his war memoirs, published posthumously as *Glücksinseln und Träume* (Islands of happiness and dreams),[5] he opted for a literary style that was carefully attuned to the perception of the soldier's body. Ratzel was captivated by the blisters on his hands that had appeared when digging trenches and by the muffled sound of cannons being wheeled through the night. He was enchanted by the shimmering beauty of ice crystals that had formed on a puddle of blood. War, in other words, was a natural spectacle.

Ratzel's student and biographer Kurt Hassert noted with admiration that the professor Friedrich Ratzel could never quite hide his military demeanor. Even in the most severe winter, the former infantryman would rarely wear an overcoat.[6] As his friend, the historian Karl Lamprecht, observed, the Leipzig geographer was quick to forge friendships with other men who had served. He would have undoubtedly been glad to have known that veterans would eventually carry his body to the grave in August of 1904.[7] During the Second World War, it was the geopolitician and Nazi sympathizer Karl Haushofer who admitted having felt the urge to model himself on this man who was "simultaneously soldier and scientist."[8] And although it is important not to read Ratzel exclusively through the eyes of his friends and disciples, it should become clear in what follows that Ratzel's necropolitical geography did emerge at the intersection of his military and scholarly identities.[9] For the young Darwinian, nature itself was in a warlike state, which meant that war was but a purer form of the natural violence that formed the background noise to all life. And as nature itself was ultimately beautiful in its immensity and exuberance, so was the possibility of dying on the battlefield.

The period from the failed revolution of 1848 until the outbreak of the First World War in 1914 was one of political conservatism and yet seismic political and socioeconomic change. Under the leadership of the powerful chancellor Otto von Bismarck (1815–1898), Prussia waged successive victorious wars against Denmark (1864), Austria (1866), and France (1870–1871), culminating in the 1871 proclamation of the Prussian monarch Wilhelm I (1797–1888) as emperor of Germany. In the 1880s, Germany also begun acquiring formal colonies and sought to assert itself as a global power. Foreign politically, this was a period of changing and often secret military alliances, although this changed after the forced resignation of Bismarck in 1890. By the time of Ratzel's death in 1904, Europe was already effectively divided into two power blocks, which would ultimately face one another on the battlefields of the First World War. In the early 1900s, the empire was also sliding into a succession of genocidal wars in its African colonies.

The period after 1871 saw high-speed industrialization, urbanization, and population growth. Whereas in the early 1880s the largest employment sector had still been agriculture, by the mid-1890s it was industry.[10] With this transformation came both significant wealth and income inequality as well as the emergence of a large and multilayered bourgeoisie. The years after unification in 1871 were marked by political competition and repression, which featured most prominently the monarchy's campaigns against the Catholic Church and social democracy. And yet the German Empire experienced the gradual legitimization of parliamentary politics, the growth of civic culture, and the introduction of social insurance legislation. These political tensions were mirrored for a while at the level of the monarchy in the conflict between the conservative emperor Wilhelm I and his son, the liberal crown prince Friedrich III (1831–1888). When Friedrich died in 1888, shortly after his father's death, Friedrich's son Wilhelm II (1859–1941) resolved this tension in favour of authoritarianism.[11] Ultimately, democratic transformations endured a fragile existence in the face of constant threats by the executive to restore a purer form of autocratic rule.[12] Parliamentary control over state policy was limited. The empire also remained socially conservative, particularly in relation to questions of the division of labor and gender roles, which was reflected too in the patterns of Ratzel's life. During the reign of Wilhelm II, Germany experienced an increasing fragmentation of its political landscape, as well as the steady rise of antisemitism, nationalism, and militarism. In the empire's eastern regions, Germanization policies were introduced. And although the last German emperor stylized himself as a nationalist, even he was increasingly overtaken on the political right by forces that congregated around the extraparliamentary Pan-German League (*Alldeutscher Verband*), of which Friedrich Ratzel was a founding member.

This larger set of political struggles functioned as the backdrop to the consolidation of the academic discipline of geography. When Ratzel entered the academy in the 1870s, the field was transforming from one of colonial exploration into one of "scientific" endeavor and imperial policy advice. This shift corresponded with a subtle change in spatial semantics. The naming of individual geographical features increasingly gave way to the production and sharpening of larger and continental-scale mental maps, many of which endure as geopolitical constructs until this day.[13] German universities had long enjoyed significant standing abroad. Because of this reputation, many foreigners came to study in Germany, taking back with them ideas and approaches. Some of Ratzel's visiting students would ascend to positions of intellectual influence in their home countries. These informal transnational scholarly networks also served as Germany's window into the world. German universities were thus crucially involved in the process of empire building, which was to some extent a form of learning from the world.[14] After the Franco-Prussian War, geography was a key driver in this modeling process by providing "a continuous and deep-seated language for German imperialism."[15] The young discipline's quest for recognition encouraged political conformism. Deemed useful in conveying imperial knowledge to the German bourgeoisie, both via secondary and tertiary education, geography thus received significant financial backing from the state. Although some faculties tried to resist its rise, the field expanded steadily. In 1871 there had only been one chair of geography (at Göttingen). In 1885, Friedrich Ratzel was one of twelve full professors. In 1910, the number had risen to twenty-three.[16]

This chapter reviews Friedrich Ratzel's biography and his most important works in the sociopolitical context sketched out in these opening pages. I begin by surveying the first half of his life, his time as a pharmacist and travel writer as well as his service in the war of 1870–1871. Subsequently, I examine his academic career. His spells at the universities in Munich (1875–1886) and Leipzig (1886–1904) saw him publish his most notable books, *Anthropogeographie* and *Politische Geographie*, which will be discussed in turn. The chapter concludes with an examination of Ratzel's political compass, although the latter will be a dominant theme throughout the chapter.

Early Life, Military Service, and Journalism (1844–1875)

Friedrich Ratzel was born on 30 August 1844 in Karlsruhe into a Protestant, lower middle-class family as the son of the chamberlain (*Kammerdiener*) of the Grand Duke of Baden. Fritz, as he was known then, spent much of his childhood in the palace gardens and library. He developed an interest in the natural world and became a collector of plant samples and fossils. A keen hiker, he enjoyed long solitary walks, especially to castle ruins. These were hobbies that lacked a social element and cultivated a love of solitude.[17] While the young Ratzel was fond of his mother and kept a curl of her hair in his private possessions, he had a more distant relationship with his father, who was often away from the family and on one occasion told the young Fritz's teacher to beat him "black and blue if he does not obey."[18]

And while his childhood memoirs paint the image of a young man who was gradually shaped by *and grew into* the idyllic rural landscape around him, it also featured experiences that brought him face to face with the question of death, disappearance, and loss. Twice he almost drowned and once he witnessed another child die. He also recounted a solar eclipse, which had made his mother cry in fear and which he interpreted as the sun's warning to all creatures on earth. This great fireball, the young Fritz had observed in awe, had made so many things blossom, only to return them back into the earth's fold.[19] Often spending days worrying about his parents' mortality, he kept the skull of a mouse next to that curl of his mother's hair.

Ratzel had three siblings: Carl (1839–1876), Ludwig (1842–1899), and Emma (1849–unknown). Carl had begun a career in architectural engineering in the 1860s and was promoted professor in 1872. He suffered from a nervous condition and died in 1876. Ludwig too struggled with his mental health. Prone to psychotic episodes (during which he was convinced that he was starving), he tried, unsuccessfully, to take his life.[20]

At school, the young Fritz was intellectually curious but appreciated his teachers' digressions more than the syllabus. He fled into a world of books, reading whatever he could get his hands on. Books enchanted him, but they also exerted pressure on him, like a "black army of letters which encircled his mind."[21] The Protestant boy learned about Catholic and Jewish religious rituals from his classmates and enjoyed being momentarily plunged into their world. Ratzel was particularly drawn to one of the grand duke's grooms, a man whom Ratzel enjoyed observing in secret. His "wonderful appearance, slim, bright-eyed" and surrounded by the sweet scent of horses gave birth in Ratzel to a general romanticization of the peasant community.[22]

Because his brother Carl had embarked on an academic career, Friedrich Ratzel was not allowed to undertake his *Abitur* and was thus initially barred from realizing his dream of becoming a student of the natural sciences. Instead, his parents sent him off to become a pharmacist in the rural Kraichgau region. He spent his evenings staring in the direction of his parental home, learning all too well the salty taste of his tears. Only after the homesickness had eventually "sucked empty his soul," was he able to return to the world.[23] But life did not get any easier. He was tormented by the unreciprocated love for Luise, a one-eyed girl whose dead eye seemed to cast "its shadow over the world around her" and almost killed himself by experimenting with some of the substances available to him in the pharmacy, an episode that may well be read as a suicide attempt.[24]

But Ratzel was dedicated to his dream of contributing to the world of science, and he spent his evenings, when not engaging in alchemy or intoxication, reading and studying for his *Abitur*. In the summer of 1866, he finally convinced his parents, who had shown little interest in him since he had left their home, to matriculate at his local university in Karlsruhe and thus begin his training in zoology, geology, and comparative anatomy. Swiftly, he moved on to Heidelberg, Jena, and Berlin. He finished his studies in 1868 with a doctoral dissertation on earthworms, titled "Anatomical and Physiological Observations of Oligochaeta," which he promptly published in the form of three articles.[25] Under the influence of Charles Darwin and Ernst Haeckel, he also wrote a popular science book, which came out in 1869 as *Sein und Werden in der organischen Welt* (Being and becoming in the organic world).[26] Although he later came to recog-

Figure 2 and 3. Friedrich Ratzel around 1870 and 1900. (Source: Leibniz-Institut für Länderkunde.)

nize that his book was overly indebted to his masters, and even admitted this embarrassedly in a letter to Haeckel himself, the book did contain, in embryonic form, many of the themes and interests that would come to dominate his thinking in later years.[27] Following the completion of his doctoral studies, Ratzel traveled to southern France to conduct fieldwork on annelids. When his most valued possession, a microscope, was stolen and his funds dried up, he approached the *Kölnische Zeitung* and was offered a position as a travel writer.[28] On commission from this major national daily newspaper, Ratzel traveled to Italy. He had begun to learn Malay in preparation for a journey to East Asia when, in the summer of 1870, the sound of an imminent war intervened.

After his unit, the 6th Baden infantry regiment (1st battalion, 4th company), had been fully mobilized, Ratzel was in August 1870 deployed in the siege of Strasbourg. The German military leadership had initially decided to shell the city and thus strike the morale of the civilian population. Strasbourg was badly prepared for any kind of siege, let alone a four-day bombardment with explosive and incendiary shells. Ratzel's unit acted as a forward post in case of a breakout of French forces. In the final stages of the siege, Ratzel fought in a suburb of Strasbourg and was involved in a skirmish that cost the life of a young girl who was caught by a stray bullet.[29] Deviating on this occasion sharply from the otherwise dry tone, his regiment's war diary would later note the following on the 1st battalion's experience of the bombardment:

> From here one could see the shells making their way in a high arc, through a fiery tail in the night sky and into the unfortunate city. With a great roar and clatter one heard the projectiles crash into the roofs and then explode in dull crashes. The residents wailed and cried for help, the garrison and the fire brigade shouted their commands. All of this was clearly audible whenever a new fire broke out. The bells rang incessantly from the brightly lit cathedral which rose high above the sea of flames. The fortress returned only weak fire. On this occasion the regiment's young soldiers became familiar for the first time with the terrible severity of a military campaign and witnessed the work of destruction necessitated by the iron laws of war.[30]

Inflicting urban destruction in this way was in 1870, as Howard notes, "a comparatively novel expedient," made possible by the advances in artillery technology, the very cannons Ratzel's unit had wheeled through the night.[31] Although many civilians were killed or made homeless and Strasbourg's most notable public buildings were left in ruins (fig. 4), the campaign was unable to break the civilian spirit and the German command moved to a more conventional siege.

Ratzel's fellow combatant Karl Stieler, a poet to whose war memoirs the Leipzig geographer contributed the foreword, was unambiguous about the traumatic effect the bombardment had had on him. But while Stieler spoke of a "bacchanalia of war" that left his "heart bleeding,"[32] Ratzel had surprisingly little to say about the destruction of Strasbourg, though he did note that burned-out houses had grinned at him like skulls.[33] Ratzel had volunteered to go to war against a country in which he had only recently traveled and about which he had written with great admiration.[34] But neither his appreciation of French culture nor the experience of losing comrades on the battlefield could dampen his enthusiasm for the war. Ratzel would conclude that war between rival states was inevitable and that his fallen comrades had lived on in him.

He enjoyed the self-discipline that long marches taught him and noted the "love" that soldiers gave one another. "How great was our first joy at the subordination into the covenant of soldiers!" he exclaimed in his war memoirs.[35] He felt attracted to other soldiers, describing one as having a "soft, friendly mouth" while also observing that military leaders were "often more beautiful in old age."[36] But most of all, he framed his war experience through a naïve nationalism that was itself infused with the language of botany and horticulture that he came to apply, in an increasingly indiscriminate fashion, to most phenomena. And so he wrote about the war of 1870–1871:

Figure 4. Strasbourg, 28 September 1870. Wiki commons

> In reality, every one of us is a tree within the forest of his *Volk*; this is how it is, will be and must be. We live with it, we die with it, we harvest the fruits of its victories and pay for its mistakes, when exuberance and recklessness bring it to fall. Today, I can feel how we were intoxicated, and how all of us, people of this *Volk* who felt unique and lonely, rustled along [*rauschten mit*], just as a neighbour rakes his leaves.[37]

Rather than focusing on the war's violence and destruction or indeed new technologies that had come to redefine the conduct of war, Ratzel thus chose to romanticize the experience, playing here on the words *rauschen* (to rustle) and *Rausch* (rush, intoxication), a move that brought out not merely his nationalism but the immersive dimension of war, the way he believed he had become but a part of a larger natural spectacle.

It was in Auxonne, near Dijon, that Ratzel was injured in November 1870 by a shot through his ear. In the days preceding his injury, his company had been patrolling villages, disarming villagers and reconnoitering the fortress of Auxonne. On the 14th of November, his unit was ordered to act as a forward post. When night fell, Ratzel and his comrades launched a surprise attack on a position from which they had repeatedly come under fire. They found it empty but were subsequently attacked from a road embankment, resulting in the "heavy injury of the musketeer Ratzel, who had already been distinguished around Strasbourg," as the regiment's war diary would note.[38] He was one of ninety-two soldiers in his regiment who were decorated with an Iron Cross second class.[39] His hearing never fully recovered.

By the time of his injury, German armies had won the major battles of Metz and Sedan, but Paris had yet to fall. Ratzel was forced to recover for five weeks in military hospitals in Lunéville and Dijon. As a trained pharmacist, he was asked to aid the nurses and doctors in treating the wounded. He recounted how he had awoken feverish in the midst of a gruesome scene of "flesh and blood, of bleeding hands cutting into other human beings, which lay pale on long tables," and which reminded him more of an abattoir than a hospital.[40] Indeed, the first object that Ratzel had glimpsed was a sawed-off human leg, placed in a bucket. He claimed to have envied the person who had lost this leg, for that soldier would have undoubtedly been given a warm bed.

Only in April 1871 was Ratzel able to resume his studies in Munich, a city whose tranquil beer gardens he adored.[41] It was at this time that he developed a close relationship with his later mentor Moritz Wagner (1813–1887), an eccentric explorer, geologist, paleontologist, and evolutionary theorist.[42] He also found the time to translate Thomas Henry Huxley, one of Charles Darwin's loudest advocates, into German.[43] Ratzel continued his travel writing for the *Kölnische Zeitung*, journeying the Austro-Hungarian Empire, Southern Italy and Sicily, before embarking, upon Wagner's suggestion, on a transatlantic voyage in 1873.[44] The journey to the United States was a considerable step up from his previous ones, as expressed not least in the newspaper's decision to grant Ratzel free rein over his route, major stops of which would include New York, Boston, Washington, Richmond, New Orleans, Chicago, Denver, San Francisco, Acapulco, Morelia, Mexico City, Veracruz, Oaxaca, and Puebla. More than Mexico, the United States left a lasting impression on him. Comparing Europe's cities to their North American counterparts, he noted:

> What a difference all this is from the enterprising life style one finds here, where modern civilization grows right out of nature like seeds in virgin soil, where it does not have to force itself with difficulty through the residue of earlier stages of development, and, restrained in a thousand ways, be satisfied with a compromised situation, with a mixture of the old and the new![45]

When he returned to Germany in 1875, he brought with him a sustained enthusiasm for the United States. The outburst of sociopolitical writing provoked by his American journey would prompt the geographer Carl Sauer to conclude a century later that it was North America that had finally made Ratzel a human geographer.[46] But the journey had clearly awakened him politically too. "I believe that I learned in America a lesson which I am permitted to apply to

Europe: The art of politics consists, to a very large extent, in leading political conflicts out of narrow spaces, where they eat away like ulcers" the Leipzig geographer would write decades later.[47]

It was on the eve of his American journey that Ratzel met his future wife, Marie Wingens (1843–1926).[48] Born in Paris into a German merchant family from Frankfurt, Wingens was financially independent and well-traveled. In 1873, she was living with relatives in London after the death of both of her parents.[49] Ratzel's letters to Wingens from the mid-1870s reveal him to be lovestruck but anxious that his admiration was not being returned. Indeed, he complained on a number of occasions that she did not write back nearly as frequently as he would have liked her to.[50] Once he did receive an answer, he vowed overeagerly to write as often as he could.[51] But it was not just the infrequent contact that made him feel uncomfortable. He had experienced much the same from his parents after leaving home.[52] In fact, he seemed to be distressed by Wingens's independence, expressed in her love of tourism and hiking. In response, he stated his desire for her "to adopt a plant-like quiet life, to take root and cocoon."[53] There it was, the language of botany and horticulture, used so often by Ratzel to categorize and discipline the world around him.

Although Ratzel's insecurities had persisted and led him in 1876 to pursue friendship instead of marriage,[54] the couple did eventually marry in 1877. They bought a small apartment in Munich near the university, where their two daughters Hedwig (1879–1965) and Lila (1881–1964) were born. While one biographer describes their marriage as perfect,[55] it is in fact unclear whether Marie or his work were the true object of Ratzel's love when they met. In the same diary entry in which Ratzel discussed his hope that Marie would finally agree to marry him, he also considered his writing, switching back and forth between the two topics, as if they were connected.[56] He even reused passages from love letters to Marie in some of his published writing. In later years, Marie did not always follow him to Leipzig during term time but preferred to stay in their Bavarian home to look after their children. His letters became increasingly practical, covering his health, common friends, their children, and the garden. When the family was united, Ratzel seemed to spend more time with his daughters than with Marie. He repeatedly noted in his journal that Marie was either unwell or melancholic.[57] In a letter to his daughter Hedwig, Ratzel complained that Marie found it difficult to enjoy life.[58] Marie would survive him by twenty-two years, continuing to correspond with many of his academic peers and assisting with the publication of various posthumous publications.

Academic Career (1876–1904)

In the summer of 1875, Ratzel decided to end his spell as a travel writer and become an academic. The reasons for this change of heart are not entirely clear, but they are likely to have been rooted in the enjoyment he received from jotting down his observations of American social life. In the same year in which he published his first study on the United States, he also completed, in astonishing speed, a habilitation on Chinese emigration at the *Polytechnische Schule München*.[59] Based entirely on secondary sources, the book showed some albeit descriptive interest in the horrors of forced migration and slavery, or what he referred to as the "coolie trade." Its main argument, however, was that Chinese migration to the United States, Australia, and elsewhere since the 1820s was in fact the epiphenomenon of a drive to colonize (*Colonisationstrieb*).[60] It was at Munich that Ratzel accepted his first academic post as a lecturer in geography, perhaps a natural fit for someone who saw himself as something of a polymath. In December 1876, he was offered a nontenured professorship and, four years later, a tenured chair.[61]

Ratzel spent much of his Munich years grappling with questions of ethnology and laying the foundation for his human geography. From 1882 to 1884, he served as the editor of the weekly ethnological journal *Das Ausland* (Foreign countries) but left the role when he felt overworked.[62] In 1886, Ratzel departed Munich to take up the post of director of the Institute of Geography at Leipzig, then known, after Berlin, as Germany's second most prestigious university.[63] Ratzel entered a city with an existing infrastructure of colonial societies, ethnological museums, and *Völkerschauen*, ethnological exhibitions or human zoos. The institute, which expanded considerably under his directorship, also offered a significant step up in financial terms, raising his salary from 5,400 to 9,000 Reichsmark. Ratzel entered the most prolific period of his life, publishing in particular the crucial second volume of *Anthropogeographie* (1891) as well as *Politische Geographie* (1897).[64] By the end of the century, he had become the most well-known of a small group of German geographers who were known beyond Germany's borders.[65] He was made a member of the Leopoldina, the German National Academy of Natural Sciences, as well as an honorary member of various geographical and anthropological institutions, including the Scottish Geographical Society in 1887 and the Royal Geographical Society in 1898. He was also appointed advisor (*Geheimer Hofrat*) to the king of Saxony. And yet the move to Leipzig was one he would come to regret, as letters to his publisher reveal.[66]

In the second half of the 1880s, Ratzel became a key member of what has come to be known as the "Leipzig circle" (*Leipziger Zirkel*), an informal gathering of like-minded Leipzig-based professors who shared a pronounced nationalism, a collective aversion to the fragmentation of the German academy into disciplines and subdisciplines, and an admiration of the Leipzig philosopher and psychologist Gustav Fechner (1801–1887).[67] Fechner had tried to combine a Romantic belief in the vitality of the universe with a nomothetic and empirical approach to science, a fusion that would become formative on Ratzel's later work. Hoping to build at Leipzig such a unified science that would span most areas of academic inquiry, the group included, at its core, the founder of experimental psychology Wilhelm Wundt (1832–1920), the historian Karl Lamprecht (1856–1915), the chemist and later Nobel laureate and president of the Monist League (*Monistenbund*) Wilhelm Ostwald (1853–1932), as well as Ratzel.[68] Although the circle's collaborative outputs were comparatively few, its common framework found its expression in Lamprecht's project of a universal history, Wundt's *Völkerpsychologie* (cultural psychology), and indeed in Ratzel's *Lebensraum* thesis.[69] The exact nature of the circle's conversations is impossible to reconstruct, but it is clear from Ratzel's diaries that he met the circle's main protagonists regularly. The interactions with Lamprecht and Wundt in particular soon rubbed off on his thinking.[70] Indeed, Ratzel noted in his journal that Wundt had assisted him in working through the methodology of his *Politische Geographie*.[71]

When not busy writing or conversing with the other members of the circle, Ratzel lectured on a wide variety of geographical and ethnological topics. Surviving lecture manuscripts are palimpsests of notes and newspaper articles, stuck together and scribbled over. They cover the regional geography of the entire globe, including the polar regions, as well as political geography, ethnography, various topics of physical geography, transport geography, the history of his discipline, travel literature, and natural philosophy. He published on each of these topics, often drawing directly from his lectures. Despite his refusal to take his students on field trips and his reputation for a monotonous delivery, which had been pointed out to him by his mentor Moritz Wagner in 1875,[72] Ratzel was a popular lecturer who sometimes used a sciopticon or appeared with a flower in his buttonhole to lighten the mood.[73] By the early 1900s, his lectures were attracting as many as three hundred students and were known as the most popular in Leipzig, drawing in a following among North American exchange students.[74] Ratzel was clearly proud of these numbers and boasted about them to his daughters.[75] While he did not write schoolbooks, he hoped that his popular science books would have an impact on school education too. Many of his students went on to become geography teachers. Ratzel was known to warn them not to rely too much on maps, which he believed struggled to capture the complex relationships between nature and society that his writings sought to unravel.[76]

Although Ratzel was often found in the departmental seminar during term time and had a wide circle of friends and colleagues, he retained a certain shyness and dislike of formal social settings.[77] In 1876, he noted that he had tried to impress the guests of a dinner party at his publisher's house with his cultural capital in order to cover up his feeling of intimidation.[78] He sometimes mentioned with some relief that he had skipped a committee meeting and repeatedly remarked on his aversion to academic conferences.[79] In a 1903 letter to his student Kurt Hassert, he lamented that his "dislike of conferences," in this case the annual convention of German geographers (*Geographentag*), had made him "such a burden on others" that he felt the strong urge to stay away, favoring instead the serenity of his summer residence on Lake Starnberg.[80] It was not that he did not take an interest in such events, just that he preferred his conflicts with other scholars to be fought out by some of his disciples.[81]

By the turn of the century, Ratzel seems to have become increasingly unhappy in Leipzig. He spoke not just of his desire to leave the city, but even to abandon academia in the hope of focusing entirely on his writing.[82] Not only was he condescending about some of his own doctoral students and his colleagues in the *Geographisches Seminar* ("lame people"), but he had become increasingly short-tempered at work.[83] He would break out into sarcasm, reprimand or rage over "some stupid answer from an ignorant scholar or about a disregard of the institute's regulations," reactions that his former student Weule speculated were rooted in a deeper frustration over his own failings.[84] Indeed, Ratzel harbored serious insecurities about his published work and doubted even his major books, including *Politische Geographie*. Although he was aware of the impact the book was having by the turn of the century,[85] he described it in private as opaque and a "formal failure."[86] Some of these insecurities may have originated in the more general eclipse of Darwinism (and its Haeckelian offshoot) at the turn of the century, a paradigmatic crisis that had prompted Ratzel to reach out in writing, though only anonymously, to some of Darwin's staunchest critics.[87]

Ratzel's most important pastime was clearly his writing. Indeed, he noted how his occasional inability to find time for pen and paper made him anxious.[88] When courting his future wife Marie, he quoted Thomas Carlyle in 1874 to warn her that only "a man who had found his work" could be truly happy.[89] After his death, Marie wrote, possibly with a sigh, that her husband had "dedicated his entire life to work and science."[90] Although Ratzel sought, from the

1880s onwards, to prioritize academic writing as an outlet for his incessant compulsion to write, this does not seem to have been sufficient.[91] He also continued to pen a large number of commentaries, the majority of which appeared anonymously in *Die Grenzboten* (literal translation: The border guards), a weekly national-liberal magazine. Frequently placed in a column called *Maßgebliches und Unmaßgebliches* (Significant and insignificant), his opinion pieces often covered recent events of national and global significance and were written with an intense passion for German colonization.

Ratzel also authored short outbursts about daily life that had a noticeably conservative, bourgeois, and moralistic undertone. He complained, for instance, that his fellow Germans had stopped wearing folkloric costumes, that they were overusing the word "brutal," or that members of the German upper classes were endangering children and the elderly by equipping their coaches with rubber tires.[92] A frequent commuter between Lake Starnberg, Munich, and Leipzig, he was disgusted by the way in which railway station pubs across the empire had become, or so he felt, a refuge for alcoholics with "tired, tobacco coloured, grey faces and goggling eyes." "All those for whom the aesthetic education of humanity is not just an empty catchphrase, should protest against this excess," he exclaimed.[93]

Unlike his brother, Ratzel does not seem to have experienced psychotic episodes. Rather, he seems to have carried an obsessional trait in him. Not only was he invaded by the aforementioned legions of letters, but his writing also exhibited crucial intellectual fixations, most importantly his convictions that there was no firm boundary between life and death and that Germany needed to expand spatially. There was also something ritualistic about writing itself. Ratzel wrote ceaselessly, and he routinely documented his progress in his journal, itself a further outlet for this compulsion to use pen and paper.[94] His drive to produce a constant stream of publications frequently drove him to submit manuscripts that were badly edited and alternated between different lines of argument. A difficulty in settling on a position, of qualifying often quite crude statements almost immediately, would later result in different readers invoking him in diametrically opposed ways. In other realms of his life, we can find hesitation, ambivalence, and procrastination. He overthought his relationship for many years, never quite found the courage to leave his academic post, and preferred to engage in academic debate via intermediaries. Later in life, Ratzel reflected on his scholarly achievements from a quasi-posthumous position. It is perhaps no surprise that he was attracted to cemeteries and ruins.

Although a keen hiker, Ratzel did not lead the healthiest of lives. His diet included strong cigars (*Virginiazigarren*), plenty of coffee, cured meats as well as the occasional breakfast of chocolate and beer, all of which he documented in his diary. When he was not hiking in the Alps (and sometimes even when he was), he was reading and taking notes, and he operated seven days a week, sometimes including Christmas. Still working on different manuscripts simultaneously almost every day by the end of his life, he suffered from insomnia, which he blamed on alcohol and intellectual overexertion.[95] His diary entries from the late 1890s and early 1900s repeatedly mention heart palpitations, a persistent cough, and later a diagnosis of asthma. In February 1904 he suffered a heart attack that briefly robbed him of his eyesight. Ratzel's last entry, on August 9, 1904, was short. It mentioned that there had been steady rain and that he had eaten gooseberry pudding for lunch. Unimpressed by a book about Italy, he had decided to leave the house for an evening stroll.[96] He was found sunk to the ground by passers-by who tried, unsuccessfully, to resuscitate him. His body was loaded onto a passing coach and returned to his home. Ratzel was later declared to have died from sudden cardiac arrest.[97] Obituaries in various newspapers and academic journals soon flooded in. A *Festschrift* to Ratzel's sixtieth birthday was swiftly rebranded as "an offering to the deceased."[98]

Major Works

In making sense of Ratzel's larger oeuvre, it is useful to follow Lamprecht and Steinmetzler in distinguishing between a zoological period, roughly from the mid-1860s until the mid-1870s, and a longer geographical period, which began in the mid-1870s and lasted until his death in 1904.[99] While the zoological period also overlapped with his travel writing, the geographical period saw him publish extensively on political and colonial questions. In the zoological period, Ratzel had familiarized himself with the philosophers of antiquity, although he remained almost untouched by medieval thought.[100] His interest in philosophy grew in the 1880s, without ever displacing the geographical thrust of his work. The final years saw a renewed interest in biogeography, and thus zoology. In February 1897, he noted in his journal that biogeography was going to be "the field on which I will harvest many a beautiful crop."[101] He daydreamt about writing an update to Alexander von Humboldt's *Cosmos*, a grand geography of eight to nine volumes "which would sell at least 10,000 copies per year."[102]

We have already noted the vast nature of Ratzel's corpus of published work. Examining only his books, it becomes apparent just how many genres and (sub)disciplines it spans. Aside from his most notable *Anthropogeographie* (Human geography, 1882 and 1891) and *Politische Geographie* (Political geography, 1897), it includes three works of popular science, *Sein und Werden in der organischen Welt* (Being and becoming in the organic world, 1869), *Deutschland: Einführung in die Heimatkunde* (Germany: Introduction to local history and geography, 1898), and *Das Meer als Quelle der Volksgröße* (The sea as a source of national greatness, 1900).[103] There are five works of travel writing and regional geography, *Wandertage eines Naturforschers* (Journeys of a natural scientist, 1873 and 1874), *Städte- und Culturbilder aus Nordamerika* (Sketches of urban and cultural life in North America, 1876), *Die Chinesische Auswanderung* (Chinese emigration, 1876), *Aus Mexico* (From Mexico, 1878) and *Die Vereinigten Staaten von Nordamerika* (The United States of America, 1878 and 1880), as well as a number of books that have an encyclopedic style but that his contemporaries nonetheless often listed among his major works: *Die Vorgeschichte des europäischen Menschen* (The prehistory of European man, 1874), *Die Erde* (The earth, 1881), *Völkerkunde* (Ethnology [English title: The History of Mankind], 1885, 1886, and 1888) and *Die Erde und das Leben* (The earth and life, 1902).[104] Ratzel also wrote a book on landscape description, *Über Naturschilderung* (The description of nature, 1904), and edited a number of collections, including one on hiking and another on the explorer Emin Pascha.[105] His memoirs were published posthumously as *Glücksinseln und Träume* but had in part already appeared in *Die Grenzboten* shortly before his death. Ratzel had a tendency to either write long books or short commentaries and book reviews. He did, however, occasionally publish longer essays, the most important of which were the 1896 "Die Gesetze des räumlichen Wachstums der Staaten" (The territorial growth of states), the 1901 "Lebensraum: Eine biogeographische Studie" (Lebensraum—a biogeographical study) and the 1904 "Nationalitäten und Rassen" (Nationalities and races).[106]

Given that the majority of his major publications were works of human geography, it is perhaps no surprise that a 1904 obituary in *Geographical Journal* mentioned his physical geography only in passing.[107] And yet Ratzel contributed to debates in physical geography in the 1880s and 1890s with work on fjords and lakes. His love for the Alps also prompted him to publish some original work on glacial drift and the snow line, and he was known too for his critique of the Kant-Laplace nebular hypothesis on the emergence of the solar system.[108] Toward the very end of his life, Ratzel returned to biogeography as he sought to develop a monist theory of all life, which rendered the human a mere effect of the natural world. Although he dealt with questions of physical geography in his two most important works, *Anthropogeographie* and *Poli-*

tische Geographie, both books treated physical geography as auxiliary to his main lines of inquiry, which were historical and political in nature.

Published nine years apart (1882, 1891), the two-volume *Anthropogeographie* sought to map the distribution of human habitation across the globe and explain regional differentiation in social organization by reference to geology and physical geography. Over almost 1,300 pages, Ratzel argued with the help of a vast array of examples that beneath the diverse and exuberant appearance of human culture lay a geographical bedrock, which had slowly and subtly shaped the way history had unfolded. Over and over again, he insisted that all life, human and nonhuman, was earthbound in that it depended on the earth for resources and was molded by climatic conditions. Ratzel wanted to offer a broad and synthetic perspective he felt other disciplines lacked. This necessitated a view, inspired by Humboldt, Herder, and Ritter, that treated both life (*halogäische Lebensauffassung*) and the earth (*halogäische Erdauffassung*) as a whole and inextricably intertwined.

> The coconut tree presupposes the coral reef, the blackbird requires the wood, the cricket presumes the meadow. Higher forms of organic life rest on a mighty base of lower organisms. The geography of our planet has most powerfully been shaped through this base, be it through vast dead deposits, such as turf or coal, or living foundations that are reefs and forests. These supplies are stored on the earth's surface, lie hard beneath it or conceal its inorganic core[109]

This was not just a totalizing view, but one that demanded a leap of faith, which was to accept that humanity was "a piece of earth" (*Die Menschheit ist ein Stück Erde*).[110] Human societies, in all their cultural diversity, were from this perspective shaped as directly by these earthly foundation as were blackbirds and crickets. In earlier years, he had often made arguments about the relationship between environment and organism in a light-hearted fashion. He had commented, for instance, that the climate in North America was "of itself so stimulating and invigorating that it probably makes alcoholic stimulants up to a certain degree unnecessary."[111] But in *Anthropogeographie* his tone changed. He now acknowledged that there was "something mysterious and sometimes frightening" in "the power of the soil," which "emerged from great depth through layers of history into contemporary life in all its richness."[112] Ratzel thus developed an approach that was interested in the limits that this earthly power placed on human agency, its ability "to annihilate the seeming freedom of man."

The book drew directly on the work he had done in his earlier *Völkerkunde* (more on which in chapter 3), paying close attention to the diffusion of socio-economic practices via trade and migration. Just as in his three-volume ethnology, he examined technologies such as bow and arrow, the plough, or the throwing knife, and he gave detailed descriptions of various customs, rituals, and superstitions. But the focus was now on the underlying material circumstances that made these artifacts and beliefs possible. Unlike *Völkerkunde*, *Anthropogeographie* was not organized according to world regions but themes. Volume 1 was titled "Fundamentals of the Application of Geography to History," while volume 2 was addressed "The Geographical Distribution of Man." They included sections on geomorphology, climate, habitation, population density, and cultural landscape, among others. The book can be read as an attempt "to overcome the Kantian distinction between history (as narration) and geography (as spatial ordering) and between geography (as the outer world of objective material conditions) and anthropology (as the inner world of subjectivities)."[113] In trying to bridge these chasms, Ratzel drew on a vast array of thinkers, from Thucydides to Machiavelli, Montesquieu, Herder, Spencer, and Darwin, as well as, most prominently, Carl Ritter. The book was popular and went through three further editions, which featured an expanded prehistory of environmental thought from ancient Greece to the late nineteenth century.[114]

Perhaps surprisingly, the style in which *Anthropogeographie* developed its arguments was rarely sweeping but delivered with nuances, caveats, and counterarguments, which sometimes made it difficult to discern exactly where Ratzel stood. He struggled to put his finger on the exact contribution his book was trying to make.[115] Only in an 1889 anonymous review article did he manage to narrow it down to the "idea of the historical development of humanity as dependent upon the natural conditions of the soil, climate as well as the animal and plant kingdoms," describing it as an approach that "searches for the limits of the demonstrability of this dependence."[116] Ratzel was quick to concede the "tremendous difficulties" of pursuing such a task, given the need "to encircle the entire earth without losing sight of the multifaceted nature of human history."[117] Despite these problems and the philosophical nature of some sections, the main body of the text was remarkably, as it were, down to earth. Ratzel conceived of his anthropogeography as a "descriptive science."[118] Good description, he believed, was not its only aim, but the most important step toward deeper knowledge. The book could thus also be read as a world history that covered different categories of *Völker* under the influence of environmental factors.

Anthropogeographie was a book about the earthly conditions of social organisms, the most important of which were *Völker*, states, and races, although there was a lack of clarity as to how these three related to one another.[119] *Naturvölker* (natural peoples) had to obey nature's forces more immediately (*sie leben unter dem Naturzwang*), but industrialized and "civilized" peoples (*Kulturvölker*), too, were dependent on natural resources, such as coal and iron, and thus on the soil.[120] The environment (*Umwelt*), understood as nature external to the organism that extended into this organism through light, air, and sustenance, either sped up or slowed down the movement of these *Völker* and variously promoted their survival, expansion, retreat, and demise.[121] The *Umwelt* acted directly onto the human body and mind, forming particular national and racial characters. More indirectly, it also shaped over time the material nature of the food, clothing, and shelter that was available to a particular *Volk*, race, or state and thus subtly formed the ideas and beliefs that were common in each of these social organisms. A *Volk* that had carved out its space in a polar region had a very different geographical imagination and spatial awareness from one located in the tropics. But climate was not the only decisive factor. Within each climatic region, the uneven distribution of particular natural resources could lead to significant differences in the technological and spiritual makeup of different tribes and *Völker*, an argument indebted to Darwin's notion of variability. Nonetheless, Ratzel concluded, climate acted to produce certain regularities. The trade winds caused nomadism, while "civilization" was more likely to emerge in temperate climates.[122] Coastlines, mountain ranges, and access to oceanic trade all impacted on national character. The human organism was stimulated by the monsoon and wilted in the tropics. "It is an old observation," he claimed, "that humans rarely become of a higher nature where nature is most generous."[123]

A second set of arguments related to migration. Ratzel understood all life as movement and all movement as earthbound. The key manifestation in humanity of earthbound movement was migration, whether in the form of traffic, colonization, or war. "The effects of nature on man," he explained, "migrate with man, and *Völker* which have been subject to such effects, carry them over large distances."[124] In some ways, this argument was almost mechanical. "Where space is available in abundance, *Völker* will pour into these openings like a liquid until they reach an obstacle," thus spreading particular traits and practices as they disperse into space.[125] But all human movement was also driven by the aim of reuniting different parts of the social organism, an idea he had taken from his mentor Moritz Wagner, to whom he dedicated the book. Rather than locating the determining principle that prompted life to evolve only in Darwin's natural selection (*Zuchtwahl*), he found it in migration (*Wanderung*) and separation (*Absonderung*). And while migration was rarely done consciously or strategically, it was the vital process by which humans were able to evade unfavourable climatic conditions. This insistence on the incessant migratory movement of humanity made Ratzel a critic of autochthony, the belief that particular regions had original inhabitants.[126]

These ideas were further developed and more explicitly politicized in his 1897 *Politische Geographie*. The term "political geography" had already been in use long before Ratzel adopted it and had been employed, for instance, by Carl Ritter, one of Ratzel's most important sources of inspiration.[127] And yet it had mainly denoted the descriptive and often dry reproduction of statistics about states and their societies. Ratzel's political geography, on the other hand, was driven by a practical ambition, to cultivate a "drive to expand, ability to colonize and innate spirit to dominate," which could already be found in many of history's greatest leaders. The book was thus an attempt to codify this "political instinct," the essence of which was "an appreciation of the geographical foundations of political power."[128] Ratzel's lecture series "General Political Geography and Political Ethnography," delivered in Leipzig from 1890 to 1894, had included a section on statecraft in which he noted the potential of his approach to predict future events in domestic and world politics.[129] Among *Politische Geographie*'s lasting contributions was not just the spawning of a new subfield that had as its subject the territorial politics of states, but the creation in its readership of a feeling that this book yielded, despite the density of its prose, useful lessons for political and particularly colonial praxis.[130]

Much like his earlier work, the book rested on environmentalist and evolutionary principles, although the latter were not discussed in significant detail.[131] The book's core argument was that existing political theory (*Staatslehre*) had lacked an understanding of the environmental and spatial conditions of statehood. In order to explain the evolution of the state, humanity's "greatest achievement," it was necessary to understand it not in juridical and institutional ways, but as an organism that was rooted in, and thus inseparable from, the soil on which it stood (*bodenständiger Organismus*).[132] Ratzel acknowledged both that he was not the first to use the organismic analogy and that social life was too complex to be collapsed neatly into the categories of biological life.[133] And yet he argued that in the same way as the divine spirit of humanity was not called into question by the observation that humans were organic beings, so did the complexities of socioeconomic organization not invalidate the view that states behaved like organisms.[134]

> If we draw the map of a negro state, then we will observe the simple image of an elementary organism: the village of the chief in the center, surrounded by small villages in gardens and fields and beyond that a liminal wildness, through which a path or two lead to neighboring areas. What a difference this is from the crowded general map of even an entirely insignificant European state with its small and large settlements, its border and capital cities, fortresses, its road, canal, and railway networks![135]

Ratzel illustrated this crude comparison with two maps, the first depicting villages and their boundaries in the Congo basin (figure 5), the second and the "area of chief Mtemi in Nyamwezi," modern-day Tanzania (figure 6). Ratzel neither discussed the maps nor the political formations they claimed to depict in any detail. What mattered was that the village communities *appeared* to have a unicellular appearance. This then performed a reversal of the earlier import of social language (e.g. "colonists," "citizens," "republics," "division of labor") into the study of cellular life in the first half of the nineteenth century.[136] The fact that the two examples were taken from the African continent was not insignificant. Ratzel believed that humanity had a single origin in Africa and had subsequently spread across the globe. He thus implied that more primitive forms of political organization had been preserved at earlier stages of history, themselves located in particular world regions. Africa's present, in other words, was Europe's past.

Despite his reservations about the organismic analogy, it drove a vast array of the book's more capillary arguments about recent European and world history. Ratzel needed his metaphor both to perform the translation of biological into political theory and to draw out the environmental conditions under which he believed political history to unfold. Over the course of seven hundred pages, Ratzel tried to provide a typology of ways in which the evolution of this social organism could be read in space. After some broad-brush strokes in the earlier chapters, the evidence was presented in the later chapters. He must have hoped that the sheer volume of case studies and illustrations would extinguish any doubt his readers might have entertained. The revised second edition of 1903 was divided into twenty-four chapters, which were organized into nine larger sections. While the first section covered the relationship of state and soil, the second and third sections addressed historical change and the growth of states. Sections 4 and 5 examined questions of geographical location and the impact of narrow and wide space, population density, and transport. The remainder of the book served the purpose of illustration, covering borders (section 6), the relationship between land and sea (section 7), oceans (section 8),

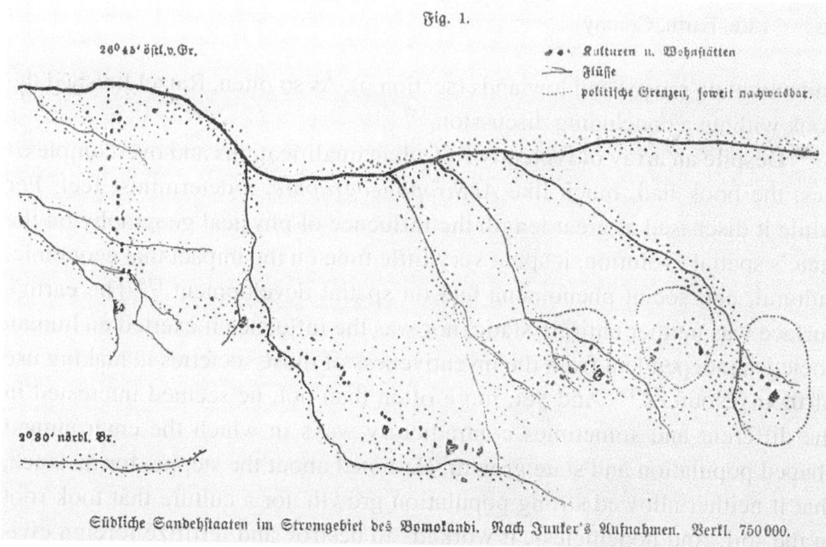

Figure 5. Southern Sandeh States in the Bomokandi River Basin. (Source: *Politische Geographie*.)

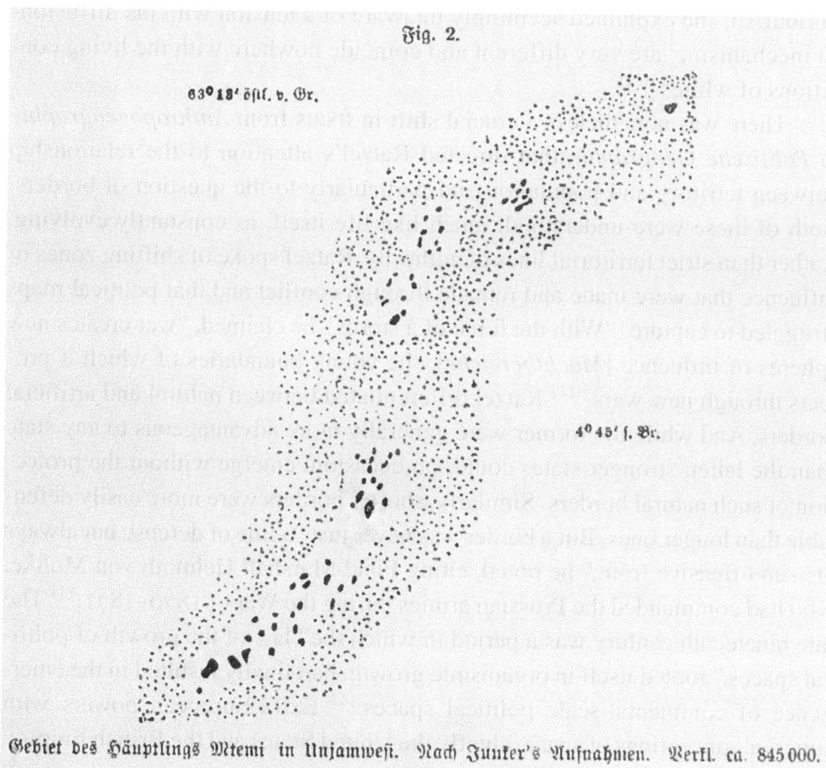

Figure 6. Territory of Chief Mtemi in Unyamwezi

and mountain ranges and lowlands (section 9). As so often, Ratzel finished the book without a concluding discussion.[137]

Despite an array of typically Ratzelian qualifications and overcomplexities, the book had, much like *Anthropogeographie*, a determinist feel. For while it discussed at great length the influence of physical geography on the state's spatial evolution, it spent very little time on the impact that economic, cultural, and social phenomena had on spatial development.[138] The earth's surface was neither entirely static, nor was the influence it exerted on human societies independent from the inventiveness of those societies in making use of their resources.[139] And yet, more often than not, he seemed interested in the different and sometimes contradictory ways in which the environment shaped population and state growth. He noted about the steppe, for instance, that it neither allowed strong population growth nor a culture that took root in the soil. And nonetheless, it worked "to destroy and fertilize foreign civilizations far beyond its area."[140] Ratzel also reasoned that it was no surprise that America's black population could be found in the South, while the Chinese had settled on the Pacific coast. "The conditions for each of these races to flourish," he explained seemingly unaware of a tension with his diffusionist mechanism, "are very different and coincide nowhere with the living conditions of whites."[141]

There was nonetheless a crucial shift in focus from *Anthropogeographie* to *Politische Geographie* that directed Ratzel's attention to the relationship between territory and population and particularly to the question of borders. Both of these were understood, much like life itself, as constantly evolving. Rather than strict territorial lines of authority, Ratzel spoke of shifting zones of influence that were made and remade through conflict and that political maps struggled to capture. "With the force of a storm," he claimed, "war creates new spheres of influence [*Machtbereiche*], the young boundaries of which it protects through new wars."[142] Ratzel differentiated between natural and artificial borders. And while the former were generally more advantageous to any state than the latter, stronger states could, on occasion, emerge without the protection of such natural borders. Similarly, shorter borders were more easily defensible than longer ones. But a border was never just "a line of defense but always also an offensive front," he noted, citing Field Marshall Helmuth von Moltke, who had commanded the Prussian armies during the War of 1870–1871.[143] The late nineteenth century was a period in which the "law of the growth of political spaces," rooted itself in organismic growth, had finally resulted in the emergence of continental-scale political spaces.[144] Extra-European powers with superior conceptions of space, chiefly the United States and the British Empire, had established a "greater-spatial" (*großräumig*) order. Germany, he believed, had to acquire just such a sense of space to compete on the international stage or it would be overrun by the forces of history.

Political Leanings

Ratzel united in him a number of professional identities, but one was curiously absent. Unlike many of the most well-known geopolitical thinkers, Ratzel was not, nor did he wish to be, a politician. He was known to harbor a deep suspicion toward political parties and wanted to be seen as an objective observer who was able to discern where and when the nation's vital needs were in jeopardy.[145] Nonetheless, Ratzel was criticized by his contemporaries precisely for his inability to distinguish between his scholarship and his journalism, for failing to draw a line between science and politics. He was known, moreover, as a loyal follower of the conservative chancellor Otto von Bismarck and "a monarchist through and through."[146] For a while, he was also aligned with the *Nationalliberale Partei* (National Liberal Party).

Ratzel had considerable respect for monarchic and aristocratic tradition. On his journey to the United States in the 1870s, he was drawn to blue-blooded travelers and later tried himself to cultivate something of an aristocratic aura, as some of his peers noted.[147] As a young man, he had felt great admiration for his father's employer, Grand Duke Friedrich I, who reigned over Baden during the second half of the nineteenth century and was known for his national liberal leanings and support for Bismarck's policies.[148] In one of his few unpublished manuscripts, Ratzel stressed the avant-garde nature of support for Bismarck in 1860s southern Germany. Only such visionaries as the Grand Duke and Ratzel himself, he implied, had been able to appreciate the historic quality of Bismarckian statesmanship.[149] And yet the old aristocratic order Ratzel admired was being challenged precisely by the rise of a bourgeois class into which Ratzel would himself ascend. In 1871, the Grand Duchy joined the German Empire.

Like many of his contemporaries, Ratzel saw in Bismarck a brilliant foreign politician. Even in dry encyclopedic entries, he would point out that Germany's status as "Europe's first continental power" had been established not just by Prussia's steady military ascent, but by its "masterly foreign policy."[150] In private too, he noted his admiration for Bismarck's system of military alliances.[151] While Ratzel did on occasion bemoan that the Prussian statesman had not prioritized settler colonialism,[152] he saw in him a "political geographer and philosopher of history" who was second to none in his understanding of Germany's geographical position.[153] Indeed, *Die Grenzboten*, the paper for which Ratzel penned many a column, was occasionally read by Bismarck himself and commonly seen as "a semiofficial organ of the Prussian foreign ministry."[154] Ratzel did not break with the iron chancellor. In a letter to his student Hassert, dating from August 1898, Ratzel spoke appreciatively of a memorial service for Bismarck he had attended.[155]

Whereas a previous generation of German geographers had often still supported German sectionalism, Ratzel was a "glowing German" who "never lost sight of the *Reich* and the bigger picture," as his student Eckert noted in 1922.[156] And although Ratzel was dedicated to Bismarck, he would become critical of those who reduced German unification to the influence of one man. The nation, this *Volk*-as-forest, was more important than the deeds of an individual statesman. Germany could only prove itself internationally, he wrote after Kaiser Wilhelm II's ascent to the throne in 1888 and Bismarck's forced resignation two years later, by "standing and walking without Bismarck."[157] Although Ratzel thus supported a further prioritization of the national interest under the new monarch, his nationalism went considerably deeper than a narrow identification with the Wilhelmine Empire. Indeed, Ratzel recorded his admiration for the famous German historian, nationalist, and antisemite Heinrich von Treitschke, "at whose feet" he had sat and from whom he had learned that "there was something great about war."[158] In 1895, he reread Treitschke, alongside Nietzsche and Spencer, while penning his *Politische Geographie*.[159]

Ratzel's nationalism was not driven by an irrefutable belief in racial superiority, but by a more fragile construction of Germany's place in the world. Although he remained unconvinced that nations constituted a "blood relation" (*Blutsverwandtschaft*), they did have their origin in "tribal consciousness."[160] But what mattered most were the geographical conditions in which a nation found itself. The geography of central Europe (*Mitteleuropa*) had favored the emergence of a significant number of great powers and thus given birth to an unstable balance of power between them. Germany's central position within this assemblage and its lack of space (*Raum*) required it to constantly renew its "sources of power," which included not just its armed forces but also its "cultural superiority."[161] Germany's status as a great power would thus need to be sustained by a combination of political will and geographical understanding.

In April 1891, Ratzel became a founding member of the Pan-German League (*Alldeutscher Verband*, initially *Allgemeiner Deutscher Verband*) and served on the organization's executive board. Although many German geographers were members of the German Colonial Society (*Kolonialverein*),

Ratzel was one of the few who were also active in the league.[162] The Pan-Germans aimed to foster nationalism internally and promote the national interest abroad as well as the interests of German nationals living in other states. The power of the league, whose membership included, for a while at least, the likes of Ernst Haeckel, Karl Lamprecht, and Max Weber, lay not just in the institutionalization of nationalist ideology, but in the informal network that spanned representatives of regional and national parliaments, colonial administrators, and public intellectuals.[163] Despite its struggle to recruit among German bankers and industrialists, the organization had a considerable political influence in the *Reich*.[164] While the Pan-Germans came from a variety of political backgrounds, including national liberals (*Nationalliberale*), free and German conservatives (*Freikonservative* und *Deutschkonservative*), and antisemites, they were united by their opposition to socialism and by their support for Bismarck.[165] In practice, the Pan-German League served to radicalize the nationalism and militarism that had become dominant during the 1870s and 1880s. Its spirit is captured well in an 1890 call entitled "Germany awake!" (*Deutschland, wach' auf!*):

> We are ready to stand in line at the call of the Kaiser and to be led silently and obediently toward enemy bullets, but we can also demand that we receive a price that is worthy of this sacrifice, and that price is: to belong to a master people [*Herrenvolk*] that takes its share of the world for itself and does not seek to receive it at the grace or benevolence of another *Volk*.[166]

The Pan-German League was headed between 1893 and 1908 by Ernst Hasse (1846–1908), a statistician and colleague of Ratzel's at Leipzig. Hasse had in 1879 founded the Leipzig Association for Commercial Geography and Colonial Policy and had in the 1880s become a leading figure in the colonial movement. A strong advocate of naval expansion, Hasse sat in the *Reichstag* from 1893 to 1903 as a National Liberal. His hobby was collecting portraits of Bismarck.[167] The league's "most influential and systematic ideologist" in the early years, Hasse drew directly on Ratzel's political geography and followed the latter in viewing German emigration to North America as a tragic failure.[168] Despite smaller disagreements that they debated in the *Kolonialzeitung*, the two men collaborated in bringing key colonial figures to speak at Leipzig.[169]

Ultimately, the advocacy of the league would bring the movement into conflict with Kaiser Wilhelm II. Its radicalization would continue apace during and after World War I and it would eventually be seen by Hannah Arendt and others as one of the most important ideological precursors of national socialism.[170] Many of the Pan-German movement's ideological constructs and political demands would indeed resurface in the national socialist program, particularly in the form of an expansionist fantasy of colonizing and emptying out territories in Europe and in a *völkisch* nationalist and racialized discourse of national victimhood.[171] In the league's insistence on Germany's need to awaken there was certainly a "palingenetic ultra-nationalist" dimension.[172]

Ratzel would embrace in the 1880s a theory of history that had a distinctly "anti-liberal" tendency, given its preoccupation with the terrestrial limitations of all politics.[173] The very idea of "freedom," which he sometimes placed in quotation marks, had led to "spiritual and physical decline," especially among the colonized; freedom thus came at the expense of others.[174] His major works insisted moreover that geography itself acted to limit the freedom of human societies. Indeed, they sought to displace the primacy of the individual as an analytical or normative category.[175] And yet Ratzel's thinking also expressed "some of the dominant optimism of nineteenth-century liberalism," as Bassin notes.[176] His friend and mentor Moritz Wagner referred to Ratzel as a "liberal," as well as a "patriot."[177] Indeed, Ratzel admired the entrepreneurial spirit of US society, cited the German philosopher Immanuel Kant authoritatively, often thought in terms of civilizational progress, used the terms individualism and cosmopolitanism in the affirmative, criticized overly patriarchal societies, and believed that women should be given jobs in education.[178]

Ratzel was generally charmed by American democracy and thought little of the beer hall strongmen of his time. He was hopeful that democratic societies would increase the morality of different classes, lessening the decadence of the upper classes while working against what he saw as the degeneracy of the lower classes.[179] In the first volume of his *Anthropogeographie*, he noted that "an equal distribution of the soil promotes a democratic society while an unequal distribution tends toward [. . .] oligarchy."[180] But Ratzel was concerned about the ability of democracies to ensure social hierarchy, which he thought was a necessary condition of any healthy society. After arriving in the United States in 1873, he observed that while order was achieved in Europe's monarchies through the privileges of birth and profession (the *Ständegesellschaft*), democracy required significant wealth inequality to re-establish such social hierarchy.[181] Ratzel was uncomfortable too with the inclusion of lower classes and different races into the US democratic process and warned of the apparent damage that democratic ideas were doing to Mexico too.[182] When reflecting on the question of military discipline and order, he noted that "many *Völker* under despotic rule, in which everyone has their place and their duty, were happy, even if a false pride in their freedom did not allow them to admit it."[183] There existed, in other words, an uneasy tension in the early Ratzel between liberal democratic sentiments on the one hand, and a monarchic or authoritarian streak on the other. In the course of his working life, this conflict was gradually resolved in favor of authoritarianism.

A similar tension can be observed in Ratzel's commitment to both cosmopolitanism and nationalism. Ratzel believed, drawing on Moritz Wagner, that the most competitive nations were those that had been subjected to migration. In a biogeographical vein, he wrote with a good degree of admiration about the brown rat that was "so cosmopolitan that it had managed to gain a foothold wherever European boats had passed, even if only fleetingly."[184] He also venerated a "cosmopolitan" explorer like Emin Pascha, to whom we will return in chapter 5, for his "dazzling blend of Judaism, Christianity and Islam."[185] And yet cosmopolitanism should not mean the rule of a "borderless world bourgeoisie, deprived of responsibilities and leanings," he cautioned, but rather a gradual influx of "fresh blood into the arteries of our organism."[186] Conversely, he was concerned, much like Adolf Hitler half a century later, about emigration. While it was beneficial to the United States, he saw the exodus of Germans to North America in the nineteenth century as harmful to Germany. This was precisely where the role of settler colonialism came in. Ratzel viewed Germany's colonial expansion as a further development of the drive for territorial unity under Bismarck and a solution to the problem of population decline through emigration. In 1882, he was one of the founding members of what he called affectionately "our *Kolonialverein*."[187] While the discourses of colonial advocacy in and around the *Kolonialverein* (later *Deutsche Kolonialgesellschaft*) did not directly prompt the early seizure of colonies by German explorers and merchants, they did run in parallel to these annexations, creating opportunities for an ideological mobilization of the middle classes, which Bismarck seized upon.[188] Ratzel used his position to provide the country's most prominent explorers and conquerors, many of whom lacked institutional support and financial stability, with doctoral degrees and speaking opportunities. It was the latter that prompted him to correspond with the likes of Carl Peters, Germany's most famous land grabber in Africa.

From 1884 onwards, Ratzel increasingly used his firsthand experience of settler colonial societies in North America and his theoretical framework to provide a running commentary on current affairs, including in the *Deutsche Kolonialzeitung*, the *Kolonialverein*'s mouthpiece, but more frequently in the *Grenzboten* and other daily newspapers. Indeed, he was known among his peers as an "agitator" and a "pioneer of the colonial idea."[189] Ratzel was not the only geographer with a colonial fetish. His predecessor in Leipzig was Ferdinand von Richthofen (1833–1905), the chairman of the *Deutsche Afrika-Gesellschaft* (the German Society for Africa).[190] The later national socialist Siegfried Passarge (1866–1958) too had led expeditions to Cameroon before being appointed chair of geography in the Hamburg based *Kolonial-Institut*.[191] But even in the context of Imperial Germany, Ratzel was something of a radical. Not only were other European geographers promoting more progressive political agendas in the last quarter of the nineteenth century.[192] But many of Ratzel's German colleagues held decidedly more moderate views than he did. The anthropologists Rudolf Virchow and Adolf Bastian, for instance, supported commercial expansion but saw settler colonialism as aggressive, unnatural, and an obstacle to free trade.[193] Ratzel disagreed strongly and engaged in a dispute with Virchow on the question.[194] As we will see in chapter 5, Ratzel was particularly well-networked within a more conservative segment of the colonial elite, which appreciated his writings as an amalgamation of the theory and practice of colonization.[195] Indeed, Ratzel became in later years something of a chronicler of the empire, a eulogist for colonial administrators and explorers.

In social and economic terms, Ratzel's views were mostly conservative. He was a believer in agrarianism, although this has been overstated in the existing literature. While he admired American cities, he thought that modern urban centers, more generally, had a "tendency for ugliness."[196] He also valued the family unit as a form of sociopolitical organization, seeing it as "to the *Volk* what a cell is to the organism."[197] And although he had questioned the existence of a Christian god in the early 1870s, he visited church regularly in later years, explaining that it had been the experience of battle that had forced his hands together for prayer.[198] Where there is recognition of Ratzel's traditionalism, this is sometimes attributed to his move into the German academy, known at the time for its conservatism.[199] But if it was primarily the university that exerted pressure on him to embrace conservative views, then it is unclear why he often chose to remain anonymous when articulating his most jingoistic opinions. One must conclude that he was well aware that his views on colonization and race, radicalized on his journey to America in the mid-1870s, were not shared by all of his peers in the German academy.

It is instructive to take a close look at where Ratzel stood not just on the German political spectrum but how his views sat in the North American political context of the late nineteenth century, a question to which we will devote much of the following chapter. Ratzel arrived in the United States an abolitionist but soon made the acquaintance of racial theorists, disillusioned Republicans, and white southerners who prompted him to rethink his position on the all-important question of race and slavery. He quickly began to blame the black population for the failures of Reconstruction, arguing that it was a mistake to have extended civil liberties to the formerly enslaved. He also ignored white supremacist violence, which had begun to haunt the United States in the years following the American Civil War. By the time of his return to Europe, Ratzel had himself embraced a white supremacist historical narrative and a segregationist vision for America. Like many of his Darwinian peers, he saw the genocidal violence and subsequent disappearance of indigenous Americans as a natural, unstoppable, and in fact cleansing process.

Ultimately, neither the opposition between liberalism and antiliberalism nor that between progressivism and conservatism are entirely helpful in making sense of Ratzel's politics. As we will see in what follows, there was, beneath his various political commitments and the tensions between them, a persistent interest in death. Ratzel's necropolitics stands precisely at the intersection of these ideological confluences, but it also points unmistakably in the direction of twentieth-century fascism. Judging by his own account, death seems to have followed Ratzel throughout the first half of his life. It was contained in various childhood experiences that featured disappearance, demise, and loss. Like many Romantics, he was mesmerized by ruins. Just as other evolutionary theorists, he was captivated by fossils. Death returned with a vengeance first in his self-induced near-death experience in the pharmacy and then again on the battlefield. Increasingly, death, or rather the absence of a distinction between life and death, became a crucial lens for Ratzel onto nature itself. On his journey through Mexico, he compared mountainous landscapes to stiff corpses, which, when opened up, revealed vibrant streams of blood.[200] In *Anthropogeographie* he spoke dismissively of "the so-called dead earth," a phrase he had taken from Haeckel.[201] In *Glücksinseln und Träume* he reiterated his belief that the boundary between life in death was fluid, this time under the influence of Fechner.[202] Death was not merely an analytical but an aesthetic category for Ratzel. A preoccupation with death was so deeply ingrained in his thinking that he could not but insert it into the very foundations of his political geography.

Chapter 3

Race and Slavery

In July 1873, the twenty-eight-year-old Friedrich Ratzel, now a journalist, boarded a steamboat in Southampton. He arrived in New York twelve days later with a naïve enthusiasm for the "New World." His voyage would last for almost two years, taking him from East to West coast and onwards to Mexico and Cuba, his final destination. Published first in the daily newspaper *Kölnische Zeitung*, his detailed and vivid descriptions quickly found a broad readership among the German Empire's bourgeoisie, whose interest in North America had been awakened not least due to a sustained wave of German emigration across the Atlantic.

Ratzel visited these places during politically turbulent times. His expedition to the United States included a long and formative visit to the southern states, which were in the aftermath of the American Civil War (1861–1865) and the abolition of slavery. While Mexico had witnessed decades of war and sociopolitical unrest, Cuba was in the midst of a war of independence from Spain. Although Ratzel was clearly shaped by the recent experience of serving in the Franco-Prussian War, he was less concerned with armed conflict than with the question of race. Over and over, he depicted in great detail the physiognomy of nonwhites as well as the various social practices and rituals he used to distinguish the status of different races in the United States and Mexico. He may not have framed it in those terms, but he was ultimately trying to grasp the role that race played in the organization of settler colonial societies.[1] An understanding of the question of race in Ratzel is thus critical in getting to grips with his broader political project at this formative stage in his life.

It should be noted at the outset that it is not always straightforward to pinpoint Ratzel's position on race, both because it was subject to change and because it exhibited at every point in time tensions and ambiguities, particularly in relation to the all-important question of slavery. Ratzel stated throughout his life that there was only *one* humanity (*ein Menschheitsgeschlecht*) and was, though decreasingly so, a critic of slavery too. Toward the end of his life, he would argue with some determination against the idea of racial purity. And yet despite his insistence on the unity of humanity, Ratzel's writing displayed a profound tendency to essentialize races and to think in terms of racial taxonomies and hierarchies. Whereas some observers have foregrounded Ratzel's rejection of the biological racism that came to dominate academic debates by the early 1900s,[2] others have argued that there is more continuity between Ratzel's constructions of race and late nineteenth-century racial science.[3] In this view, the idea of race "as the decisive agent in history" which would be radicalized in national socialist discourses of the interbellum, among others, was "already nascent" in Ratzel's work.[4]

Existing debates on this question have paid particular attention to a 1904 essay titled "Nationalitäten und Rassen" (Nationalities and races) in which Ratzel discussed the ideas of Arthur de Gobineau (1816–1882) and Houston Stewart Chamberlain (1855–1927), both of whom were widely read in fin de siècle Germany. Indeed, Gobineau and Chamberlain's biological racism and its fetishization of Aryanism is widely seen to have been formative to the development of national socialism.[5] The Frenchman Gobineau believed in the supremacy of the white race and warned against racial mixing in his 1853 *Essay on the Inequality of the Human Races*, which had been translated into German only in 1897.[6] Chamberlain's 1899 *The Foundations of the Nineteenth Century* pushed these ideas further, constructing history as a perpetual conflict between the Aryan race (white Europeans, spearheaded by the Nordic peoples) on the one hand, and nebulous Semitic forces on the other.[7]

Chamberlain had criticized Ratzel in *The Foundations of the Nineteenth Century* as someone whose work was imprecise and naïve in relation to the historical agency of Jews.[8] Ratzel in turn described Gobineau and Chamberlain as "racial fanatics" whose works were "brilliant but unscientific."[9] He took issue both with their categorization of race according to anatomical principles and with a tendency to conflate race and language, as is often done with the terms "Semitic" and "Aryan." His indebtedness to Moritz Wagner's diffusionist migration theory made him doubt not just the idea that racial homogeneity brought with it a competitive edge in the struggle for survival but also that racial purity was at all possible.[10]

Smith concludes from Ratzel's discussion that "the unit of Ratzel's analysis for humans was the *Volk*, primarily a cultural entity in which race played a relatively minor role."[11] Similarly, Bassin speaks of "Ratzel's rejection of race as a significant factor in the constitution of states," which "led him to oppose for the most part the national chauvinism and aggression" of his time.[12] Natter concurs, arguing that Ratzel "even pointedly ma[de] fun of those who indulge[d] in the proclivity of paying homage to presumed (Aryan) ancestors."[13] Dijkink speaks of "wilder allegations" of racism and proto-Nazism.[14] Even some of Ratzel's contemporaries had argued that he was "far removed from the conceit of the white race" and did not see the latter as "predestined for rule."[15]

Schultz warns against the idea that Ratzel was somehow "immune" to the racial theories of his time.[16] He takes issue too with the idea that Ratzel was guided by some kind of liberal "panhumanism."[17] Already in his earliest work, Ratzel had expressed the view that black populations constituted a lower race. In later years, he would predict race wars with the determination of someone who found it impossible to imagine a cosmopolitan social order. He noted that "as long as there are *Völker* that are aware of their own national tradition [*Volkstum*], they will collide in national struggles."[18] Black and indigenous populations in North America and elsewhere were simply "too different to integrate."[19] His rejection of the "racial fanatics" was thus a strikingly subtle one. As we will see in what follows, Ratzel very much accepted the importance of race as an object of analysis and explanatory category but was irritated by the lack of attention that biological racists gave to a consideration of environmental factors.[20] His was thus not a rejection of *Rassenlehre* (racial theory) per se but merely of *some* racial theories. And although Ratzel did indeed argue that there was no such thing as racial purity, this did not mean that he thought racial mixing to be positive.[21]

We have already encountered in the opening chapter the question of whether Ratzel's thought is fundamentally liberal and progressive or antiliberal and conservative. I argue here that the opposition constructed in existing debates is, to some degree at least, based on an unhelpful dichotomy that treats humanism and liberalism as somehow opposed to racism and, by extension, to twentieth-century fascism in its different varieties. But liberalism, as it has become increasingly recognized, was not just complicit with systems of chattel slavery, but racial hierarchy was itself intrinsic to liberal discourses, from Kant onwards.[22] It was precisely in settler colonial societies that the boundaries between frontier liberalism, nationalism, and racism were essentially blurred.

In developing this line of reasoning, this chapter does not foreground the racial and ethnographic taxonomies Ratzel developed in his scholarship, although these are discussed too, but it zooms in on his more impressionistic travel writing in the first half of the 1870s. It attempts to provide a closer look at some of the colonial and postslavery subjects whom Ratzel encountered on his journey across the United States and Mexico. Most of those he described in more detail were white, some were public intellectuals and politicians. The colonized were often only "partially visible" in Ratzel's narrative. Although

Ratzel was at times concerned about the latter's living conditions, he would come to naturalize both slavery and the slow extermination of colonized populations in the last two decades of his life through the prism of an omnipresent "struggle for space." I argue that we can already find in Ratzel's travel writing a noticeable emotional economy of repulsion in relation to race that was inextricably entangled with his settler colonial fantasies.

A New World

Already before his journey to the Americas, Ratzel had worked on commission from the *Kölnische Zeitung*, a reputable newspaper with a national readership.[23] His reports on Italy and the Austro-Hungarian Empire had found a broad readership and would be republished as a two-volume book, *Wandertage eines Naturforschers* (Journeys of a natural scientist).[24] It emerges from these accounts that an embryonic interest in the relationship between colonization and race had already been present in him before his journey to the "New World." Not only had he described the mid-twelfth-century German settlements in Siebenbürgen (Transylvania) as "colonies," but he had also framed the history of the region as a "racial struggle" (*Rassenkampf*) between Germans, Magyars, and Walachians.[25] But it is also clear that these had not yet been central preoccupations of his. Even when he arrived in the United States he was initially occupied with other questions: the merits of democracy, the emancipation of women, and the status of German emigrants. But this would change once he visited the South.

Upon his return, he would republish his travel reports with the prestigious Leipzig publishing house F. A. Brockhaus. *Städte- und Culturbilder aus Nordamerika* first appeared in 1876 (and was translated into English as *Sketches of Urban and Cultural Life in North America* in 1988). *Aus Mexico: Reiseskizzen aus den Jahren 1874 und 1875* (From Mexico: Sketches of the years 1874 and 1875) followed in 1878. While little writing came out of his four-week stay in Havana, his observations of the Chinese diaspora in California would later motivate him to embark on a larger study of the topic of Chinese migration, which served as his habilitation thesis.[26] Ratzel would continue to travel through Europe in the second half of his life but turned away from travel writing after accepting his first academic post in 1875. In 1878 and 1880 he published *Die Vereinigten Staaten von Nord-Amerika* (The United States of North America), a more scholarly follow-up to *Sketches of Urban and Cultural Life*.[27] An 1893 second edition of this two-volume book pioneered many of the ideas that would form the 1897 *Politische Geographie*. It is thus no overstatement to say that his political geography emerged organically from his travel writing.[28]

In both the United States and Mexico, Ratzel lived the life of an elite traveler. In Boston, he met the famous Harvard biologist and geologist Louis Agassiz (1807–1873) just before his death. In Washington, he spent time with the Republican senator Carl Schurz (1829–1906), a German-American 1848 revolutionary who had served as a senior officer in the Union Army. He was also acquainted with the German ambassador Kurd von Schlözer.[29] Ratzel traveled and dined in considerable luxury, describing at one point a buffet of oysters, roast salmon, and beef steak that was accompanied by no fewer than six varieties of bread.[30] In Mexico too, he lived in great comfort, staying predominantly in the houses of the wealthy, some of whom were German. Ratzel relied on servants and guides, including former soldiers, as well as horses and dogs. Using references from fellow travelers and other contacts, he found Mexicans hospitable. He ate well and drank a fair amount, commenting in detail on various spirits, from Mescal to Tequila, he enjoyed en route. Rarely did his supplies of chocolate, coffee, and cigars seem to run low.

In Mexico, Ratzel fashioned himself a Humboldtian explorer, a man who ventured up volcanoes and mountains but excelled most in the description of flora and fauna. While long passages of *Aus Mexico* are dedicated to a depiction of the natural world, *Sketches of Urban and Cultural Life* is concerned primarily with urban planning, architecture, and social rituals. In the United States, he frequently visited educational facilities and keenly compared them to their German counterparts. He strolled through city centers, wrote about restaurants and street markets, railway stations and businesses, abandoned plantations, and "so-called Indian settlements."[31] In both books, he recited recent statistics on trade and commerce, population growth, and mortality rates, and he occasionally supplied an account of historical events. At times, he paused to depict some of the individuals he encountered.

Much of Ratzel's discussion in these earlier works lacked analytical depth; his aim was less one of explaining patterns of social behavior or particular types of urban planning than one of classifying these, often according to his own aesthetic preferences: American women wore too much makeup; American men struggled to hold their drink; Aztec ruins lacked symmetry; and so forth. Nonetheless, Ratzel's travel writings revealed "a power of vivid descriptive writing," which found him admirers abroad.[32] Even a century later, the cultural geographer Carl Sauer would praise Ratzel's observations as "rich," "sharply drawn and properly representative," even approaching the quality of Alexander von Humboldt's writings.[33]

Readers of *Sketches of Urban and Cultural Life* will note that Ratzel harbored an overwhelming enthusiasm for the United States. He had prepared himself well for the trip and anxiously awaited his arrival in the New World. The following excerpt, which he had lifted from a love letter to his future wife,[34] expresses the anxious excitement he felt as his ship pulled into New York after his twelve-day voyage across the Atlantic:

> Finally the last day of our journey began. Early in the morning a stretch of the Long Island coast appeared to our right, a bright strip of sand dunes and above it a flat plateau. It was a fine day, "American weather," as my companions aboard called it with patriotic pride. Today everyone was crowded together on deck, happy because of the end of the long trip and anxious in anticipation of what was to come [. . .] To the left Sandy Hook came into sight, a higher wooded landstrip on the point of a narrow peninsula that juts out toward the entrance of New York Bay. Later on Staten Island came closer on this side and Long Island on the other, and before we had traveled through the narrows, which hem in the ocean here at the entrance to the bay, the coastlines were already so clearly perceptible that the houses of their cities and villages clearly stood out from the darkness along the water's edge. These white-walled houses, standing sometimes singly, sometimes clustered together, glistened in the morning light across the water, standing there as if thrown up by the sea like snow-white, bleached mussels and shells. [. . .] And as the city itself drew nearer, with Brooklyn on the other side and the city's southern tip on this side of one of the branches of the sea, which enters into the harbor, it looked very large indeed, and when some islands with fortification and imposing buildings on both sides appeared and the ships all around along the shorelines became increasingly more numerous, all of a sudden the field of vision turned into a single picture of magnificent, rich proportions.[35]

This image of marvelous proportions, referring here to a city of then around one million inhabitants, would stay with Ratzel as he traveled through the United States, taking on the role of an organizing frame through which he continued to read North American society even after his return. He would revisit this trope of North American greatness in his political geography and continue to admire US politicians for their "spacious thinking" (*weiträumiges Denken*), something he believed their German counterparts lacked.[36]

Ratzel was enamored by the American frontier. He had read James Fenimore Cooper's *Leatherstocking Tales* series, dating from the first half of the nineteenth century, and romanticized the individual frontiersman who headed west into unchartered territory. But it was not a boyish fantasy of adventure and masculinity that attracted him to the frontier, but the idea of progressively developing the land. He was impressed, for instance, by the way settlers had transformed California from a territory of fast-lived mining adventures into a modern agricultural society.[37] He would later hail Frederick Jackson Turner's thesis on the significance of the frontier in American history as "brilliant" and as confirming his own theory of the state as organism.[38] America's ability to colonize (*Kolonisationsfähigkeit*), he noted with admiration, was rooted in the nation's character, which was able to absorb the best characteristics of different migrants in its expansive and civilizing mission.[39] By the early 1900s, he had come to see America's expansion to Hawaii, Cuba, and the Philippines not just as a healthy form of imperialism but as the "latest tug of those countless bold and self-sacrificing moves which composed the movement to the West."[40] Ratzel had no doubt that this perpetually moving frontier was colonial in nature and that its movement, in all of its capillary forms, was a sign of America's youthfulness. Germany, on the other hand, was an old society, which could only be "rejuvenated" by constructing its own colonial frontier.[41]

Rather than being exclusively drawn to the unknown and dangerous spaces of the Wild West, Ratzel was also attracted to America's urban frontier. Indeed, he read American cities as "the centers of every important cultural development."[42] He enjoyed the "energetic" and "youthful" population of New York, whose streets and fast-flowing streams of pedestrians he saw as an "animated hustle and bustle."[43] Ratzel spoke approvingly of the cosmopolitanism he found on North American's urban settlements; he loved their broad streets, transport systems, and especially their grid plan, which he found easy to navigate. Time and time again, he noted the ability of American cities to recover from fires, which proved to his mind the vibrancy and youthfulness of this society. Occasionally, he criticized the American way of life as fast-lived and superficial, but ultimately he remained positive about the nation's enterprising character.

Ratzel traveled to the major cities of the East Coast as they had begun to prepare for the centenary of the American Revolution. He clearly admired Washington DC for its political architecture. Visible from almost anywhere, he noted in astonishment, the Capitol building was "like a beacon shining in all directions"; "in the sunlight, the marble seems to glow with a warm, yellowish glimmer, yet at night the building stands out coldly like a tower of snow against the nocturnal sky."[44] He was impressed too by a debate he witnessed in the upper chamber:

> The good impression the speeches make is even further underlined by the sight the assembled group itself offers. There are mostly older men here, and among them there are many whose entire appearance is impressive. In their faces one sees predominantly energy, keen understanding, constant awareness, and together with their white hair and well-built, generally trim figures all this adds up to making a good combination. Needless to say, the members of the Senate are recruited from the best segments of the population.[45]

Nonetheless, it is crucial that Ratzel was not necessarily expressing an unreserved admiration for American democracy. He seemed impressed only by the Senate, not by the House of Representatives. In the latter, Ratzel found "no dignified atmosphere" at all. He noted that the men here were younger and sourced from large segments of American society, "[f]rom the real Yankee of New Hampshire or Massachusetts to the Negro and the cattle breeder from the Southwest, who from his looks could have been imported straight out of Mexico, all elements are represented except for Indians and Chinese."[46] Ratzel seemed uncomfortable with the fact that American democracy had by the 1870s begun to involve not just representatives of different social classes but also of different racial groups. His critique of the House of Representatives did not prompt Ratzel to speak out in favor of autocratic forms of government or strongman dictatorships. In fact, he had little time for populist "beer hall politicians" (*Bierhauspolitiker*), but favored instead a technocratic, privileged, and white elite rule.[47] The senators were older white men who had the habitus of upper-middle-class gentlemen. With more than a hint of criticism of liberal democracy, Ratzel held that

> the democratic form of government teaches quite early the subordination of the individual to the thoughts and leadership of others; this political equality introduces among the entire populace a desire for social equality, which in turn favors the unique development of the individual less than is the case in European states with their varied, sharply differentiated societal and governmental structures.[48]

Ratzel seemed not just to express a preference here for a version of managed democracy, but also provided a reading of the relationship between democracy, equality, and individualism that favored, on this account at least, Europe's monarchies over America's republic. Even if he expressed his view in an 1873 letter to his wife that the achievements of American society ultimately outweighed its shortcomings,[49] his praise was more muted than that of other nineteenth-century European travel writers. He certainly did not share Alexis de Tocqueville's teleological vision of American democracy spreading around the globe. But we can nonetheless find in Ratzel echoes of Tocqueville, whom he had read in the French original,[50] particularly in his insistence that there was a potency and danger in mob rule.[51] Tocqueville had traveled to the United States in the 1830s and would later come to reject Gobineau's biological racism in favor of a cultural or social form of racism.[52] Like Ratzel, he was a supporter of colonialism. And yet Tocqueville had gone significantly further in his critique of slavery than Ratzel would half a century later.[53]

Ratzel was not the first German geographer to have traveled North America. Alexander von Humboldt is famous for his journeys to South America, but he had spent time in the United States too, not least to visit President Thomas Jefferson in the White House in 1804.[54] Nor was Ratzel the only German writer who was mesmerized by the American frontier and the "disappearance" of the indigenous population. As we will see in chapter 6, Ratzel was part of a transnational constellation of authors whose work was marked by such necropolitical themes. But it was particularly Germans in Ratzel's wider network, men such as his Leipzig predecessor Oscar Peschel, who modeled German colonization onto North American settler colonialism.[55] Most immediately, however, Ratzel followed in the footsteps of the geographer and 1848er Ernst Kapp, with whom he corresponded in the 1880s, and his nephew Friedrich Kapp.

When Ernst Kapp (1808–1896) saw himself forced to leave Germany in 1849 and settle in Texas, he dedicated himself to the hard labor of farming. Nonetheless, the 1848er and abolitionist was able to devote at least some time to his library and academic work.[56] He was, in other words, precisely the sort of settler, both peasant and scholar, whom Ratzel would come to valorize in the 1870s. Whereas Ernst Kapp was a Ritterian geographer who wrote books of grand theoretical ambition, on which Ratzel would draw, his nephew Friedrich (1824–1884) was more directly engaged in the study of American history and society. Indeed, he had published a number of works on the topic in the years before Ratzel's journey to America, including one on German emigration to the United States. Ratzel made extensive use of these books in his regional geography. Kapp's history of slavery, republished in 1861 with a strong abolitionist tone, tried to draw connections between European and American history.[57] But while Kapp saw the struggle against slavery as an iteration of Europe's battle with aristocracy and absolutism, Ratzel's comparison would focus on the question of North American and German settler colonization. Like Ratzel, Kapp had worked as a travel writer for the *Kölnische Zeitung* and was associated with the National Liberal Party. But whereas Ratzel became in the 1880s one of the leading proponents of German settler colonization in Africa, Kapp was known as a staunch critic thereof.[58]

Venturing South

Ratzel spent a crucial leg of his journey in the American South, then still recovering from the Civil War. He saw not only the ruins of this armed conflict, but also encountered some of the lives the war had wrecked. He arrived in the latter part of the Reconstruction period, which followed the abolition of slavery in the South and was marked by the introduction of legislation that sought to bring citizenship and suffrage to the black population of the former Confederate states. Commonly seen to have started in the middle 1860s and ending in the late 1870s, Reconstruction was a time when "[t]the slave went free; stood a brief moment in the sun; then moved back again toward slavery," as W. E. B. Du Bois would famously note in 1935.[59] It also saw the rise and spread of white supremacist violence in the southern states. Indeed, the Ku Klux Klan was founded in 1865.

Ratzel's journey to the Americas is likely to have been his first sustained encounter with people of color. His 1869 *Sein und Werden in der organischen Welt* gives us an impression of where he stood on the question of race before he arrived in the New World. Although he had clearly been captivated by the field of craniometry, which was in the nineteenth century seeking to highlight racial difference through examinations of different skull types across world regions, he was unconvinced that one could conclude from such studies that different races represented different species.[60] Resorting to Wagner's migrationist thesis, he argued that even if there had once been significant separation, the impossibility of seclusion on the globe had resulted in their mixing (members of different races were after all able to produce offspring). He would reiterate these ideas in later writings, often speaking of a "unity of mankind."[61]

Ratzel had arrived in the South something of an abolitionist and remained broadly committed to the cause after returning to Europe.[62] He praised the famous black abolitionist Frederick Douglass as an "outstanding character" and admired the German settlers in the United State for their role in fighting for the North during the Civil War.[63] In Boston, Ratzel heard the abolitionist Henry Ward Beecher speak. Before arriving in America, he had read Harriet Beecher Stowe's 1852 antislavery novel *Uncle Tom's Cabin*, though probably importing many of its racial stereotypes in doing so. He had come to the South believing that what was hampering the integration of the formerly enslaved population into American society was their illiteracy and lack of education, rather than some kind of biological incompatibility with modern society.

And yet, underlying this rejection of slavery and support for the Union, there was an economy of emotional and political condescension that surfaced whenever he encountered formerly enslaved persons face to face. Even though he admitted that black Americans could be good orators, he compared their best speakers in the Columbia legislature to "an average speaker from among the white taxpayers."[64] All these men could do was to "ape their masters pretty well." Already in the early stages of his journey he had depicted black Americans in Saratoga as lacking in virtues.

> They say the negro is a born waiter, but those at this hotel had nothing of the grace, nothing of the disposition, nothing of the hidden virtues and capabilities that make a European waiter such an interesting phenomenon for an observer of human nature. Here they work like machines. They scarcely finish what was required of them before falling back into a leaden inertia in which they remain until a new command galvanizes them into new efforts.[65]

Note the ambiguity in Ratzel's description, which stemmed from his lack of attention to questions of labor and class: the waiters were simultaneously hard working ("like machines") and lazy ("leaden inertia"). Ultimately, racial hierarchy was a convenient analytic fix. As he ventured further south, his condescension became increasingly infused with repulsion. With disgust, he spoke of "negro villages, which far surpass in dirt, idleness and demoralization [. . .] the gypsy camps on the outskirts of Hungarian and Romanian cities." In New Orleans, he described black children as having "a gusto that scarcely befits humans" and black adults as having a "more animal-like nature."[66]

How did this insistence on the fundamentally less-than-human nature of formerly enslaved persons and their children fit with his abolitionist views? It is important to understand that Ratzel did not so much describe slavery as inhumane but rather as an "economic system, which could do nothing else than promote inactivity and impede the diligence and enterprise of the majority of people for the benefit of the few."[67] It was, in other words, less the violent inequalities that Ratzel objected to than slavery's inefficiencies.[68] In *Politische Geographie* he explained that the problem with slavery was that both the enslaved and the masters were fundamentally alienated from the land.[69]

Although slavery had contributed much to the rise of feudalism, it did not establish an organic connection between soil and soul. Ratzel was thus convinced that economic productivity of the slavery era could easily be surpassed, despite the southern economy's slow recovery. Nonetheless, he wondered whether "black man's labor" was still "worth as much as when the 'overseer' had stood behind him with his whip."[70]

Ratzel was acutely aware of the South's economic problems. But rather than blaming the war itself, the activities of the Democrat "redeemers," or the emerging system of racial segregation, Ratzel saw the black population's inability to act in politically responsible ways as the prime reason. He concluded that:

> Wherever these negroes are strongly represented, the agitation, discord, and hostility that their political demands arouse, and the corruption that accompanies the racial conflict, have hindered such important cities as New Orleans and Charleston from taking part in the revitalization they seemed destined for at the end of the Civil War.[71]

Ratzel failed to grasp that despite access to political institutions, formerly enslaved persons had no capital or land; their liberation was thus fundamentally incomplete. The relative absence of political economy in his thinking and a disinterest in questions of ideology meant that he could not account for the socioeconomic conditions of the freed slaves' burdened existence. Instead, he perpetuated himself "the persistent production of blackness as abject, threatening, servile, dangerous, dependent, irrational, and infectious."[72] The period of self-government he witnessed in the 1870s, Du Bois's "brief moment in the sun," had to be no more than an intermezzo before the inevitable re-establishment of white rule. Crucially, whenever Ratzel mentioned the white supremacist violence that had erupted during the late 1860s, he denied that it was politically motivated, pointing instead to alcohol abuse and a general decline in moral standards. Murders of whites by blacks, he added, were themselves proving more common as the latter were generally known to carry knives around with them wherever they went.[73]

Before traveling to the South, Ratzel had made the acquaintance of two highly influential men, the Swiss-born glaciologist Louis Agassiz and the German-born 1848 revolutionary Carl Schurz. Agassiz had emigrated to the United States in the late 1840s, where he had become a professor of zoology and geology at Harvard. He had studied with Cuvier and von Humboldt, but was, unlike Ratzel, no Darwinian. In fact, Agassiz was one of the last "respected" scientists to deny natural selection. He believed in polygenesis, the independent creation of species, including the various human races, in separate centers of origins.[74] This process, he believed, had endowed these species with innate physical and mental differences. Although often remembered primarily for his contributions to zoology and geology, Agassiz was also a crucial figure in the rise to prominence of scientific racism. Indeed, he was foundational to what Moore describes as "a virulent slavery-rationalising science," a "skull-measuring, data-crunching, boundary-drawing racial science with its magisterium at Harvard University."[75] Agassiz would die shortly after Ratzel's visit.

Agassiz had traveled to the American South in the late 1840s and had questioned, more radically than Ratzel would three decades later, whether black Americans were in fact human. While Ratzel would still hold on to his belief in the basic unity of humanity in later life, however much he qualified this statement, Agassiz had concluded from his observations of the enslaved population that the white and black races could not have descended from the same stock.[76] Like Charles Lyell, Agassiz sympathized with the southern slave owners. He also accepted donations from the New England slave-cotton industry.[77] Crucially, he believed that racial mixing would lead to degeneration in ways that prefigured many of the arguments of late nineteenth-century eugenicists.[78] Ratzel does not describe the content of his conversations with Agassiz and had dismissed his work in his 1869 *Sein und Werden in der organischen Welt*.[79] And yet Ratzel's journals reveal that he was clearly taken by Agassiz on a personal as well as academic level and would even come to applaud his religious zeal in later years.[80] For Ratzel too, it was a journey to the American South that prompted him to radicalize his views on race.

Ratzel had also spent time with Carl Schurz in Washington immediately before traveling to the former slave-owning states.[81] Like Ernst Kapp, Schurz was a German liberal who had emigrated to the United States after the revolution of 1848. By the 1870s, he had become the first German-American with a seat in the Senate. Schurz had been active in the antislavery movement and served in the Union Army alongside Ulysses Grant, who would be elected president in 1868. The German forty-eighter had, however, broken with Grant in the years immediately before Ratzel's visit and now opposed Reconstruction. A senator from Missouri, Schurz fashioned himself as someone who understood the concerns of white southerners. He supported amnesty for all ex-Confederates and, having stopped believing that the formerly enslaved were capable of political judgment, deemed Grant's efforts to establish racial equality in the South politically misguided and doomed to failure. Crucially, Schurz downplayed Klan violence and tried to resist Grant's attempts to use the

Enforcement Acts to stamp out the KKK with military force.[82] Ratzel described Schurz in his diaries as "a great and good man" and "one of the most notable people" he had met.[83] Their conversations, which continued over the course of a number of days, covered American history among other more mundane topics. Ratzel noted that Schurz had praised the freedom and individualism that had flourished along the American frontier. The two men met immediately before Ratzel's voyage to South Carolina, Florida, Georgia, Alabama, and Louisiana.

Ever since his journey to North America, Ratzel had seen the widespread lynchings of the 1860s as no more than mere *Gewaltstreiche* (coup de mains) that were only ever responses to "negro revolts" (*Negeraufstände*).[84] In the first edition of *Die Vereinigten Staaten von Nord-Amerika*, he declared the attempts to prosecute and suppress the Klan as "equivalent to the Klan's own violent behaviour."[85] Ratzel was thus casting the sociopolitical conflicts in postabolition America as some kind of military struggle between equal sides in ways that ignored both the extent of the lynching and the power dynamics behind them. Indeed, Klansmen had burned down the very schools Ratzel had initially hoped might bring literacy to black Americans. In April 1873, and thus only a year before his visit, 150 had been murdered by white supremacists in Louisiana in what would become known as the Colfax massacre, an event that is sometimes seen to have decisively marked the breakdown of Reconstruction.[86] To not address this event seems a telling omission for someone who wrote in such detail about racial conflict in North America. Ratzel's acquaintance Carl Schurz had adopted by 1873 a stance in between the emancipationist and white supremacist visions that were competing for influence during the Reconstruction era.[87] We see elements of this compromise position in Ratzel's travel writing too, but the further he ventured south, the more he seemed to veer toward segregationist white supremacism.

In Richmond, the former capital of the Confederacy and then still partially in ruins, Ratzel noted "interesting encounters" with German settlers, one of whom he described as an "able, kindly" tradesman from Hesse. The man bemoaned the poverty that had struck Richmond after the war and the way in which the Confederacy had set back his family. With disappointment, he showed Ratzel some old Confederate bank notes, now worthless. This avid collector of weapons also offered his view on the black population, which Ratzel left uncommented on in his book:

> And when the war ended, the slaves were freed, and they made the suffering only worse, not working, and living just like cattle. I have lost all faith

> in them since I see how they use their freedom. I tell you, when I came here I had a lot of compassion for their lot and tried to do them a good turn wherever I could. But I believe there are very few of them with whom one can deal any other way than domineeringly. Give one an inch and he will take a foot; the rougher you treat him, the better he behaves. That's the way it is.[88]

Although these were the German emigrant's words, they may well have summed up Ratzel's position too. The German journalist seems to have had very few direct interactions with black Americans on his travel through the United States and was thus relying on his observations of street life and the views of white men whom he spoke to, many of whom he seemed to admire. Stehlin notes Ratzel's tendency to sympathize with the white southerners and speculates that Ratzel was in fact projecting his own southern German identity on the American South.[89] The German zoologist's acquaintance of Agassiz and Schurz so shortly before his voyage to the South sheds a different light on these moments of identification.

Ratzel felt curiously drawn to the Confederates. In Richmond, he sensed a "venerable air hovering over this former capital."[90] He sought out the house where Jefferson Davis, the Confederacy's only president, had lived, as well as those of Generals Robert E. Lee and Stonewall Jackson. In New Orleans, he visited a Confederate war memorial and admired a bust of General Lee. While Ratzel noted that the many fanatically pro-Confederate statements he encountered seemed to be in denial of the South's defeat, he did not denounce the slave owners' political struggle. "Think what one may about the South's cause," he concluded, "one has to respect the heroism of its leaders and its army."[91] Perhaps the most revealing line in *Sketches of Urban and Cultural Life* on this question concerns Ratzel's proposed solution for the South—a repatriation of black Americans:

> I constantly had one wish, which appears unworkable but which seems to me would be best for both conflicting groups, the whites and the blacks: before they are removed from office, the black administrators should decree that they, together with all their people, intend to look for a new home in the West Indies, in Central or in the warmer part of South America, or even Africa again, a homeland that will feed them more readily, a homeland where less work would be required of them than here.[92]

Upon returning to Germany, he hailed the 1876 Democratic victory in Arkansas as an opportunity to heal the wounds of Republican misgovernment. Georgia, he noted, was already "in the lucky position of having a very strong white majority within which there are almost no Republicans or negro friends."[93] In Louisiana, any political cooperation with the formerly enslaved, he argued, had proven impossible: "the Democrats or Conservatives, the former slave owner party, now encompasses all whites who have an interest in the conservation of culture in the state." In the first edition of *Die Vereinigten Staaten von Nord-Amerika*, he came to a comparatively positive assessment of the presidency of Andrew Johnson (1865–1869) and declared "carpetbaggers," northerners who had moved to the South after the war, to be "rightly hated."[94]

By 1880, he had come to argue that the Civil Rights Act of 1866, which had sought to ensure the protection of freed slaves by law, had "perhaps been a step too far."[95] Even though they were sometimes underestimated, he found that "[w]e find in the negroes the most mentally deficient [*geistig minderbemittelt*] of the races."[96] In his 1893 second edition of *Die Vereinigten Staaten von Nord-Amerika* Ratzel had come to see it as a mistake to have given America's black population the vote.[97] Abolitionists like William Lloyd Garrison had always been no more than deluded dreamers. The old slave owners, who had now returned to political dominance, united all of the South's "intelligence, property and respectability." Ratzel warned that thirty years of freedom had exposed the limitations in the view that education of the black population was the solution to the "racial question." He argued that it could "no longer be questioned that the negro had savaged where he had been left to his own devices but had remained active and useful where the whites had shown him the way."[98] "The negro carries in his character the features which define childhood rather than maturity and because of which he may never be able to be entirely autonomous," he wrote in 1892. The experience of Liberia in particular proved to him that the formerly enslaved "were not able to educate themselves" without white leadership.[99]

Ratzel predicted that the liberation of the enslaved population would soon seem no more than "an episode and an attempt." By then he had concluded that the United States was heading for segregation, that is, "a relationship between the whites and negroes that avoids slavery in form but that retains in essence a type of subordination in which society is layered according to race."[100] He noted, not disapprovingly, that marriage between whites and blacks was forbidden and argued that the very presence of the black population meant that "racial aristocracy" was becoming "unavoidable."[101] In the early 1900s, he reviewed the work of black writers like W. E. B. Du Bois and Booker T. Washington, commending them for their support of the so-called "Atlanta Compromise" with white Southern leaders.[102] He applauded Washington in particular for his insight into the supposed limitations of his own race, which made him stand out as such an "exceptional negro" (*Ausnahmeneger*).[103] Ratzel was so interested in the legal status of formerly enslaved people that he would seek advice from across the Atlantic on how insurance companies, for instance, were dealing with black clients.[104] Ratzel's student Ellen Churchill Semple eagerly reported to him that black Americans were not able to receive insurance although they were legally entitled to it and that black men who married white women were not infrequently tarred and feathered.[105] Although Ratzel did not make an explicit eugenicist argument for the control of black birth rates, he was by the early 1890s anxious about the growth of the black population, which he saw as being checked only by high mortality rates. At the dawn of the nineteenth century, Ratzel had abandoned one of his core liberal beliefs, namely the idea that education could solve America's "race problem." Schooling, he warned, could in fact serve to extend life expectancy and thereby increase the black population. What Ratzel seems to have objected to was less eugenics as a political project than its scientific pretense.[106] Indeed, Ratzel was hardly seen as an opponent of the budding eugenicist movement. When he died in 1904, the *Archiv für Rassen- und Gesellschafts-Biologie* (Archive for Racial and Social Biology), edited by the eugenicist Alfred Ploetz, published a glowing obituary for the Leipzig geographer.[107]

Although Ratzel remained an abolitionist in principle, in practice his position was far more sympathetic to white supremacist visions that circulated in the Reconstruction years. Ratzel soaked up the views of the white southerners he encountered, particularly when it came to questions of race. Underlying Ratzel's narrative was an emotional economy of repulsion: he wished the black population to disappear. By 1902 he had even come to downplay the emancipatory dimension of the American Civil War, claiming, much like many southern redeemers, that the *casus belli* had never been slavery but rather the conflict between two states.[108] Indeed, as he explained in *Politische Geographie*, the war had been rooted in an inevitable conflict between a rapidly growing population in the North, which had sought to centralize power, and its stagnant counterpart in the South, which had pursued sovereignty.[109]

Into Mexico

Although he had only began learning Spanish after arriving in the United States,[110] Ratzel agreed to continue by boat from San Francisco to Acapulco in the autumn of 1874 for a trip that would take him a further six months to complete. While his letters to Marie had for a while complained about his home-

Figure 7. Vegetation of the Mexican plateau. (Source: *Die Erde und das Leben*.)

sickness, they had still been bristling with enthusiasm for his journey. Increasingly, however, travelers' fatigue had begun to dominate, and in much of *Aus Mexico* his writing would lack motivation and drive. He found riding on horseback through rugged terrain arduous and looked forward to larger dwellings (figure 7). Increasingly impatient with his fellow travelers, he found comfort only in a narrative of bourgeois condescension targeted at Mexico's "semiculture" (*Halbcultur*).[111] When the newspaper urged him to continue onwards into Brazil and Peru, he declined.[112]

Despite Ratzel's homesickness and travel fatigue, he was still clearly driven by an interest in the natural environment and in colonial settlements. From Acapulco, he headed northwest to Petatlán, then on to Michoacán. He visited Mexico City and Puebla. He climbed the Orizaba volcano before traveling to Minatitlán by boat and continuing up the river Coatzacoalcos and crossing the Isthmus of Tehuantepec. Before reaching Oaxaca and finally Veracruz, he visited the ancient ruins of Mitla. Ratzel hardly had any contact with the indigenous population of Mexico and often spent time with other European settlers, including members of the aristocracy. He generally seemed relieved when he met fellow Germans, especially if they were bourgeois. Mexico, he believed, had been catapulted from a prehistorical darkness into the present.

Ratzel was quickly taught a colonial way of life. When staying in a tavern in Tehuantepec, run by a couple whom he described as going about their business with a "typically Mexican degree of uncleanliness and carelessness," he learned that only an artificial fit of rage would get them to obey his orders.[113] Ratzel thought of some of his indigenous servants as "parasites" who stole from his stash of chocolate and brandy. Again, he displayed a tendency to classify and judge rather than explain, often according to nebulous aesthetic principles, which he confidently applied to facial features, architecture, and the silhouettes of cacti alike. Ratzel spoke with contempt of the Aztec ruins, whose design and ornamentation he found to be lacking in symmetry.[114]

Ratzel frequently compared Mexico to the United States, concluding that while the latter was a vibrant mover of history, Mexico was stagnant and rotten, a nation that merely "vegetated."[115] He had written with such enthusiasm about the speed at which people moved through the streets of New York. Mexico City, to him, was sluggish and lazy. He found the capital's media landscape poor and the political debate sentimental, superficial, and filled with platitudes. Ratzel was more positive about Mexican women but believed they were held back by their men and the latter's patriarchal ways, not of course an uncommon stereotype in colonial discourse. He compared his experience of seeing Mexico to that of viewing someone in a fever, thrusting themselves from side to side in pointless fury, compelling in Ratzel the urge to avert his gaze from this painful spectacle.

Much of what he seemed to detest in Mexico was the country's poverty. He said the following about lower class Mexicans (*leperos*): "There is something stupefying about these wide-mouthed and goggle-eyed chaps, whose facial features reveal stupidity, badness or both."[116] He concluded that their only daily activity was idleness and even that, he bemoaned, they did with a lack of enthusiasm. Ultimately, Ratzel objected to the absence of a bourgeois class in Mexico, which he argued was rooted in the lack of a "healthy peasantry," with agriculture having been left to the indigenous population. What was missing, in other words, were white settlers to cultivate the Mexican soil with modern technology. Ratzel blamed the Spanish, whom he thought of as a lower European race that had "something weak and raddled" about it.[117] Spanish colonialism compared badly to its English counterpart. Whereas North American history had been shaped by the "gruesome struggle for annihilation between the whites and the Indians," the culturally inferior Spanish had been too lenient. In fact, the Spanish were of all European nations the one that was "most disposed for an amalgamation with the red race"; they "did not have to sink all too deep to reach the level of the half-cultivated Aztecs or Zapotecs of

Mexico."[118] Finally, he added that Mexico's geological makeup and geography, particularly the country's lack of major rivers and islands, had only allowed for primitive forms of socioeconomic organization.[119] This point may have been an afterthought for Ratzel on this occasion, but it highlights precisely where his anthropogeography would one day take him.

Just as in his description of the American South, much of Ratzel's attention was on the question of race. In order to get a better understanding of the way he approached race in Mexico, it is crucial to observe some of the interactions he had with various colonial subjects along his trail. Ratzel admired his interlocutors if they displayed the insignia of class distinction, such as career ambitions and an interest in European high culture. He also respected those who had traveled and those with a military background. He reserved a special place for "gentlemen" who were "intelligent" and "handsome."[120] With this in mind, it is instructive to contrast the way Ratzel spoke about two men he encountered on his journey from Tehuantepec to Oaxaca. Here is what Ratzel had to say about his guide Don Pepe:

> When I first met Don Pepe, in whose company I should travel, I congratulated myself, for his way of discussing and preparing our journey revealed a man whose active and understanding nature was quite unlike the many others in my wide network of Mexican acquaintances. The cut and color of his face signaled that he was related, in not too great a distance, to the sons of Ham, but his eyes did not at all display the restlessness or dimness, the lack of concentration, that characterized the gaze of negroes and negro-half breeds. He had very clear and clever eyes and his whole appearance was measured. He had enjoyed a European education in Paris, which had left an imprint on him. He had later served in the Mexican army and had moved to the isthmus where he had founded the colony Suchil [. . .]. This then really is a man with a notable biography. A pleasant sight in a country in which the capacity of men to develop and shape often ends with their first encounter with women and brandy![121]

Don Pepe was useful to Ratzel as someone who could navigate a country of which Ratzel knew very little. His guide even managed to avert trouble on one occasion after the German's jokes had angered an elderly indigenous patriarch.[122] Ratzel admitted to have facetiously asked the village elder whether he took his machete to bed with him, which the man interpreted as disrespectful. Don Pepe quickly stepped in to defuse the situation, which was resolved with the three men sharing some tobacco. In displaying his admiration for this mixed-race subordinate, Ratzel was not so much revising his views on race. Rather, he was making a crucial exception to his generally dismissive view of Mexican men (highlighted in the last sentence of the above excerpt), and it is vital to understand why he did so. Needless to say, Don Pepe's military background was a sign of distinction for Ratzel. It was less important to the German journalist on this occasion that Don Pepe was nonwhite or that he lived in a simple mud hut than that he was *cultured*. Indeed, Ratzel even favored Don Pepe to some of his fellow European travelers because he had *character*. Ratzel admired the fact that the young Mexican had occasionally taken the French capital's opera singers out for dinner. Character was a vague but crucial category in Ratzel's writing, which he used to describe those men (and some women too) who had intellect and ambitions. But what elevated Don Pepe beyond his racial background was the fact that he had himself become a successful colonist.

The story was quite different when the future geographer encountered a German, "a young unbearded man whose long, red and crooked nose protrudes his face." Having found out that they were compatriots, this nameless man had approached Ratzel with a "ladylike gentleness" and an embarrassing "servility."[123] Originally from Hamburg, he had emigrated to the United States and acquired US citizenship before traveling onwards to Mexico. He was now living off a meager pension, which barely paid for his travel expenses and daily provisions. He hoped to eventually find a sunny spot to settle down as a language teacher. Ratzel noted with some disdain that this man was traveling on foot rather than on boats, trains, and horseback. All he seemed to need in life were a few warm meals and a hammock. He was in Mexico because life was cheap. Unlike Don Pepe, this man did not have traces of a middle class or military habitus. He was *without character* and even his whiteness could not entirely save him from that. Ratzel concluded that "it is not rare that one meets such strange human existences in Mexico." "If someone is clever enough to understand the advantages of this heavenly life and intellectually lazy enough to leave civilization behind, then he is ready to become a Mexican."[124] As Stoler notes, poor whites (as well as white women) were central figures in the settler colonial imaginary as they threatened the boundaries of colonial masculinity.[125] Whereas Don Pepe was something of an honorary European in Ratzel's system of classification, the nameless German was a lesser white man who had *gone native*. The anger and disgust he felt toward this compatriot demonstrate the way this man's presence disrupted his fantasy of white colonial superiority. Ratzel found comfort in the observation that only the "most characterless of fellows" were able to fall to what he saw as the low stage of Mexican civilization.[126]

We should be careful not to conclude from a description of these two characters, in a sense an inversion of one another, that Ratzel was suggesting that racial boundaries were fluid, that race was an inadequate frame, or indeed that racial mixing was somehow positive. True, he thought it false to assume that "all products of racial mixing are absolutely bad, following the platitude that mixed-race individuals [*Mischlinge*] tend to inherit their parents' worst characteristics." "Nevertheless," he continued "it becomes very clear to those who have spent time with *Mischlinge*, that they have a tendency to become *lesser whites*."[127] He repeated this view in an essay published in the same year as *Aus Mexico*. Here, he emphasized that a state that united different *Völker* within it automatically "increased the number and diversity of its population's predispositions [*Anlagen*]."[128] He noted the role that German emigrants had played in Russian administration and agriculture, and he applauded the way the Jewish population had enhanced the Polish and Romanian economy. And yet only two pages later, he concluded, "If a white *Volk* degrades itself by mixing with negroes, Malays etc. to such a degree as the Portuguese have done in their oversees territories, then this is a mere descent from a previously higher stage and thus regrettable."[129] What may appear as a tension in Ratzel's view of race is, upon closer inspection, the superimposition of a racial boundary (white/nonwhite) above the category of *Volk*. Whereas racial mixing of *Völker* was desirable to him in a white European context, the same was not true in a colonial context. Even among those white *Völker* that had embraced "an atmosphere of great freedom," such as the populations of the United States and Switzerland, "the addition of a considerable fraction of a foreign *Volk* [was] a dangerous experiment."[130]

Ratzel's tour of Mexico may not have been as formative as his journey to the United States, but his account is nonetheless significant. It reveals a crucial question that had already occupied his mind on his journey through the Austro-Hungarian empire, but which had now begun to dominate his thinking. This was the problem of German colonization and how such a project might be governed most efficiently through racial and class hierarchies. A review of his travel reports in *Kölnische Zeitung*, which Ratzel cut out and kept among his private papers, criticized him for taking for granted that a lack of German colonies in North America was a tragedy.[131] Indeed, Ratzel stated that he would have preferred if Germans had formally colonized North America in the sixteenth and seventeenth centuries but recognized that it was now too late. *Aus Mexico* concluded by way of policy advice: the German empire should *not* seek to colonize Mexico. "Despite its harsh climate and poor soil, I would prefer parts of Canada," Ratzel noted, proposing that Mexico be left to the steady colonization by the United States.[132] The fantasy of finding space for Germans to colonize had now firmly been imprinted in his mind. Nine years after his return from America, Ratzel would call for Germany's colonization of Africa in a speech he gave at the general assembly of the rightwing *Nationalliberale Partei* in Munich.[133] We will turn to the question of colonization in more detail in chapter 5. Now I would like to outline the ways in which the "racial question" was pushed further in Ratzel's ethnological writings, which came to preoccupy him during the 1880s as his colonial fantasy was realized by the German Empire.

Ethnology and Indigeneity

Ratzel's experience of settler colonial societies in the Western Hemisphere would remain formative to his academic work. By the 1880s, he had become preoccupied with ethnology (*Völkerkunde*), the branch of anthropology that studied cultures, particularly indigenous ones, in a comparative fashion. Between 1885 and 1888, he published his most important treatise on the topic, the monumental three-volume *Völkerkunde*, translated into English a decade later as *The History of Mankind*.[134] He wanted to understand the rise and fall of civilizations and "demonstrate the cohesion of the human race."[135] Crucial for this exercise was not so much a focus on distinctions than on "points of transition" and "intimate affinities." Although the variations between human societies were "numerous," they did "not go deep."[136]

Empirically, *Völkerkunde* suffered from a lack of originality. Rather than traveling to the regions he depicted, Ratzel relied on the accounts of explorers, many of whom were British (from Cook to Livingstone), and the artifacts exhibited by various ethnographic museums and private collections, particularly German, English, and Austrian ones. Volume 1 covered the indigenous populations of Oceania, Australia, Malaysia, and Madagascar, while volume 2 addressed the indigenous Americans, Europe's Arctic peoples, and sub-Saharan Africans. The final volume continued with a discussion of sub-Saharan Africans before turning to "the cultured races" of Europe and North Africa and Asia. The three volumes featured hundreds of drawings of tools and artifacts, human portraits, huts, and ruins. With the exception of the first volume's long introduction, the tone was largely descriptive. Crucial prisms for his study were language, religion, gender relations, family and social customs, as well as the nature of habitations, tools and technology, art and artifacts, agriculture, clothing, and ornament. Particular attention was given to state formation and burial rituals. And while Ratzel did try to highlight commonalities among geographically distant *Völker*, his narrative was underwritten by a tendency to exoticize non-Western peoples.

Ratzel's travel writing had never been designed to be an entirely separate project from his physical geography. His account of life in the Americas had constantly been interrupted by geological, geomorphological, and climatological reflections. While his ethnological writings too were informed by physical geography, his anthropogeography sought to apply a geological language to the question of race. Even in his latest work, he spoke of *Naturvölker* (natural peoples) in relation to the earth system.[137] Although there were variations in tone, he used similar conceptual prisms in these different genres of writing. All life forms were subject to migration. He thus concluded in his *Lebensraum* essay that America had since the onset of European settlement experienced a "Europeanization" of its flora, fauna, and human population.[138] Neither his ethnology nor his biogeography distinguished rigidly between extinction, the passive fashion in which particular species or human races disappeared, and extermination, the active process by which one species or race actively destroyed another. When he spoke of the "extinction" of the Tasmanians in the late nineteenth century or the British bear around AD 1000, these were simply two examples of one and the same process. It was the struggle for space that led to the existence of superior and inferior peoples, races, and species. In a world in which war was the natural state of affairs, nations that refused to fight were doomed to be subdued and removed by others.

Ultimately, Ratzel's insistence on the unity of humanity collapsed under his distinction between natural and cultured peoples (*Natur-* and *Kulturvölker*). At times, he included areas of ancient civilization in Northern Africa and Asia in his definition of *Kulturvolk* and added a further layer, that of nomadism (see figure 8). But the distinction between *Natur-* and *Kulturvölker* was the one he returned to most frequently. Throughout the three volumes of *Völkerkunde* and elsewhere, he spoke of higher ("strong" and "ruling") and lower ("weak," "obeying") peoples.[139] While the former had seen progress, the latter were to a significant degree "prehistoric" or "outside of history," familiar tropes in anthropological discourse.[140] Already in his travel writing he had observed that the whites were "stronger and cleverer" than the natural races.[141] He would later conclude that the history of the United States was a history of "racial struggle" (*Rassenkampf*), "in which the former owner of the land, the red man, was exterminated [*ausgerottet*] and pushed West."[142] Ratzel interpreted practices such as cannibalism and infanticide in Oceania as signs that these peoples were self-exterminating and

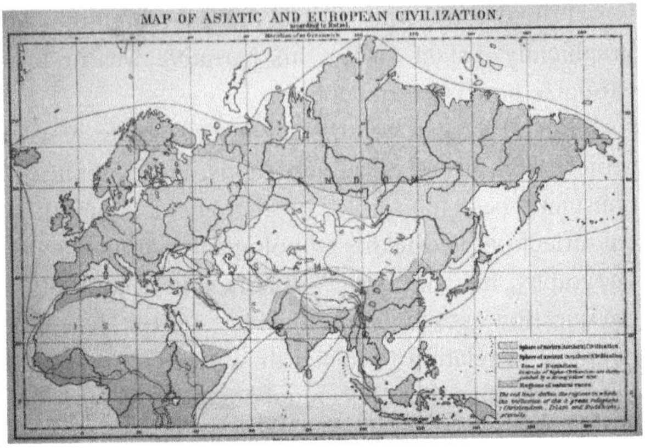

Figure 8. "Map of Asiatic and European civilisation." (Source: *The History of Mankind*.)

making way for superior white settlers. The hierarchy of races, he explained, was rooted less in "physical or intellectual power" than

> in the fact that we have labored more, acquired more, lived more rapidly, and above all, have kept what we have acquired and known how to use it. Our inheritance is larger, fuller of young life; and therefore a comparison of national positions gives us a higher standing among mankind, and indicates too how and why we have become what we are, and what road we must take in order to advance a stage further.[143]

Ratzel emphasized that the difference between "natural" and "civilized" races was about the grip on the soil that enabled the latter to develop state structures. The crucial link between humans and the earth was agriculture. "The farmer who stores his corn in a barn," he claimed, was "ultimately just as dependent on the soil of his fields as the Indian who reaps in the swamps rice he did not sow; but the former feels the dependence less, since, owing to the provision he had the wisdom to store up, the chain is longer and its pressure accordingly less severe."[144] Superior cultivation allowed for higher forms of sociopolitical organization, a process that often took place along racial lines.

We have already noted some of the tensions in Ratzel's racism. One, again not uncommon for his time, was between an occasional romanticization of the "savage" and the firm belief in the latter's inferiority. Although he argued against the idea that there could be such a thing as racial purity, he seemed to admire such purity whenever he found it among indigenous communities. He thus enjoyed hearing about the resistance with which the Spanish attempts at racial mixing had been met in Mexico.[145] In the lowlands near the city of Minatitlán, there still existed "pure Indians," which he described as "orderly people," though only because "unlike the half-breed, they had no social or political ambition, which would generally exceed their intellect and *character*."[146] Ratzel argued that mixed-race Mexicans lacked the "heeling feeling of inferiority which made the average negro or Indian useful or enjoyable people."[147] Indigenous Americans, on the other hand, could have character, but only if they continued to work in agriculture. As soon as they ceased to be peasants, shepherds, or fishermen and joined the ranks of the budding proletariat, they became inauthentic.[148]

Ratzel recognized how far the extermination of America's indigenous populations had progressed by the 1870s. He had visited California in the final years of a genocidal campaign that had decimated the indigenous population.[149] And yet he only rarely expressed any pity for a race, or species, for that matter, that had been eliminated. In *Aus Mexico*, Ratzel had called the indigenous people "half-humans" because of how he thought they mistreated their dogs. He found this trait "among their most repugnant characteristics, which quite frankly contributes to the fact that I can exclude without any regrets any philanthropic feelings when contemplating the Indian's present and future status."[150] America's native population, he argued, had not managed to move beyond a hunter-gatherer society when the first European colonists arrived. The Indians had failed to build towns and cities and exploit natural riches such as iron ore and gold. Their thinking was irrational and superstitious and they were prone to alcohol and gambling addictions.[151]

Ultimately, Ratzel came to the conclusion that the extermination of the indigenous Americans was necessary because they had proven inferior in the natural struggle for space. "One is placed in a peculiar position in regard to the displacement and annihilation of a whole race," he wrote in 1877. "As a sensitive human being, one cannot approve of the means by which this struggle is fought; but as a thinking human being, one cannot avoid feeling satisfied by its outcome."[152] The only way, he claimed, in which the indigenous Americans might find their place in American society was if they were as "scattered and few" as Europe's Sinti and Roma. He was quick to add that anyone who had witnessed the Roma in Hungary would surely recognize the "general harmfulness" that these populations exerted within a *Volk*. "Such entirely different levels of civilization as the Californians and the Indians are even less likely to coexist in the long run; the strong simply washes the weak away."[153] This, surely, was natural selection at work.

Conclusion

Race was no afterthought in Friedrich Ratzel's work, but a crucial vector from the late 1860s onwards. The history of the Americas in particular, he believed, had since 1492 taken on the form of a racial struggle. Upon his return to Europe, Ratzel remained interested in North America's new racial order and continued to update German-speaking audiences on new developments in American debate until the very end of his life. "Let us see," he wrote in 1904, what lessons might be learned from that "greatest of racial questions, the negro question in the United States of America."[154]

As we have seen, Ratzel preached the "unity of humanity" but in fact treated this unity as though it was structured in hierarchical ways. He welcomed the amalgamation of different settler populations into a new American *Volk* but was far less optimistic about the role that the colonized might play in this new people. While his diffusionism and environmentalism led him to question the existence of inborn inherited racial differences, he also argued that it was only "common sense" that there was a need to "keep clean and improve one's race" (*Reinhaltung oder Verbesserung der Rasse*).[155] In the same pivotal 1904 paper in which he attacked the racial theorists of his time, he also made the case for a "more humane racial politics" (*menschlichere Rassenpolitik*), a form of global racial segregation.[156]

> It is pointless to deny that the differences in the natural endowments of mankind's various races exclude the [possibility of] equality of achievement and aspiration. Here too, salvation thus lies only in the gradation and division of tasks, which should be combined with spatial segregation to counteract the risk of the higher race mixing [with others].[157]

Ratzel was thus ultimately torn between his academic unwillingness to commit to racial purity and a segregationist and exterminist politics he had adopted in his travel writing and political commentary. While he may have been motivated by some kind of cosmopolitan fantasy, the reality of colonialism was one that demanded racial hierarchy between the colonizers and the colonized.

It would be unsatisfactory to conclude from this discussion only that Ratzel was a racist. It is more productive to ask how his specific racism sat within the scientific and political debates during and after his lifetime. One question to tackle is thus whether Ratzel was, subliminally perhaps, already concerned with an eternal struggle between an "Aryan" and a "Jewish race." Kost certainly notes "latent antisemitic positions" in Ratzel's work, although he concedes that Ratzel "attached only marginal importance to Jews."[158] Fahlbusch and Michel suggest that it is in the idea of Jewish spacelessness that we can find an antisemitic tendency in his political geography.[159] Schultz, who has pointed to the continuities between Ratzel's geography and national socialist ideology, does not detect any such drift in Ratzel's writing.[160]

Although Ratzel rejected, as we have seen, the "racial fanatics" of the late 1890s and their preoccupation with Aryan roots, he *was* interested in Aryans, a category he defined along both linguistic and racial lines. For light-skinned Aryans, he claimed, had been among "the carriers of the highest culture over the last 2 ½ thousand years."[161] Ever the environmentalist, he wanted to point out that white skin color had developed under very different climatic and geological conditions than the present. It made little sense to search for the Aryans' "European origins," given that the European subcontinent had formed after the emergence of light-skinned peoples. Europe had only thrived culturally for millennia through "the transplantation of oriental spores on European soil."[162] The development of an Aryan race had thus been dependent on the civilizational superiority of the Near East, making it erroneous to speak of "purity" or "origins." The Ancient Greeks, he was keen to remind his readers, had been among the most thoroughly mixed races in human history.[163]

Although Ratzel would have been exposed to a significant dose of antisemitism as a member of the *Alldeutscher Verband*, his published work only touched on Jews and Judaism in passing. He recognized the power of antisemitism and believed it was a lasting phenomenon that followed Jewish migration across the globe.[164] But he also noted that "the antisemites had never been able to specify how they would extract the Jews from the economic life of European *Völker*."[165] When defending the Jewish-born explorer Emin Pascha, he noted that antisemitism "left out no opportunity for stupidity."[166] Ratzel's *Politische Geographie* emphasized what he deemed to be the Jews' weak grip on the soil, a theme we have already encountered in his discussion of indigenous peoples of the Americas. Crucially, he argued that "stateless peoples," such as the Jews, were only ever able to forge a relationship to the soil by acquiring private property and thus ultimately remained "tenants" in other host nations.[167] Ratzel was undoubtedly drawing an exclusionary distinction here. And yet there was for him many such tenancy agreements: the Spanish in late nineteenth-century Mexico held a similar position,[168] as did the German "colonists" of Siebenbürgen, whom he so admired.[169] Indeed, in the 1870s, Ratzel saw German-Jewish fellow travelers as "compatriots" and would later show understanding for the Zionist project too.[170] Ultimately, Ratzel seems to have been either ambivalent or indifferent on this matter. In short, his political geography did not class Jews as significant historical agents as long as they did not have a state of their own.[171]

It is vital to recall that while racism was integral to the ordering of privilege in colonial societies, "the *quality* and *intensity* of racism var[ied] enormously in different colonial contexts and at different historical moments in any particular colonial encounter."[172] Ratzel's racism displayed less the antisemitic and anti-Slavic sentiments[173] that had come to dominate racist discourses in Imperial Germany and the Habsburg Empire by the late nineteenth century than it did the obsessions of a typically North American frontier racism, which Ratzel had absorbed on his journey through the United States and Mexico. This is not to say that a continental European form of racism was incompatible with this North American brand but merely that the particular foci and issues that now captivated Ratzel (segregation, democracy in a multiracial society, etc.) suggest that he had adapted his thinking on this matter. Black Americans, he concluded, were incapable of democratic participation and had ideally to be resettled to other regions; chattel slavery was an inefficient economic system but necessary under certain climatic conditions; the slow extermination of the indigenous Americans was a vital and indeed natural outcome of a struggle for space that played out along the colonial frontier.

What we find in Ratzel is a racism that foregrounded cultural and economic practices as vectors of distinction (themselves always dependent on environmental factors), but which never quite managed to free itself from the biological arguments with which he had come into contact in the 1860s. He insisted in the unity of mankind and nonetheless described people of color as animal-like. Even in his latest writings he would attach some, albeit limited, importance to skull measurements.[174] He criticized the biological racists, but mainly for neglecting environmental factors. He saw members of different races as able to transcend their racial status in principle but stressed that their varying ability to acquire culture placed limits on this. He thought it wrong to condemn all racial mixing and yet was rarely positive about its outcomes. Ratzel shared with the early Du Bois the view that black Americans were not integrated into post-abolition society because they had "not reached a sufficient high grade of culture," but lacked an understanding of the ways in which it was ensured, through an intricate system of socioeconomic practices, that blacks could "not be admitted into the group life of the nation no matter what their condition might be."[175] Even though Ratzel recognized in passing the power of antisemitism, there was no room in his worldview for the role that racism itself played in stratifying society. Indeed, he criticized the anthropologist Franz Boas in 1895 precisely for emphasizing the role that racial stereotyping played in sustaining the color line.[176]

The question of (post)slavery is crucial for an understanding of where Ratzel stood on the political spectrum of the late nineteenth century. And while he was certainly not alone among those who believed that abolition had solved as many problems as it had created, it is useful to compare Ratzel's position on slavery to that of Charles Darwin. In a famous passage in his 1839 *The Voyage of the Beagle*, Darwin had described an enslaved person whom he encountered in Brazil as "uncommonly stupid" in a language Ratzel would mimic three decades later. And yet Darwin was shocked that this man had been "*trained* to a degradation lower than the slavery of the most helpless animal."[177] He thus recognized, on some level at least, that slavery was a social system designed to produce less-than-human subjects. Darwin was morally outraged by the "atrocious acts" that he witnessed and that he imagined were inherent to slave societies. Ratzel too recognized that the slave trade was a violent regime,[178] but ultimately a different story prevailed: slavery was economically inefficient rather than simply immoral. But this certainly did not mean that emancipation was the way forward. In Africa, he claimed, slavery was so pervasive that colonial powers would be well advised not to disturb the socioeconomic equilibrium by combating it.[179]

Both Darwin and Alexander von Humboldt had remained adamant abolitionists in their later life.[180] Ratzel, however, warned that "one should not praise abolition as a great sacrifice that the USA had given to humanity."[181] By the turn of the twentieth century, he had come to see slavery as the natural outcome of certain climatic conditions. Systems of forced labor, he argued, were thus likely to endure or re-emerge in the tropics.[182] In the North American context, Ratzel adopted, much like his acquaintance Carl Schurz, a stance that was closer to a segregationist and white supremacist than to an emancipatory vision, closer perhaps to Agassiz than to Darwin.[183] Just like many ex-Confederate politicians, Ratzel came to believe that the American Civil War had never really been fought for racial equality but had rather been the struggle between two states and their visions, a conclusion that coincided neatly with the lessons of his political geography. The former slave owners had got it fundamentally right: blacks were not able to act responsibly without white leadership. A new racial hierarchy based on a system of segregation, what would be known as the Jim Crow laws, was the only option. The violent forces unleashed by the KKK and other white supremacist paramilitaries during the Reconstruction era, which Ratzel chose to overlook and excuse, would ultimately give birth to an American brand of fascism.

The tension between race and space that some have identified in the Ratzelian tradition, which would come to the fore in the 1940s when prominent forces in the national socialist regime found the racial views of Ratzel's geopolitical followers too weak, thus has to be qualified in one crucial aspect. There was no fundamental tension between race and space in Ratzel's work: racial struggle was one articulation of the struggle for space. Ratzel's particular brand of racism, which lacked an Aryan and antisemitic fetish, would ultimately prove less popular in interwar Europe than those of the "racial fanatics" whom he deemed "brilliant but unscientific."

Chapter 4

Life and Earth

In 1901, and thus four years after penning his *Lebensraum* piece for *Die Umschau*, Ratzel elaborated his idea in more detail. The outcome was an eighty-seven-page essay, "Lebensraum: Eine biogeographische Studie," published as part of a *Festschrift* for the sociologist and political economist Albert Schäffle.[1] Ratzel's central argument had not changed; *Lebensraum* was an integral part of the very essence of every species or race, as it was shaped by geomorphological and climatic conditions. "The conditions under which all life develops," he reiterated in the opening sentence, were "governed by a great telluric force."[2]

> All earthly being rests on one single law: the greatest and the smallest being depend on the basic properties of the planet. Over the immense duration of its development, life in its fluidity does not find the world too big to encircle, and even a life form with the narrowest distribution is related to other forms in distant parts of the earth. The history of the particular type of mint that only occurs in a few Alpine meadows in Carinthia, the history of the insectivore that lives within the bounds of a single high valley in the Pyrenees, and that of the smallest Pygmy tribe in the African jungle—they all bear the living mark of the size and shape, movements, mass and material composition of the globe as a whole.[3]

This argument about the earthly foundations of all life chimed with Ratzel's earlier anthropogeography and ethnology and thus formed something of a summary statement of his entire oeuvre.[4] And yet there was a subtle shift in the 1901 essay toward a more dynamic and vibrant understanding of life, which was in fact a return to views he had held in the late 1860s and early 1870s. And while Ratzel depicted life as existing in some form of global harmony in which every species on earth depended on that earth as well as on one another, the processes of becoming, which he observed all around him, were hardly peaceful. Not only was there struggle for survival, but this struggle was in fact a "battle for annihilation" (*Vernichtungskampf*).[5]

Although *Lebensraum* is now inextricably bound up with Ratzel's name, he did not in fact coin the term. It was the fellow geographer and Darwinian Oscar Peschel (1826–1875), Ratzel's immediate predecessor in the chair of geography at Leipzig, who had already used it in 1874 to refer to the relation between a people and its land. The Spanish colonizers, Peschel argued, had been forced to adapt to the "tropical *Lebensraum*" they had found in Latin America and the Philippines.[6] The term had even been in circulation before Peschel. Indeed, the German writer, poet, and natural scientist Johann Wolfgang von Goethe had used it in an unrelated sense in one of his earlier novels.[7] But it was Peschel who had given it the focus on the conflictual relationship between environment and race it would have in Ratzel's writing.[8] The continuities between Ratzel's use of the term and that of the lesser-known Peschel are striking. Already in 1860, Peschel had framed the disappearance of indigenous races across the globe as a natural outcome of a general struggle for space. He too had made arguments about the origins of racial difference and the intelligence of the white race in particular.[9] As Ratzel was familiar with Peschel's *Völkerkunde*,[10] the moment in which Ratzel first spoke of *Lebensraum* cannot be considered one of inception.[11] But it is nonetheless instructive to take a closer look at his diary entries during those crucial weeks.

There is little to suggest that Ratzel deemed *Lebensraum* to be a particularly important conceptual contribution of his. As Jureit argues, he "did not present the term *Lebensraum* as the linchpin of his theory" and hardly used the essay to offer new ideas.[12] Ratzel only mentioned *Lebensraum* in his diary on a handful of occasions between 1897 and 1900 but spoke frequently about a section of the essay titled *Kampf um Raum* (Struggle for space), which, in different variations, appeared in other publications too.[13] In December 1896, he had noted in his diary that he was working in parallel on matters of political geography and biogeography.[14] By early March of the following year, he had completed *Politische Geographie* and was ready to move on to other writing projects.[15] At this stage, he still spoke of his "biogeographical essay."[16] He had contemplated using the essay as an introduction to *Politische Geographie* but discarded the idea.[17] Only on 26 March 1897, and thus five days before he sent his manuscript to *Die Umschau* from a Prague post office, did he adopt *Lebensraum* as the essay's title. He did not explain why he suddenly chose to use this term.[18]

Ratzel, in other words, added *Lebensraum* to his repertoire only comparatively late in his life and even then used it infrequently.[19] Crucially, the term cannot be found in the 1897 first edition of *Politische Geographie*, although it does appear it the opening passages of the second edition.[20] Ratzel's argument about the telluric unity of all life was no new idea of his, but one that had carefully been laid out in the 1882 first edition of his *Anthropogeographie*. What was notable about *Lebensraum* was thus its ability to sum up so well the spirit of his entire oeuvre and the thrust of his theoretical and political interventions. As he did so often, Ratzel hesitated to define his own terms. While Ratzel seemed at times to use *Lebensraum* as a synonym for the biological term "habitat," it stood in for the political concept of "territory" at other times. It was precisely the blurring of these two meanings that gave a specific sense to *Lebensraum*. To complicate matters further, Ratzel made use of "a cluster of synonyms such as *Lebenssphäre* (realm of life), *Lebensgebiet* (habitat or living area), *Lebensfläche* (surface of habitation), and *Lebensreich* (realm of life)."[21] *Lebensraum* entailed both *Wohnraum* (area of habitation) and *Ernährungsraum* (feeding area). Ratzel also had a tendency to employ *Lebensraum* interchangeably with that of the "oecumene" in both the 1897 and 1901 versions of the essay.[22] He thus used the term "oecumene" not just to refer to humanity's living space, the surface of the earth minus uninhabitable spaces such as deserts and oceans, but also to the living space of each race and people. The latter, he explained, often overlapped, thus leading to conflict, survival, and extinction. Indeed, it is in relation to the oecumene that he finally approached something of a definition. "For humans," he explained, "the great significance of living space [*Lebensraum*], which has been named the oecumene, is considered absolutely essential." "But every species of plant and animal," he continued, had its oecumene, "the space which it occupies on the earth and whose size and shape partly determine its viability."[23]

Ratzel thought there was something self-evident about all of this. Indeed, he claimed that one "need not draw on a philosophical definition of every being as a thing that takes up a space which belongs to it alone, to demonstrate the omnipotence of the need for space."[24] And yet his *Lebensraum* concept and his wider corpus did rest on a philosophical foundation. As one of his contemporaries noted, it was impossible to approach Ratzel without having been "anointed with a drop of philosophical oil."[25] In what follows, I outline the contours of this philosophy of life and earth, which he unwittingly but neatly summed up in the idea of living space. Although Ratzel often displayed a tendency, like that of many of his contemporaries, to be loose in his citational practice, he did mention his most important interlocutors by name. In order to get a sense of the particular ideas that had exerted an influence on Ratzel, it is crucial to recognize the degree to which the particular intersection of German geography and Anglo-German evolutionary biology determined the broad thrust of his arguments. A similar fusion had already been attempted by Peschel, but he had died in 1875 without completing this task.[26]

In the second half of the 1860s, Ratzel first came under the influence of what many then referred to as *Entwicklungstheorie*, a literal translation of which would be "development theory," through the work of Buffon, Cuvier, Lyell, Lamarck, Haeckel, Spencer, and Darwin, of whom Ernst Haeckel (1834–1919) was arguably the earliest and most crucial source of inspiration for Ratzel. The latter would finally settle in the mid-1870s on the idiosyncratic theory of evolution offered by his mentor Moritz Wagner (1813–1887). Among German geographers, Ratzel's most important influence was undoubtedly Carl Ritter (1779–1859), but Ratzel also drew on the explorer and polymath Alexander von Humboldt (1769–1859).[27] Beyond geography, Ratzel found inspiration in eighteenth-century enlightenment philosophers such as Montesquieu, Voltaire, and particularly Immanuel Kant (1724–1804).[28] In the 1890s, an important second philosophical streak came through in Ratzel, which took him to the esoteric writings of the panpsychist Gustav Fechner (1801–1887), as obituaries by Kittel and Lamprecht were keen to stress.[29] We will discuss Fechner's ideas about life and death in chapter 6.

Ratzel's work was consistently underwritten by a version of monism, the belief in the singleness of existence, which he took from Haeckel. It is thus that Ratzel incessantly sought to collapse binaries (life/earth, space/time, materials/ideas, human/animal), seeking to show either how intertwined these supposedly opposite phenomena were, or revealing them to express difference of degree rather than kind. Beneath this faith in the unity of all matter, there was in Ratzel's earliest and again in his latest writings a vitalist impulse to read life as an insatiable agency that drove all evolutionary change. And yet life was always in a tension with the finitude of the earth's resources; it was earthbound (*erdgebunden*). Ratzel's insistence that the biosphere, of which humanity was of course a part, was subject to geological transformations made him particularly interested in the vertical dimension. Often reduced to a basic mechanism with a conflictual logic in his earlier work, Ratzel's conception of space (and time) became more complex toward the end of his life as he turned increasingly to Kant. Although Ratzel certainly went through a determinist phase in the 1880s, which crystallized most notably in passages of *Anthropogeographie*, he later complicated his position on the relationship between life and earth, preferring to approach them as interactive (*wechselwirkend*).

Evolutionary Foundations

Ratzel repeatedly sought to decenter the human in his writing by showing how its existence was neither unique nor independent from the interplay of life and earth. To emphasize this point, he frequently anthropomorphized the animal and plant kingdoms. Different life forms were not simply able to enter into a variety of parasitic or symbiotic relationships but could themselves form societies that were not dissimilar from human ones. Ratzel spoke of animals as "citizens" or "colonists" and suggested that "overpopulation" occurred in nature as much as in society. He thought that humans and nonhumans had cooperated over time, such as in the domestication of animals. Many of Ratzel's creatures were moreover able to "act" like humans, even use technology. He was enchanted by spiders that could be found "wafting more than 100 km across the sea [] like miniature balloons: a small spider sits in the gondola, i.e. at one end of the web, thus making its journey."[30] At times, he reversed the move, drawing on a botanical and zoological language to describe cities. North American boom towns were rendered "mushroom cities," while Chinese cities were "ant hills."[31] In his 1902 *Die Erde und das Leben*, Ratzel illustrated his *Lebensraum* idea with a drawing of an ant colony defending a plumeless saw-wort, a type of thistle, against a hungry white-spotted rose chafer (figure 9). Due to the evolutionary history of the two species, the plant depended on the ant and vice versa. Some ant colonies, he explained, even cultivated aphids inside the plant. The image depicted an anthropomorphized ant standing on its hind legs, reaching out for the beetle to defend its living space.

In his 1869 *Sein und Werden in der organischen Welt*, Ratzel had already explained that there was "no fundamental difference between moss, infusoria and humans, only differences of degree."[32] Drawing on the writings of Alfred Russel Wallace (1823–1913), he had questioned whether intellect was a uniquely human ability, comparing patterns in the construction of huts by *Naturvölker* to those of bird nests.[33] Experiments with Eurasian wrens, he explained, had shown that these birds were likely to imitate other birdsongs if raised in isolation from their own. They had, in other words, to be taught how to sing like a wren. Ratzel concluded that the ways in which birds constructed their nests revealed not pure instinct but a "thinking, deliberating spirit."[34] He later backtracked and spoke more frequently of "higher" and "lower" life forms, without ever embracing a clear distinction between humans and nonhumans, animals and plants.[35] Although humans had attempted to master the earth, they were ultimately but an effect of the natural world. "The earth has

Figure 9. Ant colony defending a plumeless saw-wort (Serratula lycopifolia) against a white-spotted rose chafer (Oxythyrea funesta). (Source: *Die Erde und das Leben*.)

everything that humans have," Ratzel would later write, "because it has humans themselves."[36] While he thus held on to the notion that humans were in something of an exceptional position in comparison to other life forms, particularly due to their capacity for language and reflection, he always emphasized that they were not alien to the earth and should therefore not be treated in isolation from other life forms.

When Ratzel had first encountered the work of Charles Darwin (1809–1882), he was electrified by the elegant simplicity of the idea of natural selection. In his zoological period, he repeatedly described Darwin as a "genius" and dedicated an entire chapter of his 1869 *Sein und Werden in der organischen Welt* to him, noting in particular the importance of "variability, inheritance, the struggle for survival and natural selection."[37] By the year 1873 he had begun to casually use Darwinian terminology in his diary.[38] In the first volume of *Anthropogeographie*, he drew on Darwin again in theorizing cultural change, introducing him as "the greatest thinker" on questions of "variability and inheritance."[39] Ratzel's insistence on variability, including in his regional geography,[40] qualifies Stoddard's claim that geographers omitted probabilism when importing Darwin.[41] But the first evolutionary thinker to have shaped Ratzel was not Darwin but the German biologist and philosopher Ernst Haeckel, whose lectures Ratzel attended in Jena and whom he later corresponded with.

Known today especially for his colorful illustrations and his recapitulation theory, which claimed that the embryo developed in stages that resembled the evolution of its ancestors ("ontogeny recapitulates phylogeny"), Ernst Haeckel (alternative spelling: Häckel) was the most vocal proponent of a larger group of German evolutionary theorists. More readers are said to have encountered evolutionary ideas through the flowery writings of the Darwin enthusiast Haeckel than through his friend Darwin himself.[42] Even today, he is sometimes lauded as a "German Darwin" who successfully promoted the idea that complex life forms originated from bacteria.[43] Although Ratzel and Haeckel's contact was comparatively infrequent and the tone of their exchanges formal, Haeckel seems to have acted as Ratzel's mentor in earlier years. The two men not only exchanged their works, but Haeckel also provided references to Ratzel and in 1872 suggested him for a professorship at Graz, which Ratzel kindly declined, having not yet completed his habilitation.[44]

Haeckel's imprint on the early Ratzel was considerable.[45] A reading of the Jena zoologist's *Generelle Morphologie* (1866) drove many of Ratzel's arguments in the 1869 *Sein und Werden*, as he openly acknowledged.[46] Like Ritter, Haeckel had written about the terrestrial unity of all life. Like Darwin, he believed that all species existed in an entangled network of other creatures. In *Generelle Morphologie*, Haeckel had also developed his own brand of monism, a secular religious doctrine that "held that mind and matter were properties of a more fundamental substrate that was not to be identified with either of its salient traits."[47] It was from Haeckel, most importantly, that Ratzel took the view that there was no fundamental difference "between the organic and inorganic world,"[48] an ontology Haeckel may have himself adopted after coming to believe that his deceased first wife Anna Sethe had been resurrected in the form of a jellyfish.[49] Ratzel followed Haeckel on matter and life, which he saw as existing in a radical openness to one another. It was not just that life had once emerged from inorganic material, but rather that "a steady passage from dead to living, from inorganic to organic matter" could still be observed in the present.[50]

Haeckel was not, unlike his student Hans Driesch, a vitalist. He did not believe in a unique and untameable energy, a vital force or principle that could not be collapsed into the laws of chemistry and physics.[51] And yet his metaphysics of life saw both the organic and inorganic as governed by the same laws, which resulted in "a vitalization of matter as much as a materialization of life."[52] Whereas Ratzel had noted in *Sein und Werden* that there was no empirical evidence for a vital force, he had been preoccupied with the creative properties of life and would remain so in later years.[53] Life was an omnidirectional force that resulted in constant "change, development, deterioration."[54] "It overflows in all directions; its cause lies in the organism itself and no external factors are needed to activate its organic powers of movement," he claimed.[55] Although all life was made up of the same chemical compounds, it relentlessly produced an ever greater and more complex variety of organic materials.[56] The late Ratzel located in this vitality the very logic that made space appear:[57]

> [L]ife is a sum of internal movements that are triggered by external stimuli; life is metabolism with unvarying form: evidently, movement is expressed in all definitions of life. Now, this life is firstly an internal fact of the organism. But internal life will always generate external movement. Every increase in the organic mass, every growth, every reproduction signifies spatial movement.[58]

As we will see in chapter 6, Ratzel also came under the influence of Gustav Fechner in the mid-1880s, whose panpsychism and animism had a vitalist core too.[59] It is thus that Ratzel came to believe that life was not restricted to the earth but inhered even in the movement of celestial bodies. The ocean had a "spirit" and in "each thing of nature inhered a divine aspect."[60] The crisis of Darwinism seems to have prompted Ratzel by the early 1900s to flirt with a move into the budding neovitalist camp.[61] Indeed, he was tremendously positive in an anonymous paper about the neovitalist Johannes Reinke and even described the Darwin critic Karl Ernst von Baer as "the nineteenth century's Aristotle."[62]

Although Ratzel's view on Haeckel had by the 1890s become more critical, many of the latter's beliefs remained foundational to Ratzel's geography.[63] As Zigman rightly observes, the surviving letters Ratzel sent Haeckel between 1869 and 1904 question Steinmetzler's view that Ratzel disavowed Haeckel's *Darwinismus* in later years.[64] Ratzel had tried to play the role of a Haeckel popularizer and keenly reported back to his master about his achievements in the 1870s. And while Ratzel later abandoned this role, he remained full of admiration, variously addressing his former professor until the 1890s as his "teacher" or "master."[65] In the Leipzig geographer's final letter to Haeckel, written six months before his death, Ratzel noted that his views had "of course become more independent," but he emphasized that he would always remain hugely indebted to Haeckel.[66] Even in his latest writings, Ratzel still sought to collapse conceptual binaries.

Although the two men did not discuss politics in their private correspondence, Ratzel's views on race and conflict also overlapped to a significant extent with those of Haeckel, a fellow member of the *Alldeutscher Verband.* Like many of his contemporaries, Haeckel believed in a hierarchy of races.[67] Unlike Darwin, Haeckel insisted on the existence of lower and higher life forms.[68] So would Ratzel. Haeckel's 1868 *Natürliche Schöpfungsgeschichte*, translated into English in 1876 as *The History of Creation*, placed the "extinction" of the indigenous peoples of North America and Australia within the general struggle for the necessities of life.[69] This popular science book, which Ratzel had read, also railed against what he saw as a common misconception to construe nature as existing in some kind of peaceful harmony. "Nowhere in nature, no matter where we turn our eyes," he stated "does that idyllic peace, celebrated by the poets exist; we find everywhere a struggle and a striving to annihilate neighbors and competitors."[70] Haeckel's usage of the term "extermination" was in fact entirely congruent with that of Darwin, who had used the verb "exterminate" and the noun "extermination" more than thirty times in *On the Origin of Species*. Interestingly, Darwin had often employed "exterminated" "even where another, more neutral term such as supplanted or superseded would [have been] more accurate."[71] He had also used the notions of extinction and extermination synonymously within the same sentence. And while Darwin had not turned his thesis on natural selection into an attack on poetry, he had certainly spoken of a "war of nature."[72]

Ratzel never broke with the core principles of competition, variability, and inheritance, and he continued to read and engage with Darwin's findings in later years.[73] Interestingly however, Ratzel's Darwin remained that of the *Voyage of the Beagle* (1839) and especially of *On the Origin of Species* (1859). When the British biologist published *The Descent of Man* in 1871,[74] Ratzel barely paid any attention to it. The German geographer had read this book by the time he wrote *Völkerkunde* but only superficially engaged with its arguments. Sexual selection never became a concept of his. As Ratzel came into close contact with the Darwin critic Moritz Wagner, he began questioning some of Darwin's reasoning and adopted Wagner's position on speciation. In the early 1900s Ratzel noted that Darwin had been "an odd mixture of genius and philistine" and that "the survival of the fittest" was but a "crude hypothesis."[75] And yet he also concluded that the great clarity of Darwin's thinking had been clouded by some of his admirers.[76] The "survival of the fittest," the phrase Ratzel rejected, was of course not originally Darwin's, although Darwin did incorporate it into later editions of the *On the Origins of Species*, but rather that of Herbert Spencer (1820–1903), whom Ratzel had read attentively.

One of the most prominent intellectuals of Victorian Britain, Spencer promoted in his writings the idea that the laws that governed social and economic life ultimately reflected those of the universe. Ratzel had read both Spencer's *The Principles of Sociology* and the two-volume *Principles of Biology* and reviewed the latter favorably in the late 1870s.[77] His review of Spencer was only the second of more than six hundred book reviews Ratzel would pen over the course of his career. But it was Spencer's *Principles of Sociology* that left the biggest imprint. In this book, Spencer had developed his understanding of societies as organisms that were subject to growth and decline and whose structure became more differentiated as their size increased. Conversely, Spencer argued that a living organism was in fact a "nation of units that live individually, and have many of them considerable degrees of independence."[78] If the aggregate organism was destroyed, this did not immediately extinguish the individual units. Just as a heart kept pumping for a while after being detached from the body, so might the food production of a nation continue when the latter was defeated in war.

This kind of organicism emphasized the self-organizing capacities of larger entities, including very much the earth itself. And while organicism and vitalism are sometimes defined in opposition to one another, in Ratzel there were traces of both, alongside mechanist logics. Ratzel applied organicist metaphors to a range of entities. Crucially, he thought of organisms as having organs that could survive independently of one another and were able to separate and form new polities. Ratzel found Spencer's generalizations sweeping and lacking in broader geographical understanding. He undoubtedly had little time for the latter's expressions of anti-imperialism too.[79] And yet he noted in his journal in November 1895 that he was rereading Spencer's *Principles of Sociology* just as he was writing a crucial section on the state as organism in *Politische Geographie*.[80] Ratzel would later claim to prefer the organicism of the German sociologist and fellow Darwinian Albert Schäffle (1831–1903), who had made similar arguments to those of Spencer in the late 1870s, but he bemoaned that Schäffle had by the mid-1890s become overly critical of his own analogy.[81] The Leipzig geographer applauded the clarity with which Spencer had commented on spontaneous cooperation, despite having spent too little time on the "organic relationship between transport and state formation."[82] But this was ultimately merely an invitation for Ratzel to construct a new theory upon Spencer's "empty scaffolding."[83] This should not imply that he did not doubt his own organicist analogies. Indeed, he had in 1896 sought the view of the biologist Heinrich Ernst Ziegler on their usefulness. Ratzel ultimately decided to preserve.[84]

In Spencer, Ratzel found a thinker who had, much more than Darwin, articulated a theory of evolutionary processes that treated biological and social processes alike. Like Thomas Robert Malthus before him, Spencer was captivated by the effects of overpopulation. He spoke in terms of stronger and weaker populations. Indeed, he had described the poor as "nature's failure" and noted that "it is best they should die."[85] In *Principles of Sociology*, Spencer observed throughout human history "a continuous over-running of the less powerful or less adapted by the more powerful or more adapted, a driving of inferior varieties into undesirable habitats, and, occasionally, an extermination of inferior varieties."[86] Spencer not only used the term "extermination" precisely in the loose way Haeckel and Darwin did, but he also foregrounded the securing of habitat in the struggle for survival. For Spencer (though not for Schäffle), any form of social program to help the "weak" was a form of false benevolence that intervened into nature's course.[87] Ratzel would soon adopt a similar moral compass when writing about the decimation of indigenous populations.

Earthly Foundations

In the spring of 1872, Ratzel was posted in southern Italy to report for *Kölnische Zeitung*. He arrived in Naples shortly after the most powerful eruption of Mount Vesuvius during the nineteenth century, which had happened on 26 April 1872 (see figure 10). He decided himself to climb the volcano and peer down into its crater. Mesmerized by the occasional rumbling of the earth, he gazed at lava streams and waded through ash until his soles burned.[88] Ratzel was overwhelmed by what he saw. Language, he felt, could not capture this awesome spectacle, which he described as a *Gewalt*, a term that translates into English both as "force" and "violence."

Despite credible reports of deaths and mass panics, he was surprised to find the local inhabitants going about their daily business as if nothing had happened. They seemed to have been gripped by some kind of collective amnesia. Ratzel interpreted this as a sign of their regional character and one that was hardly coincidental.

> One would almost say that it was the hand of God which placed this light-hearted *Volk* here, where it must hang in between heaven and hell, if it did

Figure 10. The clouds of smoke above Mount Vesuvius. (Source: *Die Erde und das Leben*.)

> not appear more obvious to conclude that at least a part of its character was a product of the very conditions into which it fits so splendidly.[89]

It was this experience that convinced him that deep beneath the perceptible layer of social norms was another and more meaningful one that shaped the former. A decade later, he observed with great interest that the fear of volcanic eruptions had proven a fertile ground for social norms such as superstition in some of the earth's most geologically volatile regions.[90] His interest in the relationship between the above and below ground remained active in him until his death. "The life of humans," he noted, "is one of the plants of the flora subterranea, which thrives even in dark cellars and mining tunnels as well as it does in broad daylight."[91]

At its very foundation, Ratzel's human geography rested on the writings of Carl (alternatively Karl) Ritter. Sometimes credited alongside von Humboldt as the founder of German geography,[92] Ritter had set out to understand human existence within its larger physical setting. Often seen, and perhaps caricatured, as an armchair geographer who wrote extensively about regions he had never visited, Ritter had worked first as a private tutor for the influential Bethmann-Hollweg family and later as a professor at the University of Berlin and the nearby General War College (*Allgemeine Kriegsschule*), where he taught courses on a wide range of topics including terrain, statistics, and political geography. Ritter had decided in 1813 with a "bleeding heart," as Ratzel would later claim, not to enlist in the military campaign against Napoleon and to remain an educator.[93] And yet Ritter had subsequently shaped a generation of Prussian officers at the military academy, an institution at which the famous strategist Carl von Clausewitz lectured too. In this role, Ritter never got tired of promoting the core message of political geography, namely that the state was bound to its land. His students included not only the later field marshal Helmuth von Moltke but also the young Karl Marx.[94]

Ritter's impact on Ratzel was profound.[95] In 1879, Ratzel praised him for his attempt to unearth "the laws and conditions under whose influence the great variety of things, *Völker* and humans on earth, came into being, morphed, spread, and developed further."[96] He dedicated an entire section of *Anthropogeographie* to Ritter, applauding the latter's *Erdkunde* for uniting physical geographical and historical inquiry.[97] Ratzel did not just adopt Ritter's bombastic writing style, but also shared the latter's concern with environmental and historical transformations. He took from Ritter an interest in political geography, in oceans as connectors and in cultural distribution areas (*Kulturkreise*). Most importantly, Ritter had recognized that the planet needed to be understood as humanity's ever-transforming home, an "earth organism" (*Erdorganismus*) of which the biosphere was but a part.[98] The earth was an active force in the shaping of humanity, its "educational institution" (*Erziehungsanstalt*).[99] Life and earth thus existed in fundamental unity.[100]

Ritter had also questioned the distinction between organic and inorganic nature, noting the following on the opening page of his posthumous *Allgemeine Erdkunde* (General geography):

> Inanimate nature is often understood as so-called inorganic nature, whereas animate nature is referred to as organic nature. But this is not an absolute opposition: for in both there is progressive activity, no stasis, but in a higher, more comprehensive sense a cosmic life, the entirety of which is *one* large organism, within which so-called unorganized nature is only the basis and condition of animate nature.[101]

Ratzel too urged his readers to approach the earth as a whole, including its crust, hydrosphere, atmosphere, and biosphere. He spoke of the earth itself as organic, for he saw traces of life as scattered throughout the planet, often "mixed up with inorganic materials" to a degree that it would be impossible to separate them out.[102] At a time when exploration of deserts, oceans, and mountains were revealing new evidence of extreme life, he wrote that even the most inhospitable climatic conditions were able to host life forms. He was interested in the geologic forces and decomposition processes of small life forms that made the earth ready to carry new life. He saw this intertwining of life and earth not merely in competitive terms but as mutual cooperation, in which organic and inorganic forces conspired to make possible each other's existence. Ratzel was fascinated, for instance, by shells whose attachment to cliffs secured these landforms from erosion, by plants that were so interwoven that they helped to support the earth, and by species whose livelihood rested in the feces of other species.

Ritter's geography was naturally devoid of the scientificity later promised by Darwinian evolution. Although his vision of the earth was perhaps not quite the "harmoniously ordered whole" of von Humboldt's cosmos,[103] there was no struggle for existence or space in Ritter. While Ritter still believed that nature was progressively pushed back by civilization, Ratzel assumed that nature always already pervaded all social life. Indeed, Ratzel rejected precisely Ritter's teleology (and in earlier years the latter's religiosity too), arguing instead that evolution lacked direction. Ritter, in Ratzel's view, had merely sketched the contours of the great task of anthropogeography but had failed to provide the empirical evidence to substantiate his claims.[104] Ratzel thus saw his own mission as one of producing a more detailed geography on the canvas Ritter had provided. Ritter died in 1859, the same year Darwin published his magnum opus. It was through Moritz Wagner that Ratzel would eventually find a way to marry Darwinian evolution and Ritter's geography.

When Marie Ratzel was asked in 1901 to reflect on her husband's life, she noted that he had only ever had one notable paternal figure. This was not his biological father, but rather Moritz Wagner.[105] In her husband's eyes Wagner had indeed produced a "genius" theory that allowed both an insight into the evolution of life on earth and into human history.[106] The two men had shared a vaguely similar biography when they met in 1871.[107] Wagner had in earlier years supported himself and his research through travel writing and by collecting and selling specimens, though he would later also hold a number of posts in museums and the academy.[108] Wagner was known in the decades following the publication of Darwin's 1859 magnum opus for his insistence on the single factor of migration as an explanation for speciation. Thirty years Ratzel's senior, Wagner had still made the acquaintance of Alexander von Humboldt and is likely to have met Carl Ritter too.[109]

When Wagner first read Darwin's book, it had reportedly helped him to make sense of his observation, first noted in the early 1840s, that distinct though closely related species were frequently separated by geographical features, such as mountain ranges, deserts, and rivers. Wagner agreed with Darwin on the issue of variability but believed passionately that it was impossible for speciation to occur without a sustained separation of colonists from their conspecifics. This was due to incessant intercrossing between variant individuals within a population. Wagner was also convinced that only migration made a species or race competitive. In an 1868 essay, a version of which was translated into English in 1873, he argued that "[t]he competition of all beings for space, food, and reproduction, or the 'struggle for life' so appropriately named and illustrated by Darwin, gives the first impulse to migration."[110]

> Vegetable seeds wander by passive migration, that is to say, are transported by winds, rivers, seas, or animals. Birds, insects, and other animals generally wander of their own accord far from the station of their species; arrived in new localities, where they maintain the struggle, especially for food and reproduction, under somewhat altered conditions of life, their individual variability receives fresh stimulus; then, through uninterrupted activity of the same causes, the organic variations must accumulate during a series of years.[111]

Polar bears had, for instance, descended from albino individuals that had voyaged to ever colder regions in search of new and better habitat. Over time, their migration had induced hereditary changes because their organisms had been exposed to new climatic conditions, which included the food available in those colder regions.[112] Wagner termed his approach migration theory (*Migrationstheorie*) and initially thought of it as a corrective to Darwin's theory. He later renamed it separation theory (*Separationstheorie*), now deeming it a challenge.

When Wagner had first reached out to Darwin in 1868, the latter had been receptive and even made changes to later editions of *On the Origin of Species* to acknowledge Wagner's contribution.[113] Ultimately, however, he insisted that the individuals of a population *could* over a longer period of time evolve *within* the same territory. It was not that Darwin rejected the role of geographical isolation. In fact, he had written precisely about archipelagic settings. But he believed that isolation was, unlike natural selection, no necessary condition for speciation. In the 1870s, Darwin came to dislike the persistent Wagner, eventually writing across the cover of one of Wagner's papers the words "most wretched rubbish."[114] By foregrounding adaptation by migration and the role of environmental factors, Wagner downgraded natural selection and thus moved back onto Lamarckian territory.[115] Ratzel acknowledged Lamarck's contribution to the history of evolutionary biology and recognized too a Lamarckian core in some of Wagner's arguments.[116] And yet a firm believer in common descent, Ratzel explicitly rejected Lamarck as "naïve" and ridiculed the "inheritance of acquired characteristics."[117] While he certainly often treated environmental influences as acting in direct ways upon human history, at other times modification occurred only in an indirect fashion. Unlike Lamarck, Ratzel was a believer in slow evolution who found little use for the inheritance of acquired characteristics in his human geography but frequently emphasized that there was natural selection among polities. Ultimately, he was, like Haeckel and many other *Darwinisten*, a synthesist who saw Lamarck as a "precursor" to Darwin and used the phrase "Lamarckian-Darwinian theory" to highlight what he saw as a continuum.[118]

Many of the more than fifty letters Wagner wrote to Ratzel in the decade from 1875 to 1885 reveal an intimate personal and intellectual relationship between the two men.[119] They discussed new developments in their field and occasionally quarreled over academic matters, though this does not seem to have affected their friendship. Wagner, whose work Ratzel sometimes proofread, shared with Ratzel a dislike of socialism and of anti-imperialism.[120] Many of Wagner's letters also complained to his mentee about his declining health. Indeed, Wagner frequently compared his own aging body (he had sustained permanent injuries after jumping off a moving train) to Ratzel's youthfulness. In 1881, he noted his fear that he might not survive the next winter.[121] Four years later, he spoke of a young colleague who had suddenly become "a corpse."[122] There was then in Wagner a preoccupation with mortality, particularly his own. He would end his life in 1887.

Ratzel would be all too keen in the early 1870s to adopt Wagner's conviction that migration via geographical isolation was a necessary condition of natural selection. In the years that followed, he employed Wagner's thesis as the basis of his model of historical change. He also used Wagner's conviction that species (and not just organisms) underwent an aging process and that younger species and *Völker* were generally more competitive than older ones, a dynamic that had been mirrored on a personal level in Wagner's frequent insistence on the merits of Ratzel's youthfulness.[123] It is clear from their correspondence that Wagner saw Ratzel's main contribution as one of applying his migration theory to the history of races and *Völker*. Wagner had personally witnessed the displacement of populations in the French invasion of Algeria in 1830,[124] but only entertained a passing interest in human migration. He was drawn in the final years of his life to ethnology but ultimately recognized that it was Ratzel's anthropogeography that had performed the crucial conversion from biological to a cultural and political theory.[125]

Ratzel's critique of Darwin was largely Wagner's, and he remained committed to his mentor even after the latter's death. But because Darwin *had* considered isolation, Wagner had in fact remained much closer to the British biologist than he had been willing to acknowledge.[126] In this then, Steinmetzler in fact overstates the degree to which Ratzel too would become an "opponent" of the larger Darwinian project.[127] Despite Ratzel's support of Wagner's positions on evolution, there were important political differences between the two men. Although the eccentric Wagner had by the mid-1860s become a glowing admirer of Wilhelm I and Bismarck, he had entertained pacifist views too. The latter were of course fundamentally at odds with the beliefs of the former soldier Ratzel.[128] Despite these political differences, Ratzel was keen to apply Wagner's ideas to the human realm and repeatedly borrowed from his terminology. Not only did Wagner speak of "colonies" and "colonists" when referring to nonhumans, but he also used the phrase "struggle for space" (*Kampf um Raum*) in 1868.[129] While not a central trope in Wagner's writings, this phrase would become pivotal in Ratzel's later work. Darwin, of course, had also spoken of colonies in the sense that Wagner had, but it was Darwin's supposed inability to see that the struggle for existence was in fact a struggle for space that inspired Ratzel to venture into political geography.

Ratzel sought to develop a systematic language that could be used to understand human processes in a geologic register. The second volume of *Anthropogeographie* (1891) thus argued that language itself was both fossilized and stratified and could thus be excavated not unlike layers of sediment. He proposed that "in the same way as the geologist knows his [*sic*] index fossils, the existence of which guarantees the presence of a certain geological horizon, so can particular place names tell us about the former presence of a particular *Volk* in a region where linguistic traces have been preserved."[130] Read through a geologic lens, the great Migration Period of the fourth to the sixth century AD (*Völkerwanderungen*) had left "magma" behind.[131] This was not a mere linguistic game, for Ratzel noted how volcanic activity had impacted on place names over time. He was fascinated, moreover, by the speed at which organic life could blossom on volcanic soil after an eruption that had killed more than four thousand inhabitants of Java in 1837.

While Ratzel recognized the limitations of applying a geologic timescale to the much more recent phenomenon of human life, he was nonetheless willing to think ethnology as a form of geology. He spoke of concentric distribution areas, which featured "cultural and national layering" that resembled the "geological layering of the earth's crust."[132] Like Charles Lyell before him, Ratzel established "a geohistorical horizon [] into which questions of race and possibility [were] staged as self-evidently differentiated in time."[133] On the Crimean peninsula, for instance, he found

> an occidental and frequently blond phenotype in the steep and rugged mountains and darker and thin-bearded Tartars in the steppe [. . .], while Turkic features would give way to Mongolic ones beyond the Isthmus of Perekop. At the same time the southern population would appear as agrarian while the northern and eastern one would be revealed as nomadic.[134]

This highlights that Ratzel did not separate his racialized ethnology neatly from his (anthropo)geographical arguments, but rather sought ways to bring them together.

Geopolitics and political geography are often seen to construct a flat cartographic world of two-dimensional areas, which has neither height, depth, nor volume. And yet Ratzel's work was very much underwritten by an interest in what happened below the earth's surface. To drill into the earth's strata was to look into the planet's past. His ethnological work thus examined caves and grottos that had been used for human habitation and social rituals, while his political writing speculated about a future cultivation of the African drylands through reservoirs and drill pipes.[135] And although his analysis of the history of European sea power was mostly two-dimensional,[136] he was captivated by deep-sea exploration and the abundance of marine organisms that found living space even in the most inhospitable environments.[137]

Ratzel's biosphere was therefore a fundamentally three-dimensional space, which incorporated all life from the highest mountaintops down to the deepest seabed. His conceptual language of habitat and colonization applied to human and nonhuman life alike: the termite hill and the beehive were as much epiphenomena of the three-dimensional struggle for space as the multistory buildings humans had constructed. "On the ground, in the water, in the air, and in the different heights and depths of those elements there is room for the most diverse living," he declared.[138] Space was also dynamic. The oecumene denoted an ever evolving and temporary configuration of inhabited space. It tried to capture the relations between human movement and the earth over time, patterns that were inherently difficult to depict cartographically because of the way they had coproduced one another.[139] It is thus perhaps no surprise that Ratzel used comparatively few maps in his books and that even where he did, such visuals were rarely central to the arguments he was making.[140]

Ratzel was not the first sociopolitical thinker with an interest in geology. His *Anthropogeographie* echoed some of the writings of the seventeenth-century natural scientist and philosopher Gottfried Wilhelm Leibniz.[141] Ratzel was also not the only geographer to present arguments about geologic forces on human societies at the turn of the twentieth century. Halford Mackinder, a contemporary of Ratzel's, argued in 1902 that the "Industrial Revolution of English history could not have been accomplished without the Hercynian Revolution of geology, for the coal upon which the one change rested was preserved in the process of the other."[142] What Ratzel, however, had in mind was a more systematic vision of geography as a field that could navigate between "shallow" historical and "deep" geological time.[143] Geology, history, and geography were not just disciplines that happened to have the same object of analysis, the earth, but different perspectives onto this object of analysis.[144] The point was to carve out a niche for geography, both as a science and as a form of policy advice. For "true realpolitik," as his preferred mode of statecraft, had to have "a strong geographical core."[145]

Space and Time

Ratzel had already observed in *Anthropogegraphie* "a clear relationship between space and time," an understanding of which he saw as the objective of the geographical discipline.[146] Space was only meaningful in time. The same volcanic eruption would take longer to spread its ash over a larger planet than it would take to cover a smaller one. If the same two planets underwent an identical process of planetary cooling, the larger one would merely see its ice caps freeze, while the smaller one might be entirely engulfed by a layer of ice. Life spread over the earth in similar ways as did volcanic ash or ice. It was geography's purpose to understand the spatiotemporal nature of this diffusion. In doing so, it had to be recognized that older life forms inhabited the world next to and on top of newer ones. Evolutionary epochs coexisted within different continents and regions, even if some of the latter were shaped more profoundly by some of these periods than others. With this in mind, *Anthropogeographie* turned to the "struggle for space" (*Kampf um Raum*), arguing that the size of the space in which competition between species took place had a great impact on evolutionary processes. Wide and open spaces allowed for the evasion of an opponent; narrow spaces did not. Quickly, the Leipzig geographer arrived at an explanation of racial difference and the causes of war among humans.

Ratzel's major works were not his last word on the matter of time and space. He offered the most detailed treatise of this question in the posthumous *Space and Time in Geography and Geology*, pieced together with the help of Marie Ratzel in 1907 by the Leipzig philosopher Paul Barth out of a 1902 article and a lecture manuscript, which Ratzel had held in the months before his death.[147] Here Ratzel reiterated the arguments presented in the *Lebensraum* essays. Humans had spread around the globe, establishing an oecumene. As the population density within these habitable spaces had increased, so had the number of intersections, leading to the struggle for space. Interestingly, however, his discussion began not by highlighting the political implications of his (Wagnerian) evolutionary worldview, but by unpacking the etymology of the word *Raum*, noting its connotations of vastness and emptiness. Even in Old High German, Ratzel added, its meaning was already intertwined with that of time. The word *Zeitraum* (which translates loosely into English as "period" and more directly as "time space" or "time span") was witness to the fact that spatial awareness needed daylight. The original *Zeitraum* was the "space in which the sun shines."[148] Time revealed itself in movement across space. A carrier pigeon could orbit the earth in a mere nine days, while it would take a snail around nine hundred years to complete the same journey. But how, he wondered, did those short time spans pale in comparison to the millions of years that a particular coastline had taken to move a comparatively short distance?[149]

Ratzel had long concurred with von Humboldt that any geographical inquiry should in reality "begin with a description of the heavenly bodies," and he had thus begun a number of his books with a discussion of astronomy.[150] Ratzel noted with interest that distances were now measured in light years and that a view into the nighttime sky was always a view into the universe's past. "In space, we read time," he observed.[151] Implicit in this observation was of course that he had always been dealing with spacetime, rather than just time and space. These later writings predate not just Albert Einstein's insistence on the time and space continuum but also the mathematician Hermann Minkowski's 1908 notion of a spatiotemporal "union," which is sometimes seen as a precursor to Einstein's 1915 general theory of relativity.[152] Toward the end of his life, Ratzel also developed, as he admitted, an idiosyncratic conception of planetary history, which speculated that before the birth of planet Earth there had been earlier earths that too had supported life. He conceded that it was difficult to verify this theory but expressed his hope that previous creations might be deduced from the process of recapitulation that Haeckel had identified in embryonic development.[153]

Drawing on Immanuel Kant, Ratzel explained further that both space and time were nonempirical and thus a priori forms of intuition (*Anschauungsformen*). Although he agreed with Kant that time and space were inherent in the human mind as a ground on which experience took place, he argued that this spatiotemporal consciousness was itself an evolutionary advance that had been prompted by environmental stimuli in the distant past.[154] This was only a nuance, but a crucial one because it revealed an attempt to fuse Kant's position with that of evolutionary biology. As it was not possible for humans to step out of space, he argued, space was never empty. The very idea of empty space was an abstraction that relied on a mental emptying out of this space from objects. Space did not itself have any defining properties; it was made meaningful only by the objects and experiences that filled it. Space thus had an objective and subjective dimension. Objective space was a great and unreachable unknown, which humans could only attempt to approach by ridding themselves of their subjective notions of space. And while subjective space was ultimately brought into existence by objective space and thus in a sense subordinate as well as unapproachable, objective space left traces in the world that could be glimpsed from the perspective of subjective space. It was perceptible in the smallest sections of matter that presented themselves in the universe.[155] The earth itself was thus like a "vast clock face" that allowed humans to measure and understand movement across space.[156]

Kant's theory of space, which Ratzel found inspiring, sought to navigate a middle ground between Newton's view of space as a mere container of things (absolute space) and Leibniz's concept of space as defined by the relations between things (relative space).[157] Ratzel attempted a similar compromise, in that space for him was both a property of things in the world and something that emerged in between objects. Although the earth placed absolute limits on life's expansive tendencies in that "[e]verything that wants space on our planet Earth must draw on the finite amount of 506 million square kilometers of its surface,"[158] there was also something relational about Ratzel's spaces. "[S]pace is something that lies outside the organism, and yet each living thing is bound to its space and *connected to its space*," he argued.[159] Ultimately, Ratzel tried to fuse a Kantian concept of space with an evolutionary understanding, which he had developed from his reading of Darwin and Wagner. There remained, however, a tension between these two visions, in that the struggle for space was not implied by his Kantian assumptions.

There is another and more epistemological reason why the late Ratzel's conception of time and space deviated from a Newtonian one: his growing realization that science was a necessary but insufficient way of grasping nature, particularly the latter's aesthetic qualities. Shortly before his death, he completed *Über Naturschilderung* (The description of nature), designed as "an introduction to an artistic or aesthetic geography," which offered "a synthesis of literary and scientific representation."[160] Indeed, Ratzel argued in this book that even that most human of spheres—art—was but "the highest and most blossoming shoot of nature."[161] Nature, he observed with considerable admiration, "encloses art, invades it from all sides, grows together with it."[162] "What else is the whole history of fine art," he asked, "but a variation of motives that nature provides?" Indeed, he was even willing to argue that the artist "was himself nature," simply representing the forces of nature. In this sense, there was already "style" in landscapes that no artist had ever seen.[163]

Ratzel was keen to produce geographical knowledge that privileged direct experience. Although he occasionally spoke of "laws," he treated these as "complex, probabilistic and very un-Newtonian statements of general tendency."[164] He thought that geography had to be experienced, whether it was a volcanic eruption or the gentle rattling of chains that emanated from a nearby barn at night. In the early 1900s, Ratzel returned to the Humboldtian conclusion that science had its limitations and that "geographical description [was] not a purely scientific but to a large degree an artistic task."[165] It should be noted that Ratzel was versed in literature and fine art and produced the occasional Romantic poem. His autobiography was explicitly written as a hybrid of fiction and nonfiction. There had always been in him a creative streak and a vivid imagination that even some of his admirers referred to as a "tendency to speculation."[166] By the end of his life, Ratzel had finally come to terms with this second identity of his and wanted to extrapolate a larger meaning for his discipline: geography had to be guided by science *and* art.

Ratzel wanted geography-as-art not to remain "a slave to reality" but to bring out nature's beauty by "presenting the truth with force and wealth."[167] Observation was obviously crucial to a better understanding of time and space, but so was feeling and listening to the rhythms of nature. It was clear to him that the geographer had to "feel himself into the landscape" (*sich in die Landschaft hineinfühlen*). Indeed, he emphasized the necessity of a certain "naivety and imagination" and held that most people were too driven by reason.[168] Their felt experience (*Empfinden*), he was convinced, was far superior to their capacity for rational reflection.[169] Landscape description should seek to reproduce a certain "atmosphere that hovers over nature and the work of humans in nature."[170] Underlying all this was thus a longing to be enchanted, to be immersed by a naïve disposition of wonder. "How great is a child's reality," he pondered, "how much more does it own and master because it has the wonderful?"[171]

Conclusion

Friedrich Ratzel neither coined the term *Lebensraum*, nor did it express a new revelation of his. He used it to capture a number of ideas he had already laid down in the 1880s and 1890s. Central among these was the vision of a cosmic balance of interdependent life forms, all of which depended on the earth's resources, but between and within which there was competition and struggle. This argument rested on a number of ontological assumptions, most crucial a belief in the impossibility of separating life and earth, organic and inorganic matter. And yet the two *Lebensraum* essays marked a subtle transformation in Ratzel's thinking, which de-emphasized the iron grip of environmental agencies and foregrounded instead the creative energies of life itself.

As we have seen, Ratzel's geography largely took shape at the intersection of early nineteenth-century German geography and Anglo-German currents in evolutionary biology. At this intersection, the figure of Moritz Wagner stood out. It was Wagner who had introduced Ratzel to Ritter's work, which would ultimately provide Ratzel with a *Weltanschauung* that allowed him to see the entire planet as a structured and evolving whole. By the early 1900s, Ratzel had become keenly aware of the limits of scientific inquiry. Geography was not merely the gathering of evidence but of experiences, both of which had to be expressed through a form of writing that was enchanted by the entanglements of human and nonhuman becoming and disappearance. What thus emerged in Ratzel's later work was a concept of space that was significantly more complex than is often assumed in the existing literature. Ratzel thought of space as relational and inseparable from time. Ratzel's *Raum* was a three-dimensional and ever-evolving assemblage of competing material agencies. In what is left of this chapter, I will discuss where his materialism can be placed in relation to other materialisms, both historical and contemporary ones.

Although Ratzel is said to have participated in lively beer hall discussions about socialism during his Munich years,[172] he did not engage directly with historical materialism in his published work. He frequently mentioned the "lower classes" in his travel writing, often condescendingly, but he did not seem to have entertained a rigorous concept of social class. Ratzel did not address economic processes through the language of political economy but preferred to do so through that of botany. The Marxist Karl Wittfogel would criticize Ratzel precisely for this move in the late 1920s, at a time when the latter's ideas were gathering force. Wittfogel held that there was no room for capital or labor in Ratzel's theoretical edifice. Ratzel treated the state and other social organisms as "earthworms," which lived directly off plants and animals that inhabited the soil.[173] There was more than a hint of sarcasm in these words, for Ratzel had indeed written his dissertation on earthworms. Wittfogel implied that it was precisely Ratzel's background in the natural sciences that hindered him from recognizing that natural resources first had to be "produced and made consumable through labor." Instead of a detailed study of agricultural production, Ratzel often left his reader with a supposedly self-explanatory soil mysticism.[174] "Having [] removed the problem of the social sphere and therewith the urgency of a social question, Ratzel has only to let the economic sphere disappear and the trick is accomplished; a direct relation between state and soil is established."[175] Because there was no political economy in Ratzel, his theory was only inadvertently correct, Wittfogel argued. In fact, Ratzel's organismic metaphors were necessary precisely to conceal his theory's failure to explain historical change.

Like many scientists who pledged allegiance to nation and monarchy in Imperial Germany, Ratzel did not find it necessary to engage directly with Karl Marx or the brand of Marxism that was making itself felt in many corners of the urban bourgeoisie in the second half of the nineteenth century. Indeed, he was embarrassed that the founder of the *Sozialdemokratische Arbeiterpartei Deutschlands* (Social Democratic Workers' Party), August Bebel, cited him approvingly in his book *Die Frau und der Sozialismus* (Women under socialism).[176] But it was not just that he expressed his disregard of Marxism, but that he ignored the inner workings of a globalizing capitalist economy, limiting himself instead to a description of capitalism's symptoms, particularly urbanization and trade. Ratzel did discuss "primitive communism" in passing, but he was not one to romanticize it. "The first thing that makes a European, among the pastoral races of Africa or the hunting tribes of North America, feel that he has left the constraints of civilization behind him," he noted, "is the way in which rights of property are in some cases neglected."[177] There was, we can conclude with some confidence, not a trace of historical materialist or socialist thought in Ratzel, however much some Marxists sought, ultimately unsuccessfully, to marry his brand of geography with Marxist political economy in the decades after his death.[178]

We arrive then at the question of Ratzel's environmental determinism, a label used by his critics. The broad consensus in the existing scholarship is that despite the convoluted and diverse nature of many of Ratzel's writings, they rested on a determinist stance, sometimes implicit, that the environment shaped, if not controlled, the course of history.[179] And yet Natter suggests that Ratzel was in fact more of a possibilist, while Stogiannos even declares Ratzel's geodeterminism a mere "myth."[180] Here, a nuanced answer is called for. On the one hand, it is impossible to deny that materials overdetermined ideas in Ratzel. In fact, ideology and religion, with the exception of indigenous beliefs, remained both underexamined and unexplained in his works. On the other hand, ideas were not entirely absent. Ratzel believed that tools, for instance, were "animate" (*beseelt*). Rather than rendering them "mute matter," he treated them as "shapes inhabited by ideas."[181] By the 1890s, Ratzel was "becoming more, not less, wary about possible determinist implications in his understanding of the interaction between geographical setting and human agency."[182]

But geographical imaginations were to him always directly or indirectly produced by the *Umwelt*, so that there was, in principle at least, variety in spatial conceptions only to the degree that there were different environmental conditions. "The problem of gaining the mastery over space brings man every moment face to face with the things of nature; the result is a literal and *material tendency of thought*," he held.[183] Faced with a variety of spatial imaginaries under the same climatic conditions, he was often quick to resort to racial essentialization or declare some spatial imaginations as mere misconceptions. In short, some *Völker* had a stronger, others a weaker awareness of space. History would punish the latter, thus leading them to disappear in the long run. Another move was to declare the ways in which particular ideas were materially produced as *difficult to trace* due to the complexity of interlinking environmental factors. As Smith notes, "[c]ontradictions contained within Ratzel's theories tended to be disguised as mere complexities."[184] This enabled him to gloss over the cases about which his theory had little to say. Ultimately, ideas were second-order phenomena to Ratzel, and he dealt with them as swiftly as he could without opening himself up to the charge that he had not considered them at all.

Humans, moreover, were often rendered passive, and humanity's capacity for freedom and self-determination was frequently downplayed in Ratzel's human geography, leading some to conclude that he treated them as mere disturbances.[185] This de-emphasizing of human agency was rooted in his evolutionary perspective, which had prompted the early Ratzel to blur the boundary between animal instinct and human intellect. He later recanted, arguing that while the distinction between animal and plant was a superficial one, the human mind was a new and game-changing phenomenon from an evolutionary standpoint.[186] And yet Ratzel continued to treat the human mind and its capacity for independence with suspicion. He even speculated that the higher development of the mind might ultimately prompt the human species to "wilt earlier" than older and more primitive life forms,[187] thus implying that advanced cognition brought with it a certain evolutionary disadvantage. Most importantly, Ratzel was unconvinced that consciousness was the decisive piece of evidence that proved humanity's ability to determine its own future.

> It is true that the bright light of freedom shines over human society. But does [this society] not return to the dust from which it was taken? We do not know to which higher sphere the human spirit belongs. What we do know about humans belongs to the material, physical and evolutionary properties of the earth.[188]

It seems that Ratzel was puzzled by the human mind and thus preferred to focus his attention on the more tangible aspects of the material world which his zoological and geological training had prepared his intellect for. Although the first edition of *Anthropogeographie* conceded that the mind had liberated humanity from the shackled fate of the animal kingdom, freedom was still no more than "a gift of nature," something to be explained through evolutionary principles in ways that would ultimately reveal the limitations of human agency.[189]

But despite the determinist streak that underwrote many passages of *Anthropogeographie*, Ratzel was not always motivated by the question of environmental causation. Even in his earliest work, he had not made a simple equation between natural resource wealth and economic development. The silver and gold mines of Peru and Mexico had left these societies drained and poor.[190] He had always been concerned with human transformations of the earth and how cultural landscape (*Kulturlandschaft*) had been forged out of a natural one. His *Anthropogeographie* commented on agriculture, human settlements and ruins in detail. In *Die Erde und das Leben*, he noted the Suez canal's impact on trade routes.[191] "A *Volk* that inhabits and works the same soil over the course of millennia," he argued, "necessarily transforms this soil, giving it a new appearance."[192] Humanity's impact on vegetation and animal populations was thus potentially considerable.

Even if the power of the earth was moreover limitless, it was hardly immediate. Ratzel was skeptical, for instance, of Cuvier's idea that external events such as floods, fire, or ice ages had destroyed the entire flora and fauna at the end of each geological era.[193] "[N]o species or race," not even the dinosaurs, had "suddenly died off"; they had slowly been driven into spaces with poorer living conditions.[194] "Geologic catastrophes do not intervene profoundly into human history," he concluded.[195] Ratzel undoubtedly felt that catastrophism was no more than a scientific relic of biblical beliefs, but he was perhaps also thinking back to the Neapolitan nonchalance he had witnessed after the eruption of Mount Vesuvius in 1872. Too great was the discrepancy between what he deemed to be the *slow* processes of deep geologic time and the *rapid* transformations that could be observed in shallow historic time. The point was not that geology did not impact on human societies, but rather that it did so slowly, over the course of many generations.[196]

We can conclude from this that a certain environmentally determinist streak remained in Ratzel's later work, even as his writings increasingly gravitated towards a more vitalist ontology. While he was eager to emphasize the limits that knowledge of physical geography could bring to his political geography and warned against an assumption of simple causality between geological agency and national character,[197] his mature oeuvre still sought to understand "peculiarities that must have geological causes."[198] As so often, Ratzel's arguments were tentative: although they assumed causality, he conceded that it was often impossible to decisively demonstrate such connections. Indeed, his works were ambiguous enough to appeal to those, like Carl Sauer, who were strongly opposed to environmental determinism.[199]

Although Ratzel cultivated a theoretical persona that was "open to all sides,"[200] including to neovitalism, Christianity, and Darwinism, Ratzel's most characteristic move was the transformation of a plethora of ideas that were only implicitly political into a unified political theory. In doing so, he displayed a tendency to take from other thinkers primarily those ideas that have since emerged as their most problematic, while often showing little or no interest in their more progressive sides. From Darwin he derived the belief in the naturalness of the extermination of indigenous populations but not his abolitionism. In Spencer he found organismic struggle but neglected his anti-imperialism. From Wagner he borrowed his monocausal insistence on migration but ignored his pacifism. And although we will see in later chapters that Ratzel was read on the political left, these readers did not necessarily share his politics. It is the latter to which we will turn in the following two chapters.

Chapter 5

Colonization and War

Throughout the nineteenth century, Hamburg- and Bremen-based shipping companies had actively engaged in colonial trade. But they had only traded with the colonies of other European powers. Even after German unification in 1871, Chancellor Otto von Bismarck had remained focused on European politics and promised there would be no formal colonization so long as he remained in office. But in the spring of 1884, Germany claimed Lüderitzland, a small coastal territory in modern-day Namibia that had been named after a German tobacco merchant and out of which would soon emerge German South West Africa. The public mood swung decisively in the direction of colonization. In June, the Iron Chancellor announced a new colonial policy to the Reichstag, which, although still emphasizing his preference for economic over settler colonialism, promised to protect the commercial sovereignty of the new colonies, also known as "protectorates" (*Schutzgebiete*).[1] German colonial expansion was fast paced. In July of the same year, Togo and Cameroon were declared protectorates; East Africa followed in early 1885, making Germany one of the largest empires on the globe.

On the eve of the 1884 federal election and thus in the midst of Germany's entry into what would soon be known as "the scramble for Africa," Friedrich Ratzel gave a keynote to the inaugural convention of the National Liberal Party in Munich. The party, later termed the "imperial party" by the second president of the Pan-German League Heinrich Claß, had emphatically embraced the colonial cause and would continue to do so until its dissolution in 1918.[2] Ratzel noted his embarrassment upon having to explain to foreigners why Germany had established no colonies of its own. All great powers, with the exception of Italy, had done so, whether in India or Bosnia. He lamented the lack of popular support for Germany's settler colonial aspirations. Large segments of the German elite, he felt, had failed to see that colonies were both a crucial element of "a nation's armament" and a "natural necessity."[3] German colonists had thus neither been backed by a solid colonial strategy, nor had their attempts "to

Christianize the negroes" received the recognition they deserved.[4] Instead, they had perpetually been undermined by a group of "sterile incompetents" who did nothing but whine about the empire, the *Reichsnörgler*.[5] There was an immediacy in his tone. The question for Germany was no longer whether to acquire colonies, but where and how.

As so often, Ratzel used events in the United States to illustrate his points. German emigration to America had powerfully revealed the perils of a decentralized and haphazard colonization without state support. German settlers in North America had lost their language and culture, even though German street names still spoke of their deeds. This "bloodletting" on the German body politic had been a lasting concern of his since his visit to the United States in the mid-1870s.[6] Two hundred years earlier, he explained, it would have been wise to colonize the New World, but the opportunity had passed. With the Western Hemisphere now saturated with colonists, only Africa and the Pacific were left as viable options. Africa might be more difficult to develop agriculturally than North America, but superior German technology and farming practices, he announced, could be brought to bear even on the most inhospitable soil. "Who dares to question," he asked, "that Africa, wherever it has highlands, will absorb white settlers?"[7] Ratzel was convinced that energies, which had previously only slumbered in ordinary Germans, would be set free once these citizens had become colonists. All that was required was for the state to inject "a dose of German steel into the blood of these ethnic compatriots."[8]

Geographers like Ferdinand von Richthofen had lobbied Bismarck since the late 1860s to acquire a naval post on the East China Sea.[9] The message of Ratzel's speech was more ambitious. The same drive that had united Germany in 1871 after centuries of sectionalism (*Kleinstaaterei*) should now be channeled into the acquisition of territory in Africa. Underlying this call for colonization as a form of national expansion was the assumption that migration, war, and colonization were all epiphenomena of the same drive. This drive was fundamentally biological but expressed itself too at the level of a *Volk*'s geographical imagination. This was an idea Ratzel had already aired in earlier works and which his 1897 *Politische Geographie* would develop further.[10] Here he would suggest that the defining feature of successful colonization was a "greater conception of space" (*größere Raumauffassung*).[11] Germany desperately needed to cultivate just such a greater sense of space, for its absence would ultimately lead to national decay. There was no better way to expand one's spatial horizon than in the *school of war*.

In November 1884, two months after Ratzel's keynote, Bismarck invited the colonial powers to Berlin to formalize the division of Africa and set in stone

Germany's colonial status. Lasting more than three months, the 1884–1885 Berlin Conference, known also as the *Kongokonferenz*, redrew Africa's borders in ways that would have lasting effects on the continent until this day. Over the next decades, the German Empire became embroiled in a string of colonial wars. As part of an influential network of German colonial enthusiasts, writers, explorers, and administrators, Ratzel interacted with some of the most prominent figures of the German colonial movement, men like Friedrich Fabri, an early advocate of settler colonialism, and Carl Peters, Germany's most famous land grabber in Africa. Although Ratzel never visited Germany's colonies, he supported these men by wielding his pen and by using the institutional authority of his position.

In chapter 3, we began to outline the contours of Ratzel's settler colonial fantasy. This chapter explores this question further, examining both how he sought to theorize the colony and how he used personal connections to further the colonial cause. It situates his theory of colonization within his broader political geography and the concept of war developed therein, examining the extent to which we can find in Ratzel already a fully formed geopolitical worldview, as it would come to define the decades following the end of the First World War. We will pay particular attention not just to his scholarship, but also to his many journalistic interventions. The next chapter will delve deeper into the colonial dimension of his work, foregrounding the death motif that was so crucial to the majority of Ratzel's writings.

The Drive to Colonize

Colonization was one of Ratzel's lifelong obsessions. Already in the 1870s, he had studied German settlements in the Transylvanian highlands with great interest, framing them as colonies engaged in a racial struggle.[12] In 1880, he wrote two front-page commentaries in the major national newspaper *Allgemeine Zeitung* on the tragedy of German emigration without colonization. He lamented that German emigrants were often greeted by natives with pity or contempt. The problem, he explained, was that Germany had arrived on the global stage only after the world had been divided up. It should now look to acquire the necessary space (*Ellbogenraum*) in Argentina.[13]

In 1882, Ratzel became one of the founding members of the *Kolonialverein* (German Colonial Society), having been inspired by the early spirit of the organization, particularly the speeches of the theologian and writer Friedrich Fabri (1824–1891).[14] Fabri had in 1879 published a manifesto under the title

Bedarf Deutschland der Kolonien? (Does Germany require colonies?), often seen as a crucial early catalyst for the German colonial movement.[15] As Hell notes, the first volume of Ratzel's *Anthropogeography* appeared a mere three years after Fabri's manifesto, providing a more abstract and scholarly attempt to legitimate colonial expansion.[16] The two men continued to exchange letters in the late 1880s. Fabri reported to Ratzel on his attempts to forge a common colonial society out of a number of competing associations. He also sought to align Ratzel behind the controversial Carl Peters, seeking to dispel any doubts the geographer might have by writing to him that Peters "takes advice and shows me a special, great level of trust."[17]

Ratzel's role in the colonial movement of the early 1880s was to promote colonial enthusiasm in southern Germany.[18] He remained committed to the cause after he moved to Leipzig in 1886 and continuously wrote about German colonial affairs in his newspaper columns. Ratzel became part of a highly visible network of intellectuals who assertively promoted settlement in the southern hemisphere in the years following German unification. The Leipzig geographer used his often anonymous op-eds to comment on political developments in other European empires and assess the achievements of individual explorers and administrators. He also remarked in microscopic detail on infrastructural issues in the German *Schutzgebiete*, such as irrigation and road construction, as well as the political relations among Central African tribes.[19] His informants included the chief representative of the South West Africa Company, Georg Hartmann, a former student of his.[20]

In many ways, Ratzel's colonial enthusiasm was characteristic of the nineteenth century more generally in that it framed the colony both as a source of national prestige and a space of oriental otherness that enchanted the western mind. "Germany's colonizing presence" he held, had been "pioneered by the enterprising and capable nature of individuals," particularly of course by male explorers.[21] But Ratzel was in a more radical segment within the colonial movement and continued to fly the colonial flag even when the public began in the 1890s to become increasingly concerned about the political and financial costs of the colonial project. He found any speculation that Germany might exchange some of its unprofitable African territories for more attractive ones deeply disrespectful toward the German settlers who were already cultivating African soil.[22] Clearly he identified with these colonists, who he believed spearheaded the nation's expanding living space.

Following Fabri's steer, Ratzel began to admire the ruthless adventurer and *Reichskommissar* for the Kilimanjaro Region Carl Peters (1856–1918), a man so notorious for his violence against civilians that Hannah Arendt would later describe him as a German "Mr. Kurtz."[23] A letter dating from 1889 to his Leipzig colleague and future leader of the Pan-German League Ernst Hasse implies that Ratzel was well aware of the nature of Peters's undertakings in Africa, but nonetheless urged Hasse to support him.[24] In 1890, Ratzel wrote directly to Peters to invite him to Leipzig, praising him as a "genius and a warm-hearted patriot" and thanking him for his services to science and fatherland.[25] Whether Peters enjoyed the flattery or just hoped to forge links with academic geographers is unknown, but he certainly agreed. A surviving transcript reveals that he introduced Peters to his audience as a man around whom "wafted the magic of a historical figure," a man who had made "the dreams we had as young men come true."[26] When reviewing Peters's travel writing in 1891, Ratzel labeled him an "excellent compatriot" who should not merely be applauded for his military victories against various African tribes, but also for his intellect and writing style.[27] He saw in Peters a new colonial man who could wield both the sword and the pen. Only when the public mood had firmly turned against Peters in the late 1890s, in what came to be known as the colonial scandals (*Kolonialskandale*), did Ratzel denounce Peters.[28] But it was not the brutality with which Peters had governed or his obsession with racial superiority (*Herrenmenschen*) that Ratzel disapproved of, but rather the latter's reputation for sexual relations with African women. Although he had by 1897 come to ridicule the many "Nietzscheans" who had spotted in this man the *Übermensch*,[29] this said less about Peters than it did about Ratzel's anxiety around racial mixing.

Ratzel corresponded with a large number of colonial adventurers and cartographers, many of whom served as administrators in the *Schutzgebiete*. In fact, the list of names with whom he exchanged letters reads like a who's who of German colonialism. This group included the aforementioned Fabri and Peters as well as the colonial writer Wilhelm Hübbe-Schleiden, who unsuccessfully tried to enlist Ratzel in the effort to found a German university in the United States.[30] Ratzel was in close contact with his colleague Hans Meyer, a geographer and the first European to climb mount Kilimanjaro, and Erich von Drygalski, whose 1901–1903 expedition to Antarctica Ratzel supported.[31] As chair of the Leipzig *Verein für Erdkunde* (Society for geography), Ratzel provided an interface between explorers and the general public. His academic position also allowed him to offer institutional standing and financial stability to individual explorers. He made sure that Oscar Baumann, the Austrian explorer of Africa, received a PhD from Leipzig after Baumann had complained to him that his was a "glorious but hardly solid profession."[32] The orientalist travel writer Hugo Grothe too sought in 1903 to complete a habilitation under Ratzel's guidance but was forced to abandon his plan when Ratzel died the following year.[33] Ratzel's writing also benefited more directly from these interactions. The Austrian explorer Heinrich Klutschak reported back from polar expeditions, promising Ratzel drawings of "as yet unknown" Innuit tribes.[34] Baumann, who had personally witnessed the tyrannical rule Peters had established in the Kilimanjaro region, became Ratzel's research assistant while under contract with the German East Africa Company.[35]

Even more crucially, Ratzel exchanged letters with the later *Reichskommissar* for West Africa Gustav Nachtigal. In 1880, Nachtigal had invited Ratzel to Berlin to speak at the *Gesellschaft für Erdkunde* (Society for geography), over which he had then presided.[36] Four years later, he had laid claim to areas in Cameroon for the German Empire. When he died shortly afterwards, Ratzel praised Nachtigal's "world changing deeds," which one did "not overestimate by recognizing in them the beginning of the conclusion of the national rebirth of Germany."[37] The aforementioned chief representative of the South West Africa Company, Georg Hartmann, provided Ratzel with a detailed account of his expedition to the Kaokoveld Desert, asking the Leipzig professor to report his findings to the Leipzig *Gesellschaft für Erdkunde* and help him find a publisher.[38] In 1888, Ratzel corresponded with Ernst Hasse to organize an event with the explorer, administrator, and military commander Hermann Wißmann.[39] He found in Wißmann an excellent physical and political geographer and praised his *Im Innern Afrikas* (In Africa's interior) as "one of the most substantial and stimulating" books on the African continent.[40] Wißmann's East African *Schutztruppe*, which had been formed out of a private militia, suppressed the Abushiri revolt of 1888–1889. Ratzel was captivated by reports of the German military campaign, which came at considerable human cost.[41] Wißmann's travel companion, the doctor, anthropologist, and *Reichskommissar* for the German colony in Togo, Ludwig Wolf, was a personal acquaintance of Ratzel's with links to the Leipzig community of geographers. Upon the news of Wolf's death, Ratzel wrote two glowing obituaries, one of which was for the *Allgemeine Zeitung*. Quoting from one of Wolf's letters to him, he noted that Wolf had "succumbed to the struggle for Africa."[42]

In other words, Ratzel's role was that of an armchair cheerleader for Germany's most well-known colonists and explorers. The latter had been objects of fascination in Germany before the establishment of formal colonies. In the 1880s, there had been a marked shift to framing these adventurers as national rather than cosmopolitan heroes.[43] Ratzel's exuberant admiration for these men was not limited to Germans and Austrians and extended, for instance, to the Scottish explorer David Livingstone and the Welshman Henry Morton Stanley.[44] Ratzel did not acknowledge this of course, but he knew there was a great chasm between the kind of appreciation which someone like Stanley—or even his Leipzig colleague Hans Meyer—had received for their African expeditions and the recognition he had himself obtained for his journey to North America in the 1870s.

Ratzel harbored a particularly deep appreciation for the doctor, naturalist, and celebrity explorer Emin Pascha (1840–1892), whom he saw, like Peters, as the ideal fusion of scientist, adventurer, and colonial administrator.[45] Born into a Silesian German-Jewish family as Eduard Schnitzer, Emin had risen through the ranks of the Ottoman Empire and made it to the position of governor of Equatoria (in modern-day Sudan). When the Mahdist war broke out and Emin's troops were cut off, Ratzel supported an expedition to rescue him in the late 1880s. The Emin Pascha relief effort, which was led by Stanley and involved Peters too, and which would serve as an inspiration for Joseph Conrad's 1899 novel *Heart of Darkness*, was ultimately unsuccessful and became notorious in Germany for its violence, its collaboration with slave traders, and its high level of casualties.[46] Ratzel was so enamored with Emin Pascha that he signed a letter in 1891 to the Reichskanzler that urged him not to end Germany's collaboration with him.[47] He had also collected money for the relief expedition. Unused funds were transferred in 1890 to support the construction of a naval vessel *Dr Carl Peters*, which was intended to operate on Lake Victoria but which was never built.[48]

After initially eying territories in South America, Ratzel had in the first half of the 1880s concluded that Africa was the most attractive of all available spaces for German colonization. Although he believed that Europeans struggled with the tropical climate, he was optimistic that neither the climate nor the quality of the African soil would ultimately inhibit the establishment of large-scale German-run plantations. Africans, he explained, were ideal colonial subjects. They were neither "disappearing" like the indigenous population of Australia (more on which in the following chapter), nor were they engaged in racial mixing with white settlers to the degree that he had observed in Mexico. "Calmly, the negro allows the white man to rule over him," he claimed, a perfect precondition for the establishment of a social order based on a rigid racial hierarchy.[49] The outcome could be, he believed, a symbiosis of a "weak" and a "strong" race, similar to that which had characterized the Atlantic slave trade.[50] In order to rule over and civilize African populations, Ratzel fantasized about a new colonial man: intelligent, educated, and resilient toward any denationalizing tendencies. The Germans, he believed, were prototypical colonial masters, for they united in them everything that had made the Dutch such success-

Figure 11. Ngaoundéré, Adamawa region, Cameroon, after Curt von Morgen's description. (Source: *Die Erde und das Leben*.)

ful colonists, but with added military, bureaucratic, and scientific discipline (figure 11).

Ratzel's role model had long been the white settlers of North America and their "drive and force to colonize," which he believed had fueled first a colonial economy and subsequently the formation of an independent state.[51] His *Politische Geographie* saw a "heroic trait" in those who took it onto themselves to till new soil under great personal sacrifice.[52] Benefiting from their greater conception of space, these white settlers had purchased the indigenous territories one by one. While the natives had deemed this land worthless, the settlers had been able to appreciate that these many acquisitions were in fact slowly forming into a greater whole.[53] And yet the Americans were not the only nation with a colonial drive (*Trieb*). Germany, for instance, had only acquired overseas territories in the 1880s, but had sent settlers into its "colonial East," to Poland and Siberia for a much longer period.[54] Ratzel conceded that while Europe's great powers had thus followed different paths to colonization, the underlying reasons for their drive (*Trieb*) to colonize was uniformly environmental and spatial. Thinly populated areas attracted colonizers, as did natural resources. Wide spaces promoted a colonial spirit both because they were likely to yield good soil and because their very vastness awakened in the colonists a passion to work the soil.

Ratzel wrote that the unprecedented activity of European colonists had resulted not only in the "victorious spread of the white race," but in a global process of "Europeanization."[55] Although they dominated the world economy, European powers had come to depend on colonized areas to nourish their *Volk* and accommodate surplus population.

> As Europe's soil is unable to support 400 million inhabitants, it is necessary to import grain and meat from America, India, Egypt, and Australia, for which Europe pays with the products of its industry and its generally superior civilization. The same ships that supply these goods carry away population excess westwards and eastwards over the oceans.[56]

Colonization was not specific to modern history, Ratzel held, but had occurred since the last glacial period.[57] With only a small number of Pacific islands and polar regions newly colonized, the entire earth was thus best read as a palimpsest of previous colonization movements.[58] In fact, even the animal and plant kingdoms had been subject to similar waves of colonization. It was imperative to approach biological and social evolution with the same conceptual language. He thus called on biogeographers in 1901 to settle on using the term "colonization" instead of "migration," so as to remove what he perceived to be a source of great misunderstanding.[59] Colonization, in other words, was a bellicose logic of movement that emerged from the struggle for existence itself.

Ratzel laid out a basic typology of different forms of human colonization in chapter 6 of *Politische Geographie*. On the most basic level, there was interior (*innere*) and exterior colonization (*äußere Kolonisation*). Each state had within its existing territorial boundaries underutilized areas that could be populated and developed agriculturally. Once interior colonization was no longer possible or lucrative, states began to move toward exterior colonization through trade and conquest. Secondly, Ratzel made a cautious distinction between a maritime tradition of colonialism, perfected by the British Empire, and a more terrestrial tradition that had been the Roman, medieval German, and Russian mode of empire.[60] The former involved sending merchants into faraway lands and expanding influence through trade posts and routes. In the latter, such merchants followed peasants and craftsmen into neighboring areas. Terrestrial colonization could be further broken down, based on the relationship between the colonizing people and the land. Mining and trapping colonies only had a loose relationship to the earth. The most resilient and thus historically superior, Ratzel argued, were agricultural colonies. Only the cultivation of crops allowed a *Volk* to truly take root in the soil (*Einwurzelung*). As so often, he resorted to the language of botany, framing the "colony as a garden and its settlers as a growing plant, like an offshoot of the motherland."[61] But this *Einwurzelung* had to be carried out on an intellectual and educational level too. Ratzel was angered by the way colonial science was occasionally mocked in the German press and called on newspaper editors and book publishers to throw their weight behind the colonial effort.[62] "All colonial policy is educational policy," he claimed.[63] Immodestly, he thus saw his own armchair advocacy as vital to Germany's colonial project in Africa.

While an interest in colonization stood at the heart of Ratzel's political geography, there were other crucial concepts that gave shape to his approach. We have already examined *Lebensraum* in some detail in the previous chapter. I will now shed light on a number of further geopolitical foci, such as the land and sea power dynamic and the logics of location, greater space, and war.

Geopolitics

Friedrich Ratzel did not use the term "geopolitics." But as it was coined by one of his followers, Rudolf Kjellén, the question poses itself whether Ratzel's writings already prefigured geopolitics, a *Weltanschauung* that is commonly seen to reduce the complexities of world politics to the interplay of great powers and their clashing spheres of influence under the influence of geographical factors.[64] How deep, in other words, do the resemblances between Ratzel's political geography and twentieth-century geopolitics go?

Geopolitics is commonly broken down into a German strand, associated with Karl Haushofer (1869–1946) and his "master" Ratzel, and an Anglo-American branch, which draws on the works of Alfred Thayer Mahan (1840–1914) and Halford Mackinder (1861–1947).[65] In attending to the multiple strains and roots of the geopolitical imagination, it is thus important to examine how Ratzel's work related to that of other classical geopoliticians and to those in the tradition of political realism, which has sought to explain foreign policy, armed conflict, and alliance formation through reference to great-power competition.[66] But whereas the realist tradition of international relations has shared with geopolitics a certain state-centrism, it has not insisted on the causal influence of geographical factors—such as topography, geographical position, climate, and the distribution of natural resources—on world politics.[67]

A naval officer, president of the US Naval War College, and vocal proponent of US naval expansion, Mahan argued in his seminal 1890 *The Influence of Sea Power upon History* that global hegemony depended on the control of maritime trade flows. He was not the first military strategist to have made this point. But his book was noteworthy for its systematic attempt to understand the deeper "elements of sea power," which Mahan broke down into the geographical position, physical confirmation, extent of territory, number of population, character of the people, and nature of the government of a given state.[68] A proponent of free trade, Mahan thought of the sea as a wide common that was criss-crossed by vital trade routes. He was interested in seaboard length, the number and quality of harbors, the distance between these harbors, as well as the role of straits and canals. Mahan's book was more of a work of military history than of human geography. And while it is thus perhaps surprising that *The Influence of Sea Power upon History* is so frequently counted among the major works of geopolitics, there *was* in this book a certain spatial determinism that can be found in Ratzel too.[69]

Whereas Mahan had written in great detail about the rise of the steamboat, it was the spread of railway technology that captivated Halford Mackinder. A British academic, occasional explorer, and briefly a Member of the House of Commons, Mackinder was in fact the kind of ruthless explorer Ratzel so admired, a man who was quite literally willing to walk over dead bodies to further his career.[70] In 1904, the year of Ratzel's death, the Oxford geographer famously argued that the end of the age of exploration and the advance of railway technology was tipping the global balance from sea to land power.[71] His world was one of large continental-scale areas that were declared as being of various degrees of strategic importance: the heartland, the outer crescent, the North Atlantic.[72] Although sometimes considered a theorist of sea power, a label not entirely inaccurate given his work in this area,[73] Mackinder was in fact deeply concerned with the demise of British naval hegemony and the rise of terrestrial powers in the Eurasian heartland. Mackinder's thesis on the Eurasian heartland would have a significant impact on the formation of a school of geopolitics in interwar Germany.[74]

Although Ratzel was familiar with Mackinder's work, he only engaged with him in passing.[75] And despite his continuous interest in the United States and the popularity of *The Influence of Sea Power upon History* in Germany, he seems to have chosen to ignore Mahan's book entirely. Indeed, Ratzel's 1903 second edition of *Politische Geographie* mentions neither Mackinder nor Mahan. And yet the theoretical overlap is significant. Ratzel too offered a perspective that assumed that world politics had to be understood in a totalizing fashion, as a structured whole.[76] Like Mackinder, Mahan, and Haushofer, he was a passionate imperialist who wanted his ideas to shape the fortune of his nation. Many of his newspaper columns from the 1890s are instantly recognizable as "geopolitical."[77] He commented on the territorial struggles between European nations in the colonies, speculated about Russia's desire for warm-water ports, and listed the number of torpedo boats, turret ships, and gunboats various great powers had at their disposal. He even foreshadowed in his *Anthropogeographie* Mackinder's perhaps most famous prediction about Russia's ability to extract Siberia's riches through railway technology.[78]

Ratzel's political geography shared crucial ground with political realism too. A focus on the balance of power had already emerged in *Anthropogeographie*. "If a tribe becomes so strong that it poses an imminent threat to its neighbors," he noted, "then the latter will unite in order to exterminate it."[79] *Politische Geographie* argued that the European balance of power was no diplomatic invention, but rather the epiphenomenon of an urge for spatial expansion that was common, though to varying degrees, to all states at all levels of civilizational development.[80] Over time, *Völker* and states became functionally similar because those that were less bellicose lacked the ability to compete with those that were. In a basic sense, then, there was natural selection among states.[81] Ultimately, those states that were not adapted to their environment and the competitive nature of international sphere wilted and disappeared.[82] Ratzel commented too on alliance formation and had a keen interest in the complex web of alliances that had marked the Bismarck era, emphasizing that existing accounts had underestimated the territorial dimension of these balances.[83] He also outlined an interest-based foreign policy approach that was guided neither by the changing views of the domestic political realm nor by the aim of building up a good reputation abroad. "Germany never stood lower," he explained, "than when it had tried to mimic the liberal patterns of France, Belgium and so forth" in European politics.[84]

Ratzel's political geography did not just overlap with the writings of Mahan and Mackinder, but also prefigured those who would in the 1920s and 1930s envision a world of autarkic continental-scale spaces. While Carl Schmitt would speak of greater spaces (*Großräume*), Karl Haushofer would use the term pan-ideas (*Pan-Ideen*) to capture this idea.[85] Ratzel used neither of these terms but applied the adjective *großräumig* (greater spatial) to states, nations, politics, and political designs in *Politische Geographie*.[86] In principle, the logic of smaller and greater space was transhistorical and rooted in a gendered binary through which Ratzel occasionally framed political phenomena.

> The small-space tendencies in the life of *Völker* today appear less significant than the greater-space ones, which strive toward global transport and world politics; but they are so firmly established in human nature that they

> will only ever sleep. They will never die and could be revived at any time. As long as there are people, greater-space and small-space tendencies struggle against one another in every society and essentially in every individual soul. The women who delimit the domestic hearth, the crop farmers who depend on the clod [*Scholle*], are small-scale, the warrior caste contemplating conquest, the patriarchal communal house [*Männerhaus*], which unites the most youthful and energetic of the tribe, represent greater-spatial conceptions [*großräumige Auffassungen*]. Even at lower levels, the religious sphere has the greatest tendency to expand in all directions, even to populate or humanize the sun and stars.[87]

Großräumige Völker, such as the Romans, English, and North Americans, he argued, had produced "better practical geographers" than others, even though they had paradoxically often neglected the task of geographical theorization.[88] He explained that at the end of the nineteenth century, only the United States, Britain, Russia, and China were true "great powers" (*Großmächte*) in the sense that they were "continental powers" (*kontinentale Mächte*) who dominated spaces that extended far beyond their vast territories.[89] This was also the reason why he preferred the less precise but more common German term *Gebiet* (area or territory) to the more legal *Territorium*. Ratzel's usage of *Gebiet* was not reducible to a strictly Westphalian concept of territory as a bounded space under the control of a sovereign state. The *Lebensräume* of states, he explained, were not just subject to war and shifts in the balance of power among states and races but were always in flux due to transformations of external living conditions. The *Lebensräume* of European states in particular were not governed by state sovereignty alone. Even where the endeavors of missionaries, trading companies or emigrants were not driven by state activity, they extended the living space of a state's *Volk*. A colonizing people's *Lebensraum*, in other words, could extend far beyond the territory of its state.

Ratzel was keenly aware of America's Monroe doctrine, which had sought to ensure European nonintervention in the Western Hemisphere since the mid-nineteenth century.[90] Indeed, he saw a direct link between greater conceptions of space and this doctrine, which in some respects foreshadowed Carl Schmitt's proposal for a world order based on greater spaces (*Großraumordnung*):

> In the area of foreign policy, [the United States] operate with a greater conception of space than the middle powers of old Europe. The Monroe-Doctrine is not an isolated and even less an accidental doctrine. [. . .] We have to learn to understand this [spatial] conception in the political designs

> of these powers, even when we dispute their legitimacy. Otherwise we run into the danger of assessing these political questions from a myopic perspective.[91]

Unlike Schmitt and Haushofer, Ratzel did not argue explicitly for a German version of the Monroe Doctrine. And yet in 1904 he contemplated the possibility of a European economic union which might include at its core the "Germanic states" Germany, Belgium, Switzerland, Austria-Hungary, Luxemburg, and the Netherlands. Historical transformations, particularly in the realm of transport, were increasingly turning "Central Europe [*Mitteleuropa*] into a single economic area." Europe had carried its many territorial borders and narrow spatial horizons into this new age of "great spaces," which made it less competitive than the true continental powers.[92] Shortly before his death, Ratzel even considered the proposal for a "United States of Europe."[93] Although he concluded that such a vision was premature, he noted that the rise of the United States to the status of the world's most powerful economy and market demanded that European states enter into a customs union.[94]

Ratzel's academic works, popular-science writings, and newspaper columns also dealt in detail with the question of geographical position or location (*Lage*) in which a state found itself. "By receiving a *Lage* and thereby its land, a Volk receives itself," he stated in *Politische Geographie*.[95] *Lage* functioned as a "reservoir" into which the waves of historical change had inevitably to return. Here Ratzel was further developing arguments he had laid out in *Anthropogeographie*. In order to determine the *Lage* of a state, it was crucial to examine the latter's territorial size, the nature of its borders and neighbors, its geomorphology, climate, and its access to oceans and rivers. From these emerged crucial implications for a nation's economic potential and security. Of course, island states were sheltered from attack, but even insularity did not bring with it "complete invincibility." Each *Volk* faced the challenge of making the best out of the hand nature had dealt it.[96] On this occasion, Ratzel's position was broadly possibilist in that it emphasized human control beyond environmental constraints.

The German Empire, he explained in the late 1890s, had many powerful states in its vicinity, the most important of which were its direct neighbors France, Russia, and the Habsburg Empire. Unlike Britain, Germany did not have the comfort of an insular position and had to rely on its North Sea ports for access to the oceans. This made Germany an Atlantic power of second rank.[97] And yet the Alps had served to protect Germany's territory from the "oriental" invasions that had swept through Europe for many centuries.[98] While

Germany's climate had allowed its agriculture to flourish, its central position (*mittlere* or *zentrale Lage*) and absence of natural borders to the east and west had served both as a source of weakness and strength.[99] Before unification in 1871, Germany had been kept in a state of political slumber by the European great powers, relegated to the status of a "spiritual broker" between East and West, North and South. Ratzel did not disapprove of this role but was adamant that Germany aim higher and seek to become the continent's "politically robust and spiritually free core heartland."[100] By entering into an alliance with Italy and Austria-Hungary, he explained, it had established the basis for such a hegemonic status.[101]

Land and Sea

In 1898, Ratzel had been one of the founding members of the *Flottenverein* (Navy League) and one of the first of the German professors to ally with the cause of the naval expansion. Many of his colleagues had greeted the latter with some hesitation, but Ratzel wrote an enthusiastic pamphlet on the "naval question" in the year 1900.[102] *Das Meer als Quelle der Volksgröße* (The sea as a source of national greatness) appeared in the same year in which Kaiser Wilhelm II and Grand Admiral Alfred von Tirpitz had passed the second installment of their *Flottengesetze* (German Naval Laws), which committed Germany to the construction of a navy they hoped would rival that of the British Empire. The pamphlet's core aim was to examine the way the sea had "educated" *Völker* over the course of history. Ratzel wanted to trace the role that particular oceans, straits, and gulfs had played in dividing or uniting humans and forging empires. Both the Romans and the British, after all, had not only benefited greatly from the control of the waters, but had also been shaped by this elemental force.[103] As Santini notes, the booklet was largely a collage of existing works and included passages from both *Anthropogeographie* and *Politische Geographie*.[104] As with many of the texts Ratzel composed in later years, political praxis was very much on his mind. The pamphlet was received well in Germany's naval circles, finding him admirers such as Vice Admiral Curt von Maltzahn (1849–1930).[105]

What Ratzel really wanted his reader to understand was "that Germany must be strong at sea in order to fulfill her global vocation."[106] He was mesmerized by the substance out of which all life on earth had once emerged. Even more than the steppe, the sea was characterized by a certain monotony that molded the spatial consciousness of those who navigated it. Devoid of all geomorphological complexity, it thus brought into sharp focus both the laws of spatial growth, Ratzel claimed, and indeed the nature of "space in itself" (*Raum an sich*).[107] The sea was "purest nature," a natural force that was much more difficult for humans to wrestle with than the earth. The endless horizon injected boldness and perseverance into the seafaring nations, serving as "the carrier of historical progress."[108]

The essence of superior sea power, the booklet argued, was to intertwine political and economic expansion, allowing the projection of power from narrower into wider spaces. Land powers grew slowly. Sea powers, on the other hand, were able "to subjugate half of the world before a land power had even tried to reach out for a bordering province."[109] Because the ocean was a connected whole, sea powers always strove toward global domination. Wrapped in this argument was a subtle critique of liberalism. Sea power brought with it a distinct amalgamation of national egoism and cosmopolitanism that paired "the pettiest acquisitiveness with the broadest understanding of the interests of communality."[110] The Napoleonic Wars had proven that while England had generally been perceived to have acted as Europe's liberator, it had only served the aim of preserving its global empire. Ratzel had long been convinced that it was the British doctrine of free trade that had caused great harm to the colonized.[111]

Despite this critique of British foreign policy and free trade, Ratzel's proto-geopolitics chimed with that of many of his Anglo-American contemporaries. He shared with Mahan the assumption that land and sea influenced the character of *Volk* and state. He also concurred with him that "the silent and monotonous horizons of the sea" were in fact a space of "a martial encounter that tests and measures the greatness of a people."[112] He furthermore agreed with the American strategist that sea power "ultimately paid for itself through increased maritime trade, thus making sea power a more efficient tool of state growth."[113] Napoleon's continental empire had failed precisely because it had lacked an oceanic dimension. Like Mahan, Ratzel consequently had to advocate naval expansion. And yet Ratzel differed from Mahan and Mackinder in that he was not convinced that history divided states neatly into land and sea powers. This was because he saw the oceans as a source of political power for every *Volk* that had access to the sea and was bold enough to build a navy. Sea power was a necessary but not a sufficient condition for global hegemony. Ratzel repeatedly noted the decline of the British Empire vis-à-vis continental powers like Russia and the United States at the end of the nineteenth century. Pure sea power was limited by its reliance on long supply lines and the difficulty of projecting power inland. A state that relied solely on the global reach of its navies would inevitably cross too many interest spheres and make too many enemies in doing so. At the turn of the century, the distinction between land and sea power had thus been called into question. A true global power would have to unite both.[114] "The ancient opposition between sea nations and land nations, sea powers and land powers, is being eroded by the growth of states," he claimed.[115]

Drawing on his relational concept of space, he observed that the true power of straits was to strengthen the connection between urban centers. Carthage's dominance had depended on the control of the Strait of Gibraltar. The strait had subsequently been rendered irrelevant until the "resurrection" of the Strait of Gibraltar had led to the blossoming of Bruges and Flanders in the thirteenth century.[116] Similarly, the Suez Canal (see figure 12) underwired the ties between Bombay and London in the late nineteenth century. He was also attentive to the fact that Britain's view of Gibraltar had changed significantly over a short span of time. Even in 1712 and thus a year before the Treaty of Utrecht, London had not recognized the rock's strategic potential and had been willing to settle for a different territory. Straits, in other words, were not transhistorical assets but depended on larger political and economic connections.

Ratzel may have referred to France as "the great enemy,"[117] but ultimately England was for him and large parts of Imperial Germany's political elites the main opponent, at least from the 1890s onwards. Although he had met his future wife in London in the early 1870s, he seems to have been indifferent to the city. He had returned to Britain in the late summer of 1881 but again seemed uninterested in the politics and culture of the British Isles, noting the poor quality of shops and the anti-intellectualism of English businessmen.[118] And yet by the 1890s, this had changed. He had begun to read the *London Times* on an almost daily basis, as his diaries reveal.[119] It was at this time that he wrote a string of long and heated commentaries for *Die Grenzboten*, which sought to assess the British Empire's strengths and weaknesses on the global stage.[120] In these articles, he sharply criticized London's military campaigns in the First Boer War and First Matabele War. He called the imperialist politician and businessman Cecil Rhodes a "brutal man of violence" and praised the Boers as "admirable, loyal and freedom-loving," a *Volk* that carried its ambition of being agricultural colonists in its very name.[121] Ratzel believed English imperial policy to be characterized by a "shameless hypocrisy": Britain promised freedom and emancipation and even had the audacity to criticize the United States for its lynchings, but only ever brought conquest and bloodshed to Africa.[122]

And yet, as Ashworth observes, this was hardly an outburst of anticolonial sentiment.[123] In fact, Ratzel admired the British for their ability to construct the

Figure 12. A section of the Suez Canal. (Source: *Die Erde und das Leben*.)

"greatest state edifice of all time."[124] His dislike for Rhodes aside, he felt the man was "to be taken seriously" for he had "influence, money and vigor."[125] But what seems to have angered Ratzel most was that his own appreciation for English colonialism was answered by London with a lack of respect for Germany's interests and achievements.[126] There was clearly, in embryonic form, a narrative of national victimhood at play. It was in relation to England that Ratzel articulated most clearly and ambitiously what he saw as Germany's mission on the world stage. In 1898, he published a piece in the liberal conservative *Münchner Neueste Nachrichten* that dealt with Germany's position in international politics. All *Völker*, he explained, were either "hammer or anvil"; "[t]here will always be dominant [*herrschende*] and serving [*dienende*] nations."[127] For centuries, Germany had been pushed around by the European powers who had dominated the seas. But England was now in decline and Germany on the ascent. As he wrote in a letter to his student Kurt Hassert, he was stunned by the "demise of England" and the "deterioration of English moral qualities" that accompanied this decline.[128] History was thus opening up a window of opportunity. "Only as a world power can Germany hope to secure the soil that its people require to grow," he concluded.[129]

The tension, Ratzel believed, between the previously unsuccessful emigration of Germans and the successful wars of territorial reunification could only be resolved through the establishment of settler colonies in Africa and the construction of a navy that could challenge British hegemony. Ratzel did not call for a war with Britain, certainly not for the kind of war that would sweep across the European continent a decade after his death and that would ultimately lead to a new interest in his writings. Instead, he called for a "showdown" with England over the Boer states, the prize for which, he hoped, would be the establishment of a German farming colony in a temperate climate.[130] As for political realists, war was in Ratzel's political geography something that was always on the cards and not some kind of crime or evil that should be outlawed or policed. "Nothing angered him more," Wanklyn observes, than the periodic condemnation in the German political landscape of the war that had almost killed him.[131] Indeed, it was Ratzel's reflection on his active service that prompted him to lament shortly before his death that it was "a shame that there are not more wars."[132]

War as the School of Space

War was not a central concept in Friedrich Ratzel's work in the way that *Raum* or *Kolonisation* were. In contrast to Carl Ritter, the Leipzig geographer did not teach at a military academy. Unlike Karl Haushofer, he did not produce works of military strategy.[133] Perhaps unsurprisingly, the study of war has thus long glossed over Ratzel's writings.[134] And yet war had long been a crucial interest of Ratzel's. The first volume of his *Völkerkunde* had argued that "[m]ost of what we know of the history of the natural races is the history of their wars."[135] Here, he had first made a crucial link between war and migration, noting how "the same legend or tradition" frequently recurred in Africa and elsewhere. "A monarch sends out a band of warriors to conquer a country or town; if the enterprise fails they settle down quietly and marry the daughters of the people whom they came to overthrow."[136] In the second volume of *Anthropogeographie*, he had seen an "inevitability" of war between humans and commented in some detail on the military organization of nomadic peoples.[137] Recognizing that war among "natural races" was generally less bloody than that between European colonists and indigenous populations, he nonetheless held that the state of war was more common than peace on lower levels of state evolution.

But it was only in the 1903 second edition of his *Politische Geographie* that he sought to provide a more comprehensive and conceptual answer to the question. Indeed, he gave this edition the subtitle *Geographie der Staaten, des Verkehrs und des Krieges* (Geography of the state, transport, and war).[138] Ratzel stuck to his Wagnerian ontology of movement but added to it a focus on transport networks.

> In every historical movement of *Völker* or states, be they dislocations, expansions, reductions or condensations, there is also an element of human transport. There is never a war that does not create new traffic routes, sometimes destroying old ones, never a colonization that does not establish new roads, no rise or demise of states without the promotion or decline of such routes.[139]

While he conceded that the peaceful expansion of transport had contributed to the growth of economic spaces, his argument about transport was less one about interconnectedness and increasing global unity than about its ability to induce violent struggle between races and states that had previously been separated by vast distances.[140]

In places, a staunch militarism shone through the otherwise dry discussion: "it is war which has always remained the great school of the ability to master space."[141] And yet Ratzel did not believe that organized armed conflict, unlike colonization, was truly transhistorical. It had not played a role in the Palaeolithic, given how dispersed populations had been before the agricultural revolution. And while armed conflict had for many millennia been dominated by "wars of annihilation" (*Vernichtungskriege*) and "predatory wars" (*Raubkriege*), these had been replaced in modernity by "wars of soil" (*Kriege um Boden*), essentially bellicose attempts to unite and round off territory.[142] Even the American Civil War, he claimed, had been no more than a war for territorial unity.

The aim of modern war was thus always the same, "to wrest the control of land from the opponent," which was best done by "defeating the defensive elements of a *Volk*."[143] The history of states had at its core the logic of spatial struggle. War had a geographical, political, and societal dimension. Conceived geographically, war was "a fierce, intermittent, and violent movement of a great mass of humans from one land into another."[144] Politically, it was "the most powerful means of a continuation of state growth that had stagnated in times of peace and of clarifying convoluted relations among *Völker*." As soon as war broke out, the borders between two warring parties were dissolved and "molten into a battleground" until they were redrawn when the fighting ended. Finally, then, there was a societal dimension to war in that it brought out "masculine traits," particularly among rulers. Peace, on the other hand, fostered family life, "chaining man to woman." Here Ratzel resorted to a masculinist vision of society that classed war as the realm of men and peace as that of women. Armed conflict was, he noted crudely elsewhere, "the creator of men out of boys and wimps."[145] While war was thus a form of movement, peace was stasis.[146]

Interestingly, these views were not merely tagged on to his theory. For, as he went on to explain, there were historical and evolutionary benefits to a militarized and aggressive society. In his 1896 essay "The territorial growth of states" (*Die Gesetze des räumlichen Wachstums der Staaten*), he noted the following on aggression:

> In a stationary condition weakness and decay creep in, whereas aggression demands organization, which among the Tartar hordes, as in the ships of the Vikings and Malays, accomplished great results with small forces. The most extreme cases exist in Africa, where a martially organized people, like the Zulus, is found side by side with a people that for generations has been falling to pieces, like the Mashona. They form part of a single whole, for one people exists at the expense of the other.[147]

Ratzel's conceptualization of war was indebted to the Prussian general and military strategist Carl von Clausewitz (1780–1831), who had acknowledged the importance of geography in his posthumous *On War*, published in 1832 by his wife Marie von Brühl. Von Clausewitz had famously declared war a rational instrument of politics, a continuation of politics "by other means," and "an act of violence intended to compel our opponent to fulfill our will."[148] Like many of his contemporaries, Ratzel had read von Clausewitz closely and cited him repeatedly. But whereas the latter had reduced geography to the study of terrain and understood war as an institution shaped by intentionality and rules, Ratzel had something else in mind. War was more than a clash of wills; it was the outcome of a biological urge, common to all life on earth, to master space (*Raumbewältigung*). And while there was direction and aim in the advance of an army, namely "to penetrate the other's territory [*Gebiet*]," there was also something instinctive and nonintentional about this movement.[149] Warring parties had to be grasped as organisms that aimed to incapacitate one another by striking their vital organs, chiefly the soil and thus subsistence on which the existence of a state's population depended.[150] The social conventions that governed organized violence were thus by extension no more than shifting consequences of a permanent biological necessity. Ratzel thus added to von Clausewitz's dictum of war as the continuation of politics by other means the logic of spatial and organismic growth. Although this theory of war as the continuation of biology by other means was his own, it had a Darwinian core. Charles Darwin had begun his 1858 paper to the Linnean Society with the words "all nature is at war, one organism with another, or with external nature" and continued, invoking Malthus, to highlight the tension between biological reproduction and the earth's finite resources.[151]

Ratzel's concept of war underwent a shift between *Anthropogeographie* and *Politische Geographie*. In *Anthropogeographie*, his thinking had still been dominated by the aim of explaining conquest by reference to climatic origins, as can be seen in the following excerpt:

> The question thus poses itself as to whether it is a coincidence that the conquerors and founders of states emerged so often from the North to subdue the lands of the South? [. . .] It is impossible for stronger and weaker *Völker* to coexist next to one another for longer periods of time. One should thus expect the pressure toward the equator to originate from *Völker* that have been invigorated by the rough climate of the temperate and cold zones.[152]

In *Politische Geographie*, this *environmental* determinism was superseded, albeit not entirely replaced, by a form of *spatial* determinism.[153] Just as Darwin's struggle for existence was conducted in the animal plant kingdom as a struggle for space, he argued, "so are the battles of *Völker* frequently just battles for space."[154] War was something of an "experiment" that often resolved spatial problems between polities swiftly and decisively. History's great military commanders, such as Hannibal or Alexander, were not merely to be admired for their physical ability to conquer vast spaces, but for their superior "spatial conception" (*Raumauffassung*), which had enabled their conquests to become victorious. They "saw further and yet more sharply" than others.[155] In the late nineteenth century, the most superior *Raumauffassung* was one that thought in terms of continents and had detailed knowledge of distant places at its disposal.

If taken out of context, this emphasis on spatial conception could easily give the impression that Ratzel allowed for ideational factors to be decisive in battle, to be a force that could overrule geographical constraints.[156] More precisely, though, he was speaking of "space that had been converted into the spirit [*Geist*] of man," prompting "crippling or inspiring" effects.[157] *Geist* was thus reduced down to a binary logic of space, which was either limiting or creative and which depended on a political agent, a political or military leader, to harness it. Conceived space thus ultimately remained secondary in Ratzel to the force of space as it emerged from the struggle for existence.

> It is the struggle for space through which spatial conceptions steadily grow. War suddenly augments the theater of conflict, which has arisen on the narrowest of spaces, before proliferating and reproducing itself, dividing *Völker* and states into enemy camps.[158]

The prize of victory in modern war was space gain. Each *Volk* could be gradated according to its spatial conception, which encompassed the ways it used not just weapons and fortification but transport and communications infrastructure, such as highways, canals, railroads, bridges, and telegraphs. In this sense then, war was to Ratzel "the school of space."[159] But as war was itself the outcome of spatial conditioning, there was a certain determinism—and indeed circularity—in Ratzel's argument.

> In history books we sometimes find expressions such as "Napoleon's Russian campaign foundered on the vastness of the space of the Russian Empire," and strategy handbooks do not fail to notice the significance of wide spaces, which protect the attacked while being potentially disastrous for the attacker. But there is something much greater about wide space, something creative, I should say.[160]

The word "creative," which Carl Schmitt would later appropriate from Ratzel, should not deceive. Ultimately, war taught states, races, and nations a superior conception of space only because of the particular spatial conditions in which they had once found themselves.[161] Although Ratzel was not entirely consistent on this point, a spatial determinism certainly followed from the majority of his statements.

Conclusion

Friedrich Ratzel's political geography performed a typically late nineteenth-century militarist, masculinist, and in many ways crude reading of armed conflict. And yet there was something unusual about his move beyond an instrumental understanding of war. Rather than the continuation of diplomacy by other means, war was a force of nature, an effect of the struggle for space, which originated in the expansive and creative properties of biological life itself. War-as-evolutionary-struggle allowed organisms to progress to higher forms of military and evolutionary development, rendering armed conflict a force for good. War was the origin of superior spatial consciousness, the great school for the faculty of mastering space. But it was ultimately no different from other phenomena, such as migration or traffic. The great migration period that had brought the Western Roman Empire to fall revealed precisely this blurring of peaceful and bellicose movements.[162] These were all expressions of the same creative and expansive properties of life that could be observed in the animal and plant kingdoms: struggle, conquest, and colonization.

This chapter has re-examined some of Ratzel's most well-known ideas. As we have seen, Ratzel should be understood both as what Hell calls "an imperial theorist of the *Kaiserreich*" and something of a practitioner of empire.[163] On a theoretical level, Ratzel performed two crucial intellectual moves. Firstly, he naturalized colonization as a transhistorical process which could be found in all life forms on earth. Secondly, he argued that only agricultural colonization could establish a true grip on the soil. This "agrarian primacy," Smith notes, "led directly to Ratzel's functional definition of human *Lebensraum* as essentially farming land."[164] And yet there was more to *Lebensraum* than the mere idealization of the peasantry. The Leipzig geographer wanted German colonialism to distinguish itself from other colonial endeavors as a bourgeois and scientific enterprise. Thus, on a practical level, he sought himself to play the role of a scribe and armchair supporter for German colonizers in Africa.

In the previous chapter we saw how Ratzel's theoretical ideas were shaped directly by, and were thus ultimately indistinguishable from, his political commitments. This chapter has shown that Ratzel's concepts of colonization and war were caught up in many of his era's clichés. They were also riddled with contradictions and inconsistencies. There was a clear tension, for instance, between his insistence that greater spatial conceptions were mere epiphenomena of evolutionary processes on the one hand, and his conviction that Germany should actively *cultivate* just such a *Raumauffassung* on the other. Ratzel's theoretical argument, in other words, implied a spatial determinism that his politics did not. This is why he occasionally resorted to a possibilist position which emphasized that it was ultimately the responsibility of each individual state to harness its geographical location so as to produce the best historical outcome.

By maintaining that war, since the Neolithic Revolution at least, had been an epiphenomenon of the creative properties of life itself, Ratzel chose to ignore the basic but evolving conventions that make armed conflict possible and of which von Clausewitz had been aware.[165] In framing war as the continuation of biology by other means, Ratzel excluded the complex processes of signaling between antagonistic social groups from the phenomenon of war. When trying to classify different types of colonization and war, he was thus forced back either into simplified claims about the relationship between people and soil or an evolutionary perspective on history that distinguished between *nature* and *culture*. The sophisticated bureaucratic, economic, and emotional economies that had made modern war possible were all too often reduced by Ratzel to a clash of bodies (size of population and number of soldiers) and technology (telegraph masts, coastal forts, gunboats, etc.) as well as the ultimately nebulous agency of *Raumauffassung*. Such material reductionism has remained popular,[166] but it was never without its critics, as we will see in chapter 7.

Ratzel postulated that wherever there was demise, there was also becoming. He seemed to find comfort in a detached gaze upon political processes that saw in death the highest expression of life itself: the march of history-as-evolution, expressed in a fluctuation of territories. In *Politische Geographie*, he summarized his argument as follows:

> *Völker* are in constant internal movement, which changes into external movement when a piece of soil is newly occupied or a previously occupied one is given up. This creates the impression that these *Völker* are moving forwards or backwards like a slowly flowing mass. In the history that is known to us, it is seldom the case that such a movement spreads over an unoccupied space; as a rule, it leads to permeations or displacements, or to small regions and their peoples uniting to form larger ones without these changing their location. Likewise, these large states fall apart again, and as a result of this unification and separation, this growth and decrease, a large part of all historical movements appear politically and geographically as fluctuations of small and large areas.[167]

Politische Geographie was not Ratzel's last word on the subject of war. His posthumous *Glücksinseln und Träume* approached war through the medium of semi-fictionalized autobiography. His political geography had addressed war from a normative perspective as a force for good. It now assumed an aesthetic quality. Both works were underwritten by a concern with what happened to an organism, be it that of an individual soldier or that of a state, after it had died.

Chapter 6

Death and Necropolitics

Although he is today most famous for an argument about the relationship between life and space, Friedrich Ratzel was for most of his life preoccupied with the question of death. As a child, he had already developed an animist streak and would later observe that a child's vivid imagination could bring "all death to life."[1] In his scholarship, he came to promote the idea that even the inorganic realm was in fact alive. In chapter 2, we noted a number of formative traumatic experiences, which all entailed within them the mystery of death, disappearance, and loss: an eerie solar eclipse, the drowning child, and maternal abandonment. Although Ratzel seems to have believed that he was able to overcome these early traumatic moments, he acknowledged that his time as a teenage pharmacist was marked by trouble. His flight into a world of books raised eyebrows among his superiors, as did his interest in the many drugs that were now available to him. He recounted in his autobiography one experience of experimental intoxication:

> Why shouldn't I try it, to fly? Here it stands in a stone jar, that cherry laurel, a liquid whose prussic acid lends it a biting but elegant scent. I am not frightened by the skull above the archaically ornamented label "Aqua Laurocerasi." The hydrogen cyanide content is not too strong. Perhaps its effect will just be anaesthesia, dream and return, but perhaps it is death. What difference does it make? And here is an old glass of Bohemian crystal, the clarity of which I have long admired. How innocent that blueish venom shimmers. A long sip, a second, I can already feel my hands tremble.[2]

It was only after days of slumber that Ratzel finally awoke, his limbs feeling broken, his head numb. Upon recovering from what could well be interpreted as a suicide attempt, he claimed to have realized the value of his life to those who were close to him and felt embarrassed about the entire episode.

Years later, he wrote to his future wife that he found any talk of illness "unbearable" and "embarrassing."[3]

Ratzel mastered his early traumatic experiences, or so he came to believe, with a degree of emotional detachment that would come to guide him in his later encounters with the dead. As a soldier in the Franco-Prussian War, during which he lay for hours in the muddy soil about which his writings would obsess, chopped off limbs and disfigured faces became a familiar sight when he was forced to recover from an injury in a military hospital. Ratzel sought to practice viewing dismembered bodies and concluded that some soldiers had understood that it was better that they die. When he came to reflect on these experiences in his posthumous *Glücksinseln und Träume*, he would resort to a morbid Romanticism. Death, he noted, was "the most beautiful thing that can be thought."[4] At the point of writing, Ratzel had come under the sustained influence of the physicist, philosopher, and experimental psychologist Gustav Fechner (1801–1887), who prompted Ratzel to think of death as an awakening to the universe.

The dead frequently performed important roles in Ratzel's writing, ranging from the ways in which fossils and skeletons were invoked as evidence, to speculations about "bloated animal corpses," which he believed to have inspired humans to build inflatable boats.[5] More crucially, Ratzel's Darwinian preoccupation with the logic of survival meant that he was always also writing about survival's opposites: disappearance and extinction. And whereas much of Ratzel's writing examined extinction in the animal and plant kingdom, he was also mesmerized by the ruins of fallen civilizations and by the disappearance of indigenous peoples in the regions colonized by white settlers. As we will see below, Ratzel had an obsession for cannibalism, human sacrifice, head hunting, and infanticide among what he referred to as *Naturvölker* and deemed these to be symptoms of peoples who had begun to self-annihilate even before white settlers set foot on their territory. Ratzel went on to operationalize death and decay in his political geography in ways that have remained underrecognized. His concept of the state-as-organism relied on a necropolitical reading of history, which was concerned with the health and cleansing of the body politic. In the final stages of his life, the Leipzig geographer became engrossed in the aesthetics of death itself, particularly in relation to armed violence and its aftermath. It is in these later writings that we find a necropolitical dimension that points in the direction of twentieth-century fascism as much as it expresses the violence of nineteenth-century colonialism.

This chapter offers an examination of Ratzel as a thinker of death and a figure in the history of modern biopolitics. We have already seen how this question has most commonly been addressed through the work of Giorgio Agamben and Michel Foucault. Although there are crucial differences between these two thinkers' approaches, they agree that we can find within the superficially protective practice of biopolitics, the concern for the health and growth of populations, a much less benign policy of excluding forms of life that are deemed dangerous and unworthy.[6] Both theorists find the purest form of biopolitics in the national socialist policy of encampment and extermination. Roberto Esposito agrees with this assessment, adding that it was "the coincidence of homicide and suicide" that turned the violence first inflicted by Nazi Germany on others finally upon itself.[7] Whereas Foucault holds that it was only in the nineteenth century that biopolitical regimes began systematically to depend on racist orderings, Achille Mbembe points to a longer colonial trajectory. It was not just in the camp in which vast populations were "subjected to living conditions that confer[red] upon them the status of the *living dead*," but crucially in the plantation and the colony, both biopolitical spaces of experimentation in their own right.[8] While Arendt hints at the way race and death were fundamentally united in a continuum of colonial and fascist sovereignty,[9] Mbembe takes Arendt one step further by highlighting the complicity of Western humanism and liberal democracy with slavery and racism, suggesting thus that necropolitics has not been neatly confined to the worst excesses of Western civilization but is inherent, as a "nocturnal body," in its very progressive ideals. Mbembe argues moreover that the shared myth at the heart of both colonialism and fascism is *the West* "as a natural living body" with "marrow and soul."[10]

The attractiveness of Mbembe's insistence that "the ultimate expression of sovereignty resides, to a large degree, in the power and the capacity to dictate who may live and who must die" lies perhaps in its applicability to the present condition. And yet his understanding of necropolitics as the underside of Western sovereignty can also help us to make sense of Ratzel. The latter's only half-hearted rejection of slavery and support for a new and violent regime of racial segregation in post–Civil War America points the way. His political geography saw polities as living organisms and Western states as superior to "lower" forms of political organization. The very *jus publicum Europeaum*, which his political geography tacitly celebrated, was in fact a racial world order which distinguished between spaces that could not be colonized and those that could. It drew a line between a first zone in which war was fought between sovereign states, which recognized one another as enemies but between which peace could be struck, and a second, in which limitless violence could be unleashed against irregular combatants and *Naturvölker* in the name of *Kultur*.

The Aesthetics of Death

There is no doubt that Ratzel was gripped by the question of death throughout his life. By the time of his own funeral in 1904, he had written a staggering two hundred obituaries, an achievement for which he was applauded by his contemporaries.[11] He deemed the practice of writing obituaries, or necrologies as he preferred to call them, so important that in 1894 he found himself forced to complain that his "dilettante" contemporaries were no longer taking the practice of necrology seriously enough.[12] Only in military journals could one still find obituaries that were appropriate to the occasion, he claimed.[13] There was in him a profound urge to render death meaningful and historically significant.

A preoccupation with the aesthetic properties of death followed Ratzel on his journey to the United States and Mexico in the 1870s. He was especially enamored by the sepulchers, garden layouts, and above-ground mausoleums of the American South, which he found more attractive than those elsewhere in North America.[14] In Jesup, Georgia, he observed a young family that had just sat down for breakfast in a railroad restaurant. While Ratzel described the mother as "a very healthy enterprising-looking woman," her husband was stricken by illness and "corpse-like."[15] He continued, "She later brought in their lovely, cheerful baby whose joyful young life seemed to be of more concern to her than the hopelessly disintegrating one of her husband."[16] The same logic of demise and creation was detectable in Ratzel's descriptions of urban life. On a number of occasions on this same journey he saw how quickly the urban fabric had recovered after fires in cities such as Boston and Chicago. Destruction and death, he observed, so effortlessly gave way to life in the New World.

When writing about matters of biogeography in the early 1880s, he urged his readers to learn from the inevitable end of organic beings, "which return to the very dust from which they were born."[17] This cyclical understanding of time, which coexisted uneasily with his occasional insistence on civilizational progress, came to the fore in the spatiotemporal figure of the ruin. Urban ruins were to Ratzel "works of art." Rural landscapes which still exhibited the traces of former human cultivation were more precious than those that had not been touched by humans.[18] Wherever he went, the man who had been himself involved in the destruction of Strasbourg in 1870 commented on the aesthetic quality of ruins. He was unimpressed by Mexico's ancient sites but valued the ways in which ruined castles had been preserved by the English and the Welsh.[19] The German geographer also had a keen interest in the dilapidated settlements that had been left behind along the railway lines of the American frontier. Wandering through one such ghost town, he was surprised to find occupants, a couple who had remained childless and whose austere, monosyllabic, emotionless, and asexual nature he found captivating. It was this experience that prompted him to ponder that "[l]ife is indeed not stronger than death."[20] Again, there was a beauty to such ruin landscapes:

> How the field of our thoughts and feelings expands, how everything seems to go on more peacefully in this place of quiet decay, where matter naturally strives to return to the old earth again. Here nothing is really hurried, here nothing forces anything else. It has been a long time since these objects were inhabited and used by humans, they are already going downhill and have nothing to offer us no matter how attractive they look, but are instead completely immersed in the process of becoming dust and ashes.[21]

Ratzel was impressed by the speed at which nature had reclaimed these American ghost towns. "What will be there in a couple of decades other than debris inundated by green and brown vegetation and the flowering beauty of creeping and climbing plants?" he asked.[22] A Romantic, he generally preferred the medieval ruins of Europe to those of North America but could imagine that even modern buildings might ruin beautifully. Nevertheless, he speculated that before that were to happen America's black population was likely to steal the bricks, thus denying a white European aesthetics of decay.[23] The question of race was always on his mind.

Anthropogeographie too was mesmerized by "ruin lands" (*Ruinenländer*) where "a whole culture ha[d] been annihilated without any new life blossoming from the ruins."[24] Not able to conceal his admiration for the more monumental traces of civilizational decline, he argued that a nation "that leaves humanity with witnesses of its existence, lives on in its works."[25] Not all peoples had left notable ruins. Ratzel observed in an 1884 encyclopedic entry that the peoples of Central Africa in particular had not produced ruins that could serve as witnesses to their civilizational deeds.[26] Meaningful ruins, in other words, were only achievable by advanced civilizations and white settler colonial societies in particular, though even those were threatened by racial struggle. By the time he published the second volume of *Anthropogeographie*, he had begun to frame his interest in decay in a more metaphysical language.

Despite his background in the natural sciences, Ratzel had always had a keen interest in philosophy. And while we learn from his diaries that he was spending his evenings reading Friedrich Nietzsche while writing *Politische*

Geographie, it is not obvious that Nietzsche left a significant mark on his thinking.[27] As we have seen in chapter 4, Ratzel was directly influenced by the biophilosophy of Spencer and Haeckel, Immanuel Kant's theory of time and space, and the earth mysticism of Carl Ritter. In the last two decades of Ratzel's life, the name Gustav Fechner moreover increasingly appeared in Ratzel's published and unpublished writing.[28] Fechner had originally held a chair in physics at Leipzig but had turned to questions of psychology and philosophy after successive experiments had almost left him blind. Often seen, alongside Ratzel's friend Wundt, as foundational to the emergence of experimental psychology, Fechner sought to find mathematical relations between bodily stimulus and mental sensation. And while he is remembered as an advocate of quantification and science, he was in fact "a multifaceted figure" who published works of a satirical and indeed esoteric nature, including an essay on the comparative anatomy of angels.[29] Most importantly, it was through Fechner that Ratzel was introduced to panpsychism, the view that the universe is "besouled" (*beseelt*).

Like Haeckel, Fechner rejected dualistic accounts of nature and society, life and death, religion and science. But rather than approaching body and mind as a union in a transcendent substance, he read them as existing in a functional relation. One of Fechner's hallmarks was the use of analogy. While recognizing that both the parts and the wholes of any two phenomena could be highly dissimilar, there could nonetheless be similarity in "the functional relations between parts and the respective parts and wholes."[30] He inferred the existence of sensation and mind from otherwise very different phenomena such as plant movements and animal reflexes. He thus likened, as Woodward observes, "the nervous system of animals to the vascular system of plants; in fact, he went so far as to liken the nervous system to the societal system, the system of conscious and unconscious minds of the earth."[31] Ratzel felt that Fechner's philosophy could help him to navigate some of the issues about which Lamarck, Darwin, and Wagner had only very little to say.[32] And although Fechner's panpsychism was antimaterialist, it was nonetheless compatible with Ratzel's human geography. German Darwinism (*Darwinismus*) was after all less materialist than its Anglo-American counterpart. Indeed, Ratzel was thus not the only German Darwinian with panpsychist leanings. Ernst Haeckel too had by the 1890s endorsed this worldview.[33] By rendering human consciousness as part of earth-consciousness, Fechner helped Ratzel to sidestep the question of consciousness altogether: it had become so all-encompassing that it could be neglected as an analytical category.

Most crucially, Ratzel found in Fechner a theorist who could help him move his preoccupation with death and dying onto a higher philosophical plane.

Fechner expanded in some detail on the blurred boundary between organic and inorganic molecules.[34] There was "hardly a page" of Fechner's writing that was not linked to the promotion of his "daylight view," which saw the entire material universe not as dead, but as "inwardly alive and consciously animated."[35] Ratzel not only endorsed the daylight view but also Fechner's panpsychist reading of death. Fechner had written about near-death experiences, spoken of death as a sanctification, and embraced eternity in ways that were neither incompatible with nor collapsible into a Christian worldview.[36] Fechner's 1836 *Das Büchlein vom Leben nach dem Tode* (Little book of life after death) argued that death was a form of awakening to the consciousness of the world. In Fechner's world, death was thus "only a second birth into a freer existence, in which the spirit breaks through its slender covering and abandons inaction and sloth, as the child does in its first birth."[37] The demise of a human body did not destroy what the deceased had contributed to the world. "A Goethe, a Schiller, a Napoleon, a Luther, still live among us, thinking and acting in us, as awakened creative individuals, more highly developed than at their death – each no longer restrained by the limitations of the body, but poured forth upon the world," he claimed.[38] Although Fechner's panpsychist theory of death had generally been met in the scientific community with a deafening silence, it certainly had its followers among the members of the Leipzig circle.[39]

Although it is unlikely that the two men ever met (Fechner died shortly after Ratzel's move to Leipzig), Ratzel devoured many of Fechner's writings and invoked him in 1896 when conceptualizing the state as organism.[40] In 1901, he proclaimed himself a follower of Fechner's worldview which "sees God in the world and the world in God and which dares to believe without giving up anything science knows and is yet to learn."[41] Ratzel's 1904 *Über Naturschilderung* and his 1905 *Glücksinseln und Träume* would reveal the spirit of Fechner's panpsychist and animist philosophy.[42]

> I see all life forms in a landscape, the many trees and plants, which are scattered in groups across it; with the idea in mind that life is movement, I see them as wandering, striving together and dispersing. That mountain there seems to me alive, its trees seem to be moving, some of them striving upwards, others descending. Some rear their heads upwards, others creep away into the earth.[43]

When describing the damp blackness of fallen leaves in late autumn, Ratzel observed that although they were not quite as obviously striking as the brown and yellow of previous weeks, their beauty lay in constituting the natural precondition for the earth's fertilization and renewal in spring.[44] In this sense, natural forces were always simultaneously destructive and creative.

It is perhaps surprising to see that some observers have spotted an "ecological note" in Ratzel's musing on decay and ruination, allowing perhaps for an interpretation of him as an environmentalist, an early thinker of the Anthropocene.[45] Although Ratzel did muse about the power of nature to reclaim abandoned human settlements,[46] it is important not to be deceived. When Ratzel spoke of the human effects on the environment, such as "the annihilation of the forests, the drying up of swamps, artificial irrigation, the disappearance of deadly animals,"[47] he did so not as an environmentalist of a contemporary ilk but as someone who was simply trying to assess whether the human species had been able to exert more impact on the environment than vice versa. Ultimately, nature—with all its destructive and creative forces—always won, not least because nature itself pervaded all human activity, grew over and into it. In any way, destruction and demise were only the fertile ground for new forms of becoming. Fechner had mentioned wars and revolutions in passing, but his statement on death had not been intended as a political theory. It was Ratzel who performed this move, fusing Fechner's philosophy with Darwinian evolution and the ethnological discourses to which he had been exposed.

Necropolitical Geography

The most critical way in which Ratzel's concern with finitude was worked through in his major writings can be found in his interest in extinction, a notion he used both in his biogeography and human geography. Indeed, the question of extinction had already loomed large over his 1869 *Sein und Werden in der organischen Welt*, which included lengthy discussions on the fossilization of various species.[48] But his focus then was still on the animal and plant kingdoms. This was to change in the late 1870s. Now he had become preoccupied with *Völker* that lived on in others, such as the Romans had in the Italians, and those that had disappeared entirely, such that "their traces can be found, if at all, in libraries."[49] It is in *Völkerkunde* that we get a clear sense of Ratzel's concept of time. Here Ratzel argued that "all that we call progress of civilization, may better be compared with the upward shoot of a plant than with the unconfined flight of a bird; we remain ever bound to the earth and the twig can only grow on the stem." "Human nature," he continued, could "raise its head aloft in pure ether, but its feet must ever rest on the ground, and the dust must return to the dust."[50] It is in this work that the dead took center stage, as Ratzel wrote at great length about burial rituals among the *Naturvölker*. As already noted, Ratzel was obsessed with human sacrifice, head hunting, infanticide, and cannibalism, and he described these practices, for many of which the empirical evidence was sparse, with the thoroughness of a pathologist conducting an autopsy. The following described death rites in Fiji and New Guinea:

> Burial within the hut is customary in Fiji. Among the Motus of Port Moresby the only sign of mourning is the incessant beating of drums for three days. When this is over, the grave is dug in front of the house, the dead body laid in a mat, and a little hut is built over the grave. After some time, the grave is opened, the corpse taken out and smeared on the elbows and knees with red ochre, while the widow smears herself with the decaying flesh. Then the dead man is put by again, and the little sepulchral house is gradually pulled to pieces, so that no trace of the grave is left.[51]

Ratzel depicted similar rituals on the Gilbert Islands, where the widow slept under the same mat with the corpse of her husband "until the head falls off the body," after which the skull was cleansed and the widow carried it with her for the rest of her life.[52] A few pages later, he casually discussed the way infanticide in Australia was committed, by "thrusting a stick through the ears and into the skull" of the infant.[53] Ratzel's observations of these grisly scenes were continuously presented in a curiously descriptive manner, often without drawing any wider conclusions. And yet the repeated visual depiction of burial sites, coffins, human skulls, and mummies conveyed to his readers the subliminal message that there hung over at least some of these non-Western peoples the shadow of death (see figures 13, 14, and 15).

Ratzel was not alone in telling this story of the extinction of the "savage races." Indeed, a powerful "extinction discourse" had taken root in the Western world from the eighteenth century.[54] Much like orientalism and other racist discourses, extinction narratives were transgenre and transdisciplinary; they were told by missionaries, evolutionary scientists, explorers, ethnologists, novelists, and geographers alike. They were as much tied up with British colonialism as they were with America's manifest destiny, revealing a potentially, though not automatically, genocidal "logic of elimination" at the heart of settler colonialism itself.[55] Among the scientists who promoted extinction discourse, there were a number of Ratzel's intellectual heroes, including Darwin and Lyell. Traces could already be found in the work of Thomas Robert Malthus, another of Ratzel's sources of inspiration. Ratzel himself thought that he had taken the extinction thesis from Humboldt.[56]

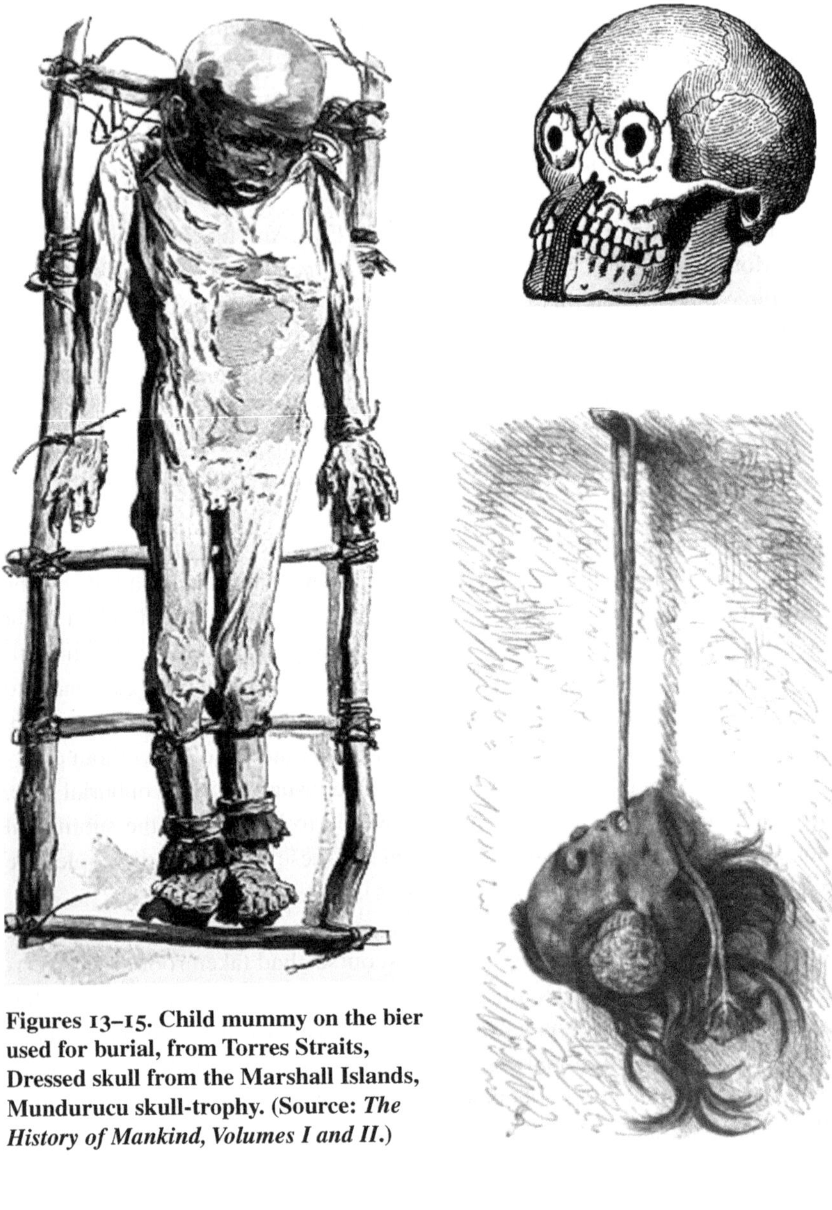

Figures 13–15. Child mummy on the bier used for burial, from Torres Straits, Dressed skull from the Marshall Islands, Mundurucu skull-trophy. (Source: *The History of Mankind, Volumes I and II.*)

Even though there were clearly earlier necropolitical thinkers, Ratzel's insistence on self-annihilation was felt by some of his contemporaries to have been original to his thinking.[57]

There was a level of disagreement among these necropolitical thinkers as to how much the struggle for survival could and should be intervened in, but there was "a basic agreement about the disappearance of some or all primitive races."[58] "When civilised nations come into contact with barbarians," Darwin would argue in his 1871 *The Descent of Man*, "the struggle is short, except where a deadly climate gives its aid to the native race."[59] Darwin had already spoken of a "war of extermination" in *The Voyage of the Beagle*. He had been captured by the observation that "no fact in the long history of the world is so startling as the wide and repeated exterminations of its inhabitants."[60] He had also observed a "mysterious agency" that ensured that "[w]herever the European has trod, death seems to pursue the aboriginal."[61] In *On the Origin of Species*, he had stressed the inevitability for new life forms to emerge through the extinction of old ones, an idea Ratzel was keen to adopt. He had also noted that some "barbarians" valued their dogs more than their old women, the latter of which they preferred to devour during famines.[62] In *The Descent of Man*, Darwin used cannibalism and infanticide among a host of other factors to explain the demise of indigenous races and stressed the inevitable decline in a population's reproductive health that accompanied contact with an economically superior race.[63] Darwin blurred the boundary between the human and nonhuman worlds. He concluded that "at some future period, not very distant as measured by centuries, the civilised races of man will almost certainly exterminate, and replace, the savage races throughout the world." "At the same time," he continued, "the anthropomorphous apes [. . .] will no doubt be exterminated."[64]

Ratzel too saw the process of self-extermination as rooted in a myriad of interlinking causes and practices, but he singled out infanticide, human sacrifice, head hunting, and cannibalism as symptoms of "lower races." He was keen to emphasize the connection between these deadly practices, arguing, for instance, that where one found head hunting and human sacrifice, one also found cannibalism. "Whether it is the devouring of a still twitching heart or the slurping of fresh brain," he suggested, there always existed an intimate connection between the quest for enemy scalps and "the desire to eat their owners."[65] Both Darwin and Ratzel made reference to Malthus and also cited the 1868 essay *Ueber das Aussterben der Naturvölker* (On the extinction of natural peoples) by the German ethnologist and geographer Georg Gerland.[66]

Motivations for cannibalism were not just bellicose, but economic. "For where it has once got a footing," Ratzel argued, "its attraction increases, till we find races among whom human flesh is an article of trade, and funerals are almost unknown."[67]

> [C]annibalism is closely involved in the whole network of cannibal customs; embracing first human sacrifice, then the employment of portions of the human frame in the ritual of consecrations and witchcraft, and lastly the preservation and use of human remains, skulls for drinking cups, bones for daggers, teeth for necklaces. This playing with human flesh and bones would be the first step to overcoming a natural disgust.[68]

Even where white settlers had arrived and cannibalistic practices had disappeared on the social surface, the natives could still be found to drink from human skulls if left unsupervised by their colonizers.

Cannibalism had captured the European mind for centuries before Ratzel wrote about it. The Victorians, in particular, had developed a sensationalist fetish around the consumption of human flesh among the colonized.[69] And yet what distinguished Ratzel from other observers was the way he used cannibalism and other deathly practices to develop a biologically driven political theory. Ratzel noted on occasion that the extinction of indigenous populations saddened him but stated the exact opposite on other occasions. His aim was not one of moral judgment but merely one of understanding the underlying political reasons for these extinctions. His conservatism did not allow for conservationist intervention: the weak had to make room for the strong. The "last Tasmanians" played a crucial role for him as an example of a recent extinction. Ratzel's discussion was brief here but harrowing images of the last Tasmanian woman Truganini (1812–1876) (see figure 16) accompanied his text.[70] And although Ratzel applauded the Tasmanians' cheerful and good-humored nature, he also described them as particularly low in culture and development, thus implying an inevitability and moral acceptability to their autogenocide. The extinction of the Tasmanians was not sad from a moral but only from a scholarly point of view; Ratzel lamented that it was no longer possible to study the Tasmanians and thus learn about world history from this "prehistoric" *Volk*.[71] And yet races never entirely disappeared: "When ancient races fall to pieces new ones quickly form themselves out of the fragments."[72] Extinction, in other words, was not a final ending, only the end of a chapter in humanity's evolutionary history.

Ratzel dedicated two chapters of his *Anthropogeographie* to the ques-

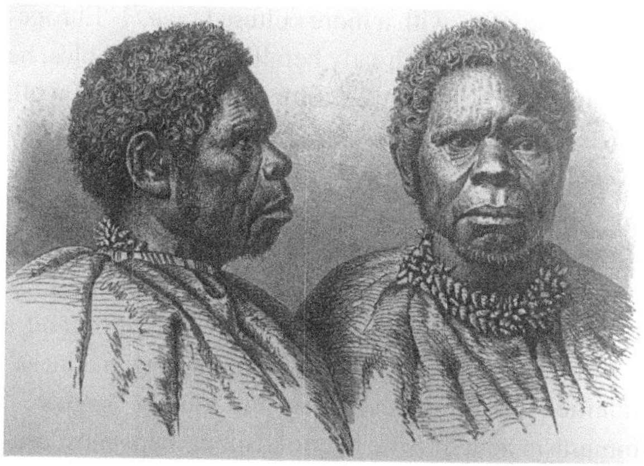

Figure 16. Truganini, the last Tasmanian woman. (Source: *The History of Mankind, Vol. 1.*)

tion of the disappearance of indigenous peoples in the colonies. And while the annihilation of the indigenous Americans took center stage in his narrative, he saw the same process everywhere, from the Arctic North to the shores of Pacific islands. Mostly Ratzel described racial extinction as a slow and steady process, but sometimes it took up speed. "The death struggle" of the Tasmanians, he noted with astonishment, had only lasted seven decades.[73] Ratzel argued that this mass extinction was not occurring primarily because of the influence the white race was exerting, but because of an internal instability that was predisposing *Naturvölker* toward self-annihilation, even before white settlers had arrived:

> They are physically and morally inferior to the civilized peoples. They deal with human life in a thoughtless and cruel way, and treat [population] growth as dangerous and burdensome. They therefore do not share with us the concept of the value of life. Illness, an unhealthy lifestyle in terms of clothing, habitation and nutrition, infanticide, the killing of the young, unnatural vices, polygamy, hunger and water scarcity, war, abduction, and finally cannibalism form a complex of facts, which work against population growth. But what does not reproduce, will be forced back, because other *Völker*, which are growing, will strive to take the space that the weaker are not able to fill.[74]

While Ratzel recognized that "manhunt, great and small, and slavery were phenomena that were closely linked with the encounter of two races of unequal power," he argued that it was not primarily war or slavery that had decimated these populations, but the simple contact with a more cultured race.[75] "Europeans have only exacerbated the evils that had already befallen" these peoples, he argued against those who blamed the barbarism of colonialism itself.[76] If at all, white civilization was proving harmful "without wanting to do so," simply because higher and lower levels of civilization were irreconcilable.[77]

In line with his thesis on the influence of the earth system on history, Ratzel argued both that it was the weak grip on the soil that proved lethal for indigenous peoples and that *Völker* went extinct because they had not acclimatized. But there were other reasons too. Ratzel pointed to superstition and a general unpredictability of life as generating "a self-annihilating madness" among indigenous communities that often descended into violent "orgies."[78] Hunger and primitive communism added to the picture. But perhaps most crucially, his *Anthropogeographie* contained the seed of a biopolitical reading of history that located the seeds of demise in population growth and density, unhealthy lifestyles, low sexual morals, and certain medical practices. Population growth was not only "the source of national vitality and the visible indicator of a sound state organism, it was the motor of natural state expansion through colonization."[79] And yet the reverse was also true: "[t]he very thinness of the population is a cause of their decay; their smallness of their numbers are more readily brought to the point of dwindling or vanishing."[80]

History itself, the former pharmacist claimed, was driven by "health and illness" in that "a *Volk* whose individuals live longer, will also live longer as a *Volk*."[81] He had an interest not only in the many diseases that had decimated colonized peoples since the sixteenth century, but also classed certain medical procedures as signs of racial decay: castration and sterilization rituals, such as the extirpation of the testicles in Polynesia and Micronesia, the slitting of the urethra in Australia, or the complete emasculation in the Americas.[82] In Central Africa, he claimed, human meat was so popular that even the sick were eaten. To capture this supposed blend of structural and psychological causes that made *Naturvölker* disappear, Ratzel spoke of a "pathology of world history," a choice of words that was applauded by some of his contemporaries.[83] But ultimately death, within this historical framing, was only a beginning and not an end. Some traces of a race would always survive through racial mixing.[84] Even Tasmanian blood still flowed in the veins of those with whom the Tasmanians had reproduced. In the course of the extinction process, destruction and renewal thus combined "but in the way that at first the former becomes operative, after which the latter tills the field upon the soil, which is [still] strewn with ruins."[85] The demise of one people meant the birth or expansion of another. This idea would become a central theme when Ratzel turned more systematically to the question of interstate politics.

While the Leipzig geographer had displayed a tendency to use the terms "race", "*Volk*," and "state" interchangeably in his earlier work,[86] *Politische Geographie* focused more consistently on the territorial state, which he felt showed similar regenerative powers to races. Ratzel thus operationalized the death motif by transposing the logic of rise, decline, and rebirth that he had described in his ethnology and human geography onto the interstate system. "The fragments of a decaying state live on and form new states," he argued.[87] Just as races were subject to inner decay, so were states. At lower stages of state development, decline set in faster than at higher ones. But at higher stages of state organization too there was regression, particularly when the spatial consciousness of a *Volk* had not caught up with the spatial extent of its territory.

The idea of decay and reorganization was thus at the very core of Ratzel's organismic theory of the state. He made a crucial distinction here between organisms and organs:

> Individual parts of the organism are connected more closely with the life of its entirety than others; these are the vital and always also most geographically precious parts of different states. One has to know their place within the organism so as to understand their political value. Every state has provinces and counties, the loss of which would bring about its death, as well as others that can be lost without danger.[88]

The state-as-organism's most significant vital organs were access to water and transport routes. If Serbia, for instance, lost the Danube it would be "untreatably mutilated."[89] But major economic regions functioned as vital organs too, particularly those that hosted natural resources and agriculturally rich land. Borders were peripheral organs. Ratzel added a layer of complexity to this model by arguing that organisms could themselves degenerate back into organs and vice versa. As the state-as-organism struggled against other states, it also had to fight to remain an organism.[90]

Ratzel returned to his discussion on the extermination of *Naturvölker*, already discussed in *Völkerkunde* and *Anthropogeographie*. When writing *Politische Geographie*, he again took a keen interest in the fate of the Tasmanians, reading a biography of the Dutch explorer Abel Tasman and noting in his diaries that he had learned about the Aboriginal Protection Society from the London *Times*.[91] He observed that the English colonizers of the Americas had been more "successful" at diminishing indigenous populations than their French and Spanish counterparts, noting that the "Germanic" Anglo-Saxons tended to colonize in family units while the "Romanic" *Völker* preferred fast-lived colonization through the endeavors of young men. The French had established trapper colonies that had relied on indigenous populations and that did not necessitate settlement. They were thus known, he claimed, to have entertained better relations with the indigenous populations, but this was in fact a mere symptom of a weaker grasp of the relationship between people and soil. Only after the French had lost their territory had they realized their error.[92]

Cannibalism too made a reappearance in *Politische Geographie*, though now as a phenomenon that could shed light on political forms of organization. Ratzel reminded his readers of the spiritual relationship between soil and state and how the dead had always been the true landowners. He mentioned not just the sacred respect with which burial sites were approached in many cultures, but how land rights could be impacted by death. The Maori often laid claim to land on which their relatives had died, "even if they had been killed or eaten."[93] Ratzel argued moreover that "states on a lower civilizational stage" were "small-spaced" (*kleinräumig*) and thus likely to experience overpopulation. In order not to lose their inner cohesion and be divided up by others, these states practiced self-extermination and cannibalism. "The *Völker*-destroying 'work' of the European conquerors and colonists," he announced triumphantly, "now appears in a different light." For the very same smallness in numbers that was necessary for indigenous American political organization to function were "an equally inevitable cause of political weakness when they encountered Europeans."[94]

While Ratzel reinforced his earlier observation that decay set in more quickly at lower stages of state development, his argument now extended beyond *Naturvölker* and their supposedly unicellular states. Indeed, Ratzel used the example of Austria-Hungary to show how these processes of decay could occur even among the great powers of the Old World. Germany too had grown naturally out of the moribund mosaic that the Holy Roman Empire had left behind. When a state declined, its nucleus could remain active:

> Such a country is surrounded by the rubble of its former greatness, just as there are mighty screes and seas of rocks at the foot of mountains, which tell the story of peaks which were once higher but which collapsed in the struggle of the elements. Germany, Italy, Austria, those are all cores, which have remained while pieces fell off their periphery. [. . .] The German Reich is surrounded by Switzerland, the Netherlands, the Baltic

> provinces, Bohemia, Austria, all of which were once parts or shoots of an earlier Reich. And on what historical sea of rubble does Ancient Rome glance![95]

Ratzel revealed again a cyclical understanding of history, in which the struggle over space could only produce unstable balances of power and temporary territorial configurations, before the latter would themselves collapse and give way to new patterns. The death of states was to be understood not simply as a downfall but as a waning that was immediately superseded by a reshaping, much in the same way as the extinction of *Naturvölker* was only the beginning of something new. Ratzel found a certain comfort in this. "A bigger state dies, smaller ones emerge in its place," he noted, which made it "mistaken to assume that the growth of states will have reached the end once one state has come to encompass the entire globe." "Should this ever happen," he continued, "there would be, long before this great moment, a multitude of growth processes already under way, which would cause decay and new formation."[96]

Ratzel returned to the question of decay in his popular science book *Deutschland: eine Einführung in die Heimatkunde* (Germany: An introduction to local history and geography), published in 1898 and thus only a year after *Politische Geographie*. It was here that he depicted in detail the geomorphology of German lands, along with their animal, plant, and human life. But while he was primarily preoccupied with the spatial dynamics of life and its relationship to the earth, his narrative also had a strikingly morbid undertone. Ratzel was struck by the coastal erosion along "the clayey coasts of Farther Pomerania" on the German Baltic and vividly described the case of a graveyard that had been eroded by the sea so that "coffin planks and bones" were sticking out of the ground.[97] This observation should perhaps be read as more than a side note, for Ratzel thought of the burial of the dead as serving a reproductive purpose that connected the nation to the soil on which it grew. For him, in other words, the earth was not just a vital natural resource on which the state-as-organism would sprout; it was the very material into the folds of which it would eventually return. The violence that erupted among and within states was, in other words, ontologically similar to the geologic and climatic forces which molded the earth's surface.

As we have seen, Ratzel was generally keen to disassociate his ideas from those of other evolutionary theorists, including Darwin himself, by giving them a spatial twist. Ratzel's spatial logic of extinction crystallized in its most abstract form in his 1901 *Lebensraum* essay, where he argued that "[c]learly, the limitation of living space on earth demands that an old species vacate the space that a new one needs to develop."[98] "In this sense," he continued, "new creation and progress presuppose retreat and demise. It is the same as with the death and birth of individuals."[99] The very idea of survival always already assumed the possibility of its opposite. His was thus a spatial theory of life and death that applied to nature and culture alike. Nations, much like species of primroses or caterpillars, were bound to disappear if they failed to assimilate to their environment.

Ratzel distinguished ontologically between flexible and inflexible organisms but not, for instance, between the extinction of humans and nonhumans. We get a sense of this in a discussion of the relation between the earth and the spatial boundary:

> In all realms of life, life's grip on the soil becomes weaker in the border regions. Speedy reduction and early extinction are the fate of plants, animals, and peoples that live near the borders of the oecumene. [. . .] The invasion of the French in Canada and Nova Scotia, and the English in Virginia, initiated the retreat of the North-American Indians at the easternmost edge of their area of distribution, progressing slowly across the entire continent, like an organ or carrier of all influences that were ruinous to that little people (*Völkchen*). The retreat of the bison, the elk, the Canadian moose, and other animals happened in a similar manner.[100]

Again, there was that slippage between extinction and extermination. Indeed, he spoke of the "dying out or extermination of large animals" without distinguishing between the two.[101] "The losers," he held, were the indigenous Americans "who had but a weak grip on the soil." There was little room in Ratzel's intellectual universe for compassion for those who had gone extinct, for as one species or tribe disappeared its soil was "immediately occupied by another, as if that other had already been waiting for an expansion of its living space."[102] Ratzel was not a fantasist of "empty space." Even self-annihilation was not enough to empty out space; "the land had to be taken" from indigenous populations "before it could be cultivated and settled by whites."[103] His Wagnerian migration theory and his rejection of the concept of indigeneity forbade him to treat the land as belonging to anyone other than the only-ever-temporary victors of evolutionary struggle.

Ratzel was willing to accept that advanced agriculture, forestry, and hunting techniques led to the extinction of certain animal species and individual human races. But he did not seriously contemplate the extinction of humanity itself, certainly not in a sudden way. Even the dinosaurs' living space, he was convinced, had been shrunk over longer periods of time by more adaptable mammal species. Extinction and extermination were not so much an end product as a new beginning. "A complete extermination of all life across a wide area cannot [] be accepted even in the case of the Ice Age." "While no more than a few decades ago," he continued, "no one doubted that fire, water, or ice had intermittently destroyed all life across vast regions, after which new life emerged out of the dead soil," today it was crucial to reject any catastrophism in evolutionary theorizing.[104] And yet Ratzel also noted that many animals were predestined to destroy the viability of humanity. He mentioned not just animals of prey that killed and devoured humans, but also parasites or microorganisms that entered the human body to destroy its functioning. This struggle between human and nonhuman life did pose to him the question of "the limits of human existence."[105]

War and the Necropolitical Impulse

We have already encountered Ratzel's concept of war in the previous chapter, but it is worth returning here not just to how he framed war in his political geography but how he reflected on the corporeal experience of armed conflict. Although his personal experience of war had occurred in the first half of his life, he was only to reflect on it in writing shortly before his death, at a time when he was outlining in the most panoramic terms a geography of all life, human and nonhuman. His autobiographical reflections thus assume, through the medium of personal experience, a culmination of his thought on the question of life and death.

Ratzel's account interlaced immersive descriptions with a detailed depiction of the French landscape. He wanted his readers to hear, smell, feel, and taste combat with his protagonist. He described how nighttime marches caused visions among worn-out soldiers or how darkness amplified sounds that blew away in the daytime: a soldier moving his rifle from shoulder to shoulder and the stumble of a tired horse. He treated war as a medium through which nature, that looming thunderstorm, imposed itself onto behaviour and consciousness. The masses who "followed the orders of a genius statesman" (Bismarck) were led by an "elemental force."[106] Bullets swirled and chirped like birds and civilians were struck dead by lightning, discipline "took root" in the young soldier, and the rifle drill movement allowed for the technology of the weapon itself "to be absorbed by the human body and psyche."[107] In winter, even the stars shivered with the soldiers, shimmering and dancing to keep themselves warm.[108]

War emerged thus as a form of hybrid nature, in which nonhuman agencies acted upon and shaped the body and mind of the combatant. Nature itself seemed to be at war.

Ratzel's narrative was marked by a notably detached depiction of the horrors of war, which treated the latter with medical detail but without overt emotional or indeed political undertones. In this, Ratzel's war narrative prefigured the experiential account of World War I trench warfare given in Ernst Jünger's 1920 bestseller *Storm of Steel*. Both accounts, published a mere fifteen years apart, are graphic in nature and place the human subject within a larger technological and environmental assemblage in which mud, shrapnel, bloodied uniforms, and torn-off limbs come together in a more-than-human dance that performs the spectacle of war. Jünger, much like Ratzel before him, described war as a natural rather than a social force. Although Ratzel was immersed in a significantly older technological landscape, his trench world too was one of colorful flowers, insects, and elementary forces, of "hurricanes of fire" and "hails of shells."[109] There are significant stylistic differences between Jünger's machine-gun-length sentences and Ratzel's Romantic prose, and yet both Jünger and Ratzel aestheticized war, including its horrors. Both writers interpreted the soldier not just as a final arbiter in conflict but as a subject whose metamorphosis from naïve and confused civilian to masculinized warrior was completed only in the experience of war.

And while the experience of combat itself featured centrally in Ratzel's account, it was in fact the crucial biopolitical space of the military hospital that exerted the most profound influence on his thinking.[110] As we have already seen, Ratzel had joined the Franco-Prussian War with great enthusiasm and remained overwhelmingly positive about his military experience throughout his life. When he was forced to recover in a military hospital, he turned his attention from ruined landscapes and towns to ruined bodies. Captivated by the well-oiled machine of the Prussian medical service, he watched the nurses as they tirelessly tended to the wounded, communicating in short commands, just as if they were themselves combatants on the battlefield. He recounted the repeated bandaging of a fellow soldier who had a gunshot wound through his face that had "ripped out both eyes and his upper nasal bone."[111] Ratzel saw this horrific experience as an opportunity to "practice viewing" such wounds, a skill that prompted him to muse on the "aesthetics of ugliness" (*Ästhetik des Häßlichen*).[112]

Ratzel's memories of this time, from encounters with French civilians to the description of near-death experiences among his fellow soldiers, reveal him as an often strangely unemotional observer. Buttmann sees in Ratzel's account "the shock at the suffering and dying on the battlefield."[113] But the latter is not obvious. Indeed, Ratzel claimed that his experience of the military had enriched him and suggested that his fallen comrades had lived on in him. Reflecting on mass graves, he noted that it was painful for a soldier's parents not to be able to pray at their son's grave. And yet, "future generations will be able to see the mound under which the dust of heroes rots, and a wide-branched tree will rustle and murmur above it."[114] This was of course the same man who passionately believed that only military obituaries were able to pay an appropriate level of respect to the dead and that only war brought out the best in a man.

Ratzel paused to discuss some of the medical practices that had been established in the Prussian military by the 1870s, including that of triage. Army doctors habitually abandoned cases which they considered too difficult to treat by hanging a sign around the neck of these soldiers which read "moribund." Ratzel depicted a scene in which a soldier himself asked to be classified in this way and thanked the doctors for doing so. By concluding that this man had "gladly" consigned himself to death,[115] Ratzel unknowingly put his finger on the central dynamic of Foucaultian biopolitics: the simultaneous fostering and abandoning of life and the state's medicalizing grasp at the physical bodies of its population.

It was also Ratzel's time in the military hospital that prompted him to ponder the ontological "easiness" of death and the inability of separating it from life. Clearly under the influence of Fechner, he concluded that there was "no mountain range or wall" between life and death. Indeed, "the way into the great and dark gateway" was really "rather flat." Ratzel, who in later life had returned to his Christian faith, described the experience of watching another human die, in this case a comrade, as a "sanctification."[116] Once a logic of survival has been accepted, Mbembe points out, "one's horror at the sight of death turns into satisfaction that it is someone else who is dead."[117] Mbembe is concerned here with the emotional response to the killing of an enemy, but it accounts too for Ratzel's experiences of the military hospital, where "[i]t is the death of the other, his or her physical presence as a corpse, that makes the survivor feel unique."[118] This was as true for the wounded soldier as it was for the eliminated Tasmanians, whose eyes glanced back at the reader of Ratzel's *Völkerkunde*, allowing for a detached and voyeuristic enjoyment of disappearance.

Much like Foucault would in the 1970s, Ratzel lamented that death had become the ultimate taboo.[119] Humans, the incessant writer of necrologies felt, should embrace death if they wanted to find eternity. He concluded that "the more life is prepared for death, the more beautiful it becomes."[120] Rather than approaching survival and extinction only through the lens of scientific detachment, Ratzel came to deal with death thus as an aesthetic category in ways that chime with the death cult introduced and institutionalized by national socialism three decades later. While at first sight Ratzel's biopolitics, discussed here through the prism of war, may seem different from that articulated by Foucault, and Ratzel certainly never developed his theory of life as a critical conceptual tool, it is certainly productive to read Ratzel as a biopolitical thinker.[121] Ratzel was aware of the two scales Foucault would later designate as crucial to his understanding of biopolitics—the population and the human body—and developed an interest in the conditions of life, such as light, air, warmth, and moisture on both the human body and on whole nations and their desire to migrate.[122] He thought of some populations and their activities as an evil on the state-as-organism precisely because they were primitive or inferior.

> The Indians were pushed inland from the Atlantic Ocean and therewith began the worsening of their living conditions, which continues until this day and from which arises a decrease of their number in never ending struggles, which are a chronic malady in the body of the state [*ein chronisches Übel am Körper des Staates*].[123]

Ratzel admired North America for its love of statistics and its tendency in the late nineteenth century to compile and publicize data on public health, living conditions, and mortality rates. On his travels to the United States he had noted in appreciation the way North American society had managed to control smallpox through inoculation and was fascinated by American architects' and urban planners' concern for ventilation and cleanliness.[124] He had a sense of how bodies became the subject of governmental practices, particularly in times of war. Upon his entry into the Prussian army, Ratzel had become aware of the way modern war required him to cast away his identities as a civilian and scientist and subject his body to the necropolitics of the war machine. "The fatherland", he wrote, "needs nothing other than our naked bodies; just as we stand in front of the army doctor, this is how one wants us: legs for marching and arms to shoot and punch, a body that holds these together and a head with the right senses, nothing else is required."[125] Ratzel thus reduced the social function of the German soldier to that of a body which had to have the "right" visceral experience of war.

By the time Ratzel's war memoires were collected and published in 1905, the German Empire was at war in the colonies. After participating in the Boxer War in China (1900–1901), the empire would fight the Maji Maji War in German East Africa (1905–1908) and conduct a genocidal military campaign against the Herero and Namaqua in Southwest Africa (1904–1907). Formal German colonization had from the start relied on forms of violent oppression. In a truly Ratzelian fashion, the Germans had publicly celebrated the de jure abolition of slavery in its colonies while tolerating its de facto continuation, particularly in German East Africa.[126] The murder of up to 100,000 Herero and Nama in German South West Africa by starvation, dehydration, shooting, and systematic encampment stands out as the most cataclysmic form of German colonial violence. While these colonial wars represented a form of violence that was pioneered in the colonies only to boomerang back into Europe in the first half of the twentieth century,[127] they also saw a radicalization of military practices and doctrines that had their roots in Bismarck's wars of unification. The harsh treatment of civilians in particular, unleashed to detrimental effect in Southwest Africa, had already been adopted in the Franco-Prussian War, during which civilians were executed with and without trial for a whole range of minor offenses.[128] As the Prussian military had developed a high degree of autonomy from civilian control, its military doctrine and procedures had moreover become governed by the ideal of a war of annihilation (*Vernichtungskrieg*). In order for a decisive victory to emerge, it was believed, the army had to aim at the annihilation of the enemy in ways that risked its own destruction. It was this strategic principle that was unleashed with superior weaponry and full force on the bodies of the colonized.

Although some of Ratzel's most popular books could be found in the shelves of military enthusiasts across Germany, it is important to keep in mind that he was only one of many colonial and racial theorists of his time. Crucially, however, the very language in which the war against the Herero and Nama was fought resonated with the logic of extinction/extermination developed and popularized in Ratzel's scientific discourse and colonial activism. The German geographer had moreover been among the first public intellectuals to suggest that German colonialism in Africa should emulate the westward expansion of the American frontier and its struggle against the indigenous Americans, an idea that was still reverberating powerfully through Germany during and after the genocidal campaign against the Herero.[129] The way the General Staff had legitimated the establishment of German settlements in Southwest Africa by reference to the Herero's inability to cultivate their own land invoked a Ratzelian logic too.[130]

Ratzel did, moreover, have admirers in the colonial administration who were directly involved in the empire's military campaigns. Eduard von Liebert, the governor of German East Africa, for instance, spoke with admiration of "the great geographer Friedrich Ratzel" in a 1905 speech to the *Kolonialkongress*.[131] Von Liebert was considered a Carl Peters–wing radical of the early colonial movement who had pioneered scorched-earth tactics in the 1890s.[132] He had been a right-wing critic of post-Bismarck-era governments and would in 1929 become a member of the National Socialist Workers' Party. Georg Hartmann (1865–1946), Ratzel's former student and later chief representative of the South West Africa Company, told the *Kolonialkongress* in 1910 that Ratzel had shown unambiguously that the tropical climate had a negative effect on a population's ability to civilize (*Kulturfähigkeit*) and thus generally brought with it a "negrification" (*Vernegerung*) of those who migrated there.[133] Coating his language in the "struggle for existence" and reminding his listeners that humans were "of the earth and would remain so," he insisted that only once colonial subjects had been subjugated to German civilization, whether voluntarily "or by force of arms," were they to be treated as equals. He quickly added that the colonial administration should use mixed race subjects strategically as a layer in between white and black races, while ensuring the "purity of our race" (*Reinhaltung unserer Rasse*).[134]

Hartmann was hardly a detached observer. Only four years after publishing the thesis on drift ice he had completed at Leipzig, he actively planned a war against the Herero in South West Africa. As a representative of the South West Africa Company he worked closely with Theodor Leutwein, the commander of the *Schutztruppe* and later governor of German South West Africa. Hartmann argued in 1895 for a war against the Damara, claiming that only "when these arrogant tribes [. . .] lie smashed upon the ground will the *Schutzgebiet* experience a truly peaceful development." He called for a "sudden and totally surprising attack" and expressed his conviction that a few hundred men could easily "sweep the area under question all the way to Windhoek."[135] A year later, Leutwein put Hartmann's plan into action, one of a number of military campaigns he led as commander of the *Schutztruppe*. Leutwein was still in command in 1904 during the early stages of the Herero Wars. Lothar von Trotha, who took over from Leutwein, believed he was fighting a "racial war" in which anything other than total victory was unacceptable.[136] When in October 1904 he would issue his infamous "extermination order" (*Vernichtungsbefehl*) against the Herero, masses of unarmed men, women, and children were already being massacred.

Ratzel did not reflect on the German Empire's worst atrocities before his death in August 1904. And while he had expressed his dislike of the brutal conduct of Cecil Rhodes's troops in the 1893–1894 First Matabele War, this was hardly a statement made on ethical grounds.[137] Ratzel had never imagined the German colonization of Africa to be peaceful. In 1892, he had written about the need to pacify "the lawless hottentots and basters" in Southwest Africa by force, if need be.[138] These then are crucial links between Ratzel's military biography, the Darwinian ideas that underwrote his political geography, a necropolitical concept of war that naturalized annihilation and extermination as inherent characteristics of humanity's struggle on earth, and the racial violence inflicted by the German Empire upon its colonies. As we have already seen in previous chapters, Ratzel was not aware that his necropolitical notion of *Lebensraum*, which summarized so well the spirit of his geography, would become his most lasting contribution. In the following chapter, we will look at the ways his political geography traveled, was inflected, and resisted in the century after his death.

Conclusion

A preoccupation with death and a concern with the boundary between life and death had already taken root in Ratzel's adolescence. As he came under the influence of Darwinian ideas in the 1860s, his work began to refocus this interest first on the extinction/extermination of nonhumans before he turned to the extinction/extermination of colonized races. He was so engrossed by the question of death and dying that he felt the need to tell his readers, when reviewing the British eugenicist Karl Pearson's *The Chances of Death and Other Studies in Evolution* in 1898, that it contained little that was new to him.[139] A year earlier, Ratzel had made the idea of death as a new form of becoming the ontological basis of his political geography. There was no "aura of meaninglessness about death" for Ratzel, nor was death a fundamental rupture, but rather a "ripening" and a "fullest expression of life."[140] Throughout his life, the Leipzig geographer was drawn to decaying architecture and bodies. Before his own death, he reflected on the biopolitics of war, although he did not refer to it in those terms. I have used the term "necropolitics" to denote this cluster of ways in which he fused death, nature, and politics.

While neither his concept of extinction, his fascination with cannibalism, nor his ruin gaze alone were atypical for a nineteenth-century subject that had been influenced, to varying degrees, by German Romanticism, Christianity, Darwinism, and colonial discourses, it is the combination of these that points to something more extraordinary. Ratzel's preoccupation with the rise and fall of races, nations, and states and his aestheticization of death reveal a system of thought that needs to be inserted in a more prominent place in the history of bio- and necropolitics, which has thus far foregrounded the mid-twentieth century's most deadly regimes. Ratzel's reading of the state as an organism and his insistence that political geographers had to be pathologists of history were clearly products of his background as a zoologist and pharmacist. And yet they also expressed the profoundly necropolitical space of the military hospital that was so formative to him. By coating his discussion of life and death in the scientific language of his day and by amalgamating, much like Darwin did, the process of extinction with the violent act of extermination, Ratzel's work served to naturalize the elimination of indigenous populations. Ratzel was not the only one to do so, but his geography laid the intellectual and scientific foundations for a twentieth-century approach to politics that saw extermination as a legitimate strategy to ensure a nation's survival. The self-annihilating *Naturvolk* was not merely a euphemism for conquest and domination, but a self-fulfilling prophecy.

As we will see in the following chapter, Ratzel's morbid preoccupation with the disappearance of races, peoples, and states did not die with him in 1904. The American geographer Ellen Semple, Ratzel's popularizer in the Anglo-American world, incorporated the extinction thesis in her 1911 *Influences of Geographic Environment*, even though she did not give it as much attention as Ratzel had.[141] The Swedish political scientist Rudolf Kjellén picked up the death motif during the First World War when he compared territorial partitions in European history to "executions." Kjellén imagined his own nation, Sweden, in the graveyard of history, accepting that there was a beauty in a gaze from a nation's future grave onto the present.[142] The Nazi sympathizer Karl Haushofer argued in the 1930s that the many maps he used in his writing were in fact X-rays, which were meant to discern where a state was so old that its skeleton was already shining through its skin.[143] Ratzel's writing on death and ruins must have left an impression on Haushofer because he chose to include these passages in his 1941 collection of Ratzel's most notable work, *Erdenmacht und Völkerschicksal* (The power of the earth and the destiny of nations). Similarly, Richard Hennig, a contemporary of Haushofer's, explained in his 1935 *Einführung in die Geopolitik* (Introduction to geopolitics) that while young states tended to have a strong impulse to grow in space, more mature states typically started to stagnate or found colonies. "Old states," he claimed, "shrink in much the same way as elderly humans do." "[V]ery old states," he added, "sometimes have to be incapacitated" if they have become "unfit for life."[144] Ratzel's necropolitical geography, in other words, was in this respect less radicalized than operationalized by his followers in the 1930s and 1940s. It is perhaps no surprise that Ratzel's ruin aesthetic proved to be largely congruent with the theory of ruin value propagated after the war by Adolf Hitler's architect Albert Speer.[145]

The close relationship between fascism and death has been pointed out by many an observer.[146] "It is not by chance," Umberto Eco notes, that *Viva la Muerte* (Long live death!) served as a motto of the Falangists.[147] Whereas in other modern societies, death was often rendered an unpleasant inconvenience, the fascist "craves heroic death, advertised as the best reward for a heroic life." It is precisely in this impatience to die that fascism sends other people to their deaths so easily. This directedness toward death would find expression not merely in an ultimately suicidal warrior cult and the extermination of enemies of the *Volk*, but in the realms of architecture and philosophy too. Arendt notes about de Gobineau that "[d]octrines of decay seem to have some very intimate connection with race-thinking."[148] She could easily have made a similar argument about Ratzel. It is perhaps precisely because she tries to absolve Darwinism of any responsibility for the rise of fascism that she discusses neither Ratzel's political geography nor the wider *Lebensraum* discourses he helped to spawn. And yet Arendt was onto something. We should be careful not to imply a teleology from Darwin to Ratzel. Ratzel's *Lebensraum*, with its deathly underside, was only one mutation of Darwinism, albeit a particularly influential one, as we will see in the next chapter.

CHAPTER 7

Ratzel after Ratzel

Friedrich Ratzel took a keen interest in his reception. Keen to document how his peers had commented on his works, he collected reviews of his books, including negative ones. In June 1904, he told his publisher triumphantly that *Politische Geographie* had now been favorably reviewed in France and Norway.[1] Nine weeks later, Ratzel was dead. The months that followed saw the publication of more than seventy obituaries for the deceased geographer, a number that would have undoubtedly brought the dedicated necrologist some satisfaction. The canon of these eulogies was predictably laudatory. There was a sense among scholars, admirals, eugenicists, and Alpine enthusiasts that someone important had been lost.[2] Geographers were particularly exuberant in their praise. While Vidal de la Blache claimed that the death of the German geographer had left a "great void," Ellen Semple felt that geography had lost its very own Adam Smith.[3] More ominously, the Anglo-German geographer Ernst Georg Ravenstein expressed his hope that politicians might soon find inspiration in Ratzel's political geography.[4]

Perhaps unsurprisingly, this outpouring of admiration for the departed glossed over the fact that Ratzel's reception in the academic community had in fact been mixed. Those who knew his works well had long observed that his tendency to produce manuscripts out of a mosaic of other texts had led to argumentative tensions. But there had been other and more damning charges. *Völkerkunde* had in 1886 been criticized for its superficiality and misinterpretations, including the book's reading of circumcision as a form of human sacrifice.[5] In 1891, the geographer Hermann Wagner, a nephew of Ratzel's mentor Moritz Wagner, had slated the second volume of *Anthropogeographie* as polemic and sophistic. Unconvinced by the book's evidence base, Wagner had spotted in Ratzel "the bravery of a publicist not afraid to utter trivialities."[6] In a second scathing review he criticized Ratzel for constructing straw man arguments and for being sloppy with his citations.[7] Ratzel was deeply offended by Wagner's attack and responded immediately, though quickly getting carried away by a convoluted explanation as to why he had alternated between kilometers and miles in his statistical analysis.[8] What seemed to have angered Ratzel most was that this fellow geographer had dismissed him as a mere "journalist."[9] Hermann Wagner was not the only one to grumble. Even two years after Ratzel's death, his widow Marie bemoaned that her husband was still being attacked.[10]

Tracing the bearing of a thinker beyond the grave is no easy task. Citations may provide crucial evidence about transmission in scholarly discourse, but they are rarely present in nonacademic writing. Ideas, moreover, are refracted by critics and disciples alike, grouped together with those of other thinkers in ways that may transform them beyond recognition. Different meanings of the same concept often circulate simultaneously. At times, one may be dealing with a theoretical or ideological compatibility, rather than with direct relations of influence. In the discussion that follows, I have kept these issues in mind. Rather than provide a comprehensive account of Ratzel's reception, which would go far beyond the scope of a single chapter, I have sought to strike a balance. I have endeavored to account for those who were most directly influenced by Ratzel but who were sometimes less historically noteworthy, and those on whom Ratzel's impact was less significant but who played a more important role in shaping twentieth-century history. A focus on his posthumous reception among German geographers, significant as the latter undoubtedly were, will thus make way for a more transdisciplinary and transnational examination. Much of what we will encounter has to do, in some way or other, with geopolitics. But in addressing Ratzel's place within this larger and multifaceted tradition, I have concentrated on direct engagements with Ratzel and his works rather than providing a more general history of geopolitics, which has already been surveyed by others.[11]

We will see that while Ratzel was embraced in North America and Europe, it was not, as it were, the same Ratzel who circulated in these different contexts. Although he was certainly an influential scholar and public intellectual during his lifetime, it was only during the First World War that his ideas began to develop their full impetus. And while Ratzel's necropolitics certainly left the biggest imprint on the period from 1920 to 1940, it can still be traced into the Cold War period. It is significantly more difficult to unearth the Leipzig geographer's impact after 1945, both because his writings were superseded by those of his disciples and, more importantly, because of the stigma surrounding his name. Even in the most recent period, Ratzel has kept reappearing.

Ratzel in Europe before "Versailles"

At the time of his death, Friedrich Ratzel had gathered admirers, interlocutors, and opponents in different German academic and political circles. He had also established himself as a public intellectual, particularly through his journalism and founding membership of nationalist and colonial advocacy groups. His public science books were read by a broad segment of the German bourgeoisie. In the academy, his readership extended far beyond the field of geography. Perhaps most obviously, there were the members of the Leipzig circle, men like the Ostwald, Wundt, and Lamprecht. But others too had come into contact with his ideas, including the philosopher Friedrich Nietzsche. Nietzsche's annotated copy of the first edition of *Anthropogeographie*, which has endured in his private library, reveals that the philosopher was particularly taken in the 1880s by Ratzel's insistence on the geographical base of historical processes.[12] Despite a personal aversion to Ratzel's mentor Moritz Wagner, which Nietzsche confessed in private correspondence alongside an interest in Ratzel's book, the philosopher seems to have been intrigued by the implications of the *Migrationstheorie* for human history. Nietzsche had also highlighted passages in *Anthropogeographie* in which Ratzel had questioned the degree to which racial difference could be reduced to phenotype.

An interest in Ratzel's writings extended beyond Germany and the Anglosphere. The Polish-born sociologist and thinker of *Rassenkampf* (racial struggle) Ludwig Gumplowicz, a professor at the University of Graz, drew heavily on Ratzel's American writings and deemed his political geography "epoch-making."[13] The French founder of sociology, Émile Durkheim, who corresponded with Ratzel in the late 1890s,[14] found the German geographer's approach "messy" but "extremely thought-provoking."[15] Indeed, much French geographical writing from the 1890s until the 1930s has been read as a response to Ratzel's work.[16] The anarchist geographer Élisée Reclus, for instance, noted that Ratzel's *Die Vereinigten Staaten von Nord-Amerika* was "a work of the first order" and, despite its didactic style, "perhaps the best and most complete treatise on the United States in any language."[17] Paul Vidal de la Blache, whose possibilist ideas would ultimately give birth to a distinct French "school" of geopolitics, keenly reviewed and built on Ratzel's major works in the late 1890s.[18] Ratzel was also read and translated by Italian geographers, including some with socialist inclinations.[19] Through Bruno Adler, one of Ratzel's students, his ideas found entry into Russian geography during his lifetime.[20] Indeed, Ratzel's import into late imperial and early Soviet Russia was part of a larger transfer of German geographers to the east.[21] The influential Russian philosopher Georgi Plekhanov, in particular, sought to tease out semblances between Ratzel and Karl Marx's writings at the turn of the twentieth century.[22] Ratzel's popularity was thus not limited, initially at least, to conservative circles.

The Leipzig geographer may have received a transnational interest from distinguished geographers, sociologists, and philosophers, but he was not uncontested. Even among his former students, reservations about the grand master emerged soon after his death. Kurt Hassert noted that Ratzel had often settled for "hypothetical connections instead of irrefutable laws."[23] Ratzel's former assistant in Leipzig Alfred Hettner observed in 1907 that many of Ratzel's books were so general that they bordered on the trivial.[24] Later, Max Eckert would accuse Ratzel of using empirical examples only to confirm his preconceived views.[25] Although some of Ratzel's works were republished in the years following his death, including *Anthropogeographie* in 1909, interest in him had begun to wane, especially outside of the geographical discipline. Kelly has suggested that social Darwinism struggled more generally in the period before the First World War to find supporters in Prussian military circles.[26] Secretary of state in the German Imperial Naval Office, Admiral von Tirpitz, for instance, was Mahanian in his thinking.[27] Indeed, he had removed Ratzel's follower Vice Admiral Curt von Maltzahn from his position as head of the Naval Academy in 1903.[28]

Darwinian-infused arguments about international politics did not, of course, necessarily make reference to Ratzel. The bestselling books by Friedrich von Bernhardi, Heinrich Claß, and Friedrich Naumann had a Darwinian undertone and certainly chimed with the rampant militarism and nationalism to which Ratzel had contributed, but they were neither applications nor further developments of the Leipzig geographer's ideas.[29] A blueprint for a more aggressive German foreign policy, General Friedrich von Bernhardi's *Deutschland und der nächste Krieg* (Germany and the next war) termed war "a biological necessity of the first importance" and made reference to Darwin but invoked neither Ratzel nor his *Lebensraum* concept.[30] In fact, the book was not a study of geography at all, but a blend of military strategic manual and political pamphlet. Heinrich Claß, Hasse's successor as president of the Pan-German league, published in 1912 the widely read pamphlet *Wenn ich der Kaiser wär'* (If I were Kaiser) under the pseudonym Daniel Frymann.[31] Claß may have used a Ratzelian language of inner and outer colonization and geographical *Lage*, but he prioritized the acquisition of territories in southeastern Europe rather than overseas. The book's hypernationalist, antiliberal, and antisemitic content made an impression on a young Adolf Hitler, but it was not in a dialogue with

Ratzel. Similarly, while Naumann's vision for a *Mitteleuropa* (central Europe) under German hegemony was perhaps broadly in sync with Ratzel's latest writings, the book was neither anthropogeographical in tone nor did any of Ratzel's works appear in its lengthy bibliography. There was a general ideological compatibility between these three bestselling texts and Ratzel's political geography, but it would be difficult to trace the spirit of the former back to the latter.

If we follow Smith in reading *Lebensraum* more broadly as an ideology of agricultural colonization, then we can certainly detect traces thereof in the ongoing debate surrounding the German war aims, although the arguments made here rarely featured the intricacies of Ratzel's geography.[32] Claß's Pan-German league lobbied aggressively for the Germanization of annexed territories during the war, but its status was that of a powerful pressure group that struggled to establish a new consensus.[33] In February 1917, the Kaiser passed an essay to the General Staff to be distributed among the German soldiers in the hope that it would reignite their will to victory.[34] But it was not an article written by Ratzel but rather by one of his opponents, the racial theorist Houston Stewart Chamberlain. Ratzel had established neither a direct connection to the monarch nor to Theobald von Bethmann-Hollweg, who would become chancellor in 1909 and remain so during much of the war.

None of this should imply that Ratzel had fallen into oblivion. Geographers began to produce works that played with Ratzelian motifs, keenly promoting imperialist visions of *Mitteleuropa* that had been gaining traction in popular discourse.[35] Alfred Hettner, now a professor in Heidelberg, claimed in 1917 "[p]recisely because we are in the center and the other states are pressing on us from all sides, we perceive our *Lebensraum* as cramped and must at least expand it in some places to remove the hostile pressure from our central organs."[36] Hettner would call the postwar settlement a "terrible peace" that had emerged from a unique combination of French hatred, English egoism, and American deceit.[37] "The German nation," he promised, would "rise again and build a new, more beautiful fatherland from the rubble and ashes of its previous national greatness."[38]

Even more prominent among wartime *Lebensraum* enthusiasts was the geomorphologist Albrecht Penck (1858–1945). A fellow colonial enthusiast, Penck had declined an invitation to apply for the chair at Leipzig to which Ratzel would in 1886 be appointed.[39] He had begun to engage with Ratzel's glaciologist writings in the 1890s and later corresponded with him about issues of academic governance.[40] Until the war, Penck's writings had primarily been concerned with questions of physical geography. But in 1914, a journey to

Britain had coincided with the outbreak of the war. Accused of espionage, he was held under police observation in a West London hotel before being allowed back to Germany.[41] When Germany lost its colonies early on in the war, Penck turned his attention to eastern Europe, where he found political boundaries to be in flux, a space that in his view was waiting to be colonized by Germans.[42] In 1916, he visited the newly occupied territories in Poland and practiced a colonial gaze upon the "primitive" forms of sociopolitical organization he found there.[43]

Although Penck had previously noted in private that Ratzel's works were too abstract to be useful for the purpose of actual military occupation,[44] he resorted to Ratzel's terminology in his September 1917 inaugural lecture as rector of the University of Berlin. Like Ratzel, Penck accepted that the struggle for space was rooted in biological principles. While natural borders might provide historically favorable conditions, they could also be limiting for a state that was growing organically. Like Ratzel, Penck was caught in a tension between thinking of politics as the mere extension of geographical influences and as a realm that could be shaped by superior geographical knowledge. Within the natural confines within which each state was required to act, Penck claimed, "far-sighted rulers" were able to "cautiously steer the development of the state towards a certain *Lebensraum* that [could] provide the necessary foundations for the state to flourish."[45] Fusing the personal and the political, he coated his use of Ratzel's concept in a narrative of national victimhood. Germany, he claimed, had only sought peaceful expansion and had thus been unfairly punished by the confiscation of its colonial acquisitions and the imprisonment and expropriation of its expatriates. Any peace treaty would have to hand Berlin back its colonial empire and grant it some of its newly conquered territories in Europe that were "necessary for our German *Volk*."[46] Inevitably, Penck would become a staunch critic of the terms of the Versailles Treaty.

In Penck's and Hettner's wartime writings, *Lebensraum* was transformed from a catchphrase that gave expression to the fantasy of further colonization in the southern hemisphere into one that stood for territorial compensation in eastern Europe.[47] Although Penck had little trouble in adapting his anti-Slavic worldview to the ideological aims of the national socialist movement, his background in geomorphology made him increasingly skeptical of a new and rapidly growing school of *Geopolitik*, which was de-emphasizing the importance of natural science as the basis for political analysis and prediction.[48] This new discourse was to a large degree sparked by the writings not of a German geographer but by those of a Swedish political scientist.[49] A Germanophile and later

Conservative member of the Swedish parliament, Rudolf Kjellén (1864–1920) had in 1891 accepted a post at Gothenburg University that included responsibilities in teaching geography alongside political science. Immersing himself in the field of German geography, he swiftly integrated geographical ideas into his thinking.

When Europe was plunged into war, Kjellén wrote a series of pamphlets that provided geographical-infused answers to questions of state organization and interaction. The German translations of these wartime writings soon attracted a sizeable following on the political right. Germany, he claimed in 1915, stood for conservative political values—order, hierarchy, and nationalism—which he saw as locked into a mortal struggle with those of democracy, parliamentarianism, and cosmopolitanism. The German values of 1914, he predicted, would be victorious over those of 1789.[50] In an attempt to produce a Linnean system of political science, Kjellén would coin a large number of concepts, including that of "biopolitics."[51] But it was "geopolitics," defined as the "doctrine of the state as a geographical organism or as a phenomenon in space,"[52] which would catch on. Kjellén had first used the term in an 1899 Swedish-language article and thus during Ratzel's lifetime, but his conceptual innovation had been reviewed negatively by German geographers.[53] Only during the First World War did the term develop its full force among its German audience.[54]

Despite drawing on a variety of German thinkers, Kjellén's main source of inspiration was undoubtedly Ratzel.[55] He had used the German geographer's anthropogeography in his earlier work, which addressed the dependence of human societies on the earth system. When he turned to the study of war and interstate politics, he brought with him a typically Ratzelian tool kit, a focus on the location and territorial size of states. He followed Ratzel in his reading of war as an *experiment* and predicted a great war that would tilt the balance of power in Germany's favor.[56] He pledged allegiance to Ratzel's geography again in his most notable work, the 1916 *Staten som lifsform* (The state as life form), translated a year later into German.[57] Kjellén noted minor terminological disagreements with Ratzel but concurred with him on other more crucial issues. He accepted the state-as-organism analogy, the view of humanity as a piece of earth, and the importance of vital geographical organs to state survival. He noted too a dangerous oblivion to the soil among seafaring and trading nations. Kjellén also accepted Ratzel's necropolitics, highlighting the state's ability to be resurrected in other states. "At such a grave, one must not shed tears but find gratification," he claimed. For "to dissolve in a higher form of life" was for states as much as individuals a way of "taking the sting out of death."[58]

Ratzel, Kjellén felt, had always been "more of a collector than an organizer" and had to be transformed into a more systematic thinker.[59] The Swede stripped away not just the philosophical dimensions in Ratzel's writings but the biogeographical latticework too, though without disposing of their bellicose logic. Those who followed in Kjellén's footsteps would perform much the same move. The Swede's most lasting achievement was thus the replacement of Ratzel's convoluted prose with a neater conceptual matrix that could more readily be mobilized by others.[60] He had found a level of abstraction in between Ratzel's political journalism of the 1890s and the highly codified account the Leipzig geographer had provided in *Politische Geographie*. When Kjellén died in 1922, he was at the height of his popularity in Germany. In his native Sweden, however, he had remained almost unknown.[61]

A final avenue in which Ratzel's geography continued to echo was in neovitalist biology. Although Ratzel had flirted with a move from the broad school of *Darwinismus* to the neovitalist camp in the early 1900s, he had died without making the decisive step.[62] Others were less hesitant. The Baltic German aristocrat, biophilosopher, and neovitalist thinker of *Umwelt* Jakob von Uexküll (1864–1944) had, like Ratzel, begun as a follower of Darwin and had developed his ideas in a dialogue with Ernst Haeckel. But he had abandoned the neo-Darwinian camp and its mechanistic interpretations in the 1890s and increasingly favored a vitalist subjectivism. Uexküll understood *Umwelt* as the lifeworld of an individual animal in which it perceived and communicated in ways that were specific to its species, an idea that further developed Ratzel's discussion of *Umwelt* in *Anthropogeographie*, as well as the latter's concepts of *Lebensraum* and *Raumauffassung*.

Von Uexküll lost his sizable estate in the 1905 Russian Revolution and First World War. A friend of Houston Stewart Chamberlain's, he began to see himself as the counterpart of a Jewish and socialist conspiracy.[63] In 1920, Uexküll published an influential popular science pamphlet under the title *Staatsbiologie* (State biology). Heavily illustrated with diagrams, *Staatsbiologie* read the state not as an assemblage of legal contracts but as one of ever-mobile bodies and materials. There were crucial differences to Ratzel's political geography. Von Uexküll was not primarily concerned with developing a theory of interstate politics, although he recognized that it was the requirement of each *Volk* to defend its state "to the utmost."[64] But he too saw the state as an organism, the smallest component of which was the family unit. Like Ratzel, who had in 1904 acknowledged the need to cultivate a state of racial belonging, Uexküll called for a "pathology" of the state. He emphasized the need to attend to the "illnesses" and "parasites" that could infest a state organism.

Even though von Uexküll did not engage with Ratzel directly and could have developed his ideas independently from a reading of Haeckel and Spencer, Mildenberger rightly notes that the latter is highly unlikely, given the status Ratzel held in the German academy at the time of his death.[65] On this occasion, the overlap was not merely ideological but conceptual. Incidentally, von Uexküll too had spent time at the Zoological Station in Naples and had discussed in some detail the influence Mount Vesuvius exerted on the character of those who lived at its feet.[66] A number of lesser-known German biologists would in the 1930s and 1940s develop, on the foundations of von Uexküll's *Staatsbiologie*, approaches similar to those of a later wave of Anglophone sociobiology. Von Uexküll's biosemiotics would moreover find entry into different strands of continental philosophy.[67] This then was the last trace of Ratzel's biogeography.[68] All others who drew on Ratzel in the years following the First World War would discard this dimension.

Transatlantic Voyages

Meanwhile, a different Ratzel had come into circulation. As we have seen, Ratzel had spent a highly formative period of his life in America and had later fashioned the United States into a role model for German colonization. But he had also developed a following in North America. The most important of his disciples was the geographer and historian Ellen Churchill Semple (1863–1932). Born in Kentucky during the American Civil War, she had graduated from Vassar College in 1882, an institution to which she would return in 1891 to pursue postgraduate work on slavery.[69] She developed an interest in questions of geography in the process and decided, after encountering Ratzel's writings, to study with him at Leipzig in the academic year of 1891–1892.[70] She returned in 1895.

Back in the United States, Semple became a prolific scholar, publishing two high-profile books which made her an internationally recognized figure.[71] Teaching in a number of North American and British universities before settling at Clark, she was also on the margins of a circle of public intellectuals who promoted US imperialism, a group that included Frederick Jackson Turner, Josiah Strong, Brooks Adams, and Alfred Thayer Mahan.[72] In 1917, she was invited by the geographer Isaiah Bowman to advise the Inquiry, a research group established by Woodrow Wilson to hammer out the conditions for a peace agreement. Semple's reports, which were read by the US president, proposed boundary settlements in Turkey, the Middle East, and in the Austro-Italian borderlands, using political geography to guide her conclusions.[73] But although she had clearly exerted herself for the American war effort, Semple was not able to attend the Paris Peace Conference because of her gender.[74] In 1921, she became the president of the Association of American Geographers. Her writings would have a lasting intellectual influence on a number of leading geographers, including Ellsworth Huntington, Derwent Whittlesey, and Isaiah Bowman.

Semple and Ratzel's relationship may not initially have been one of parity, but it was certainly one of intellectual exchange. Ratzel not only found in the American a translator of his writings.[75] He also drew on her in the early 1890s as a research assistant for the second edition of *Die Vereinigten Staaten von Nord-Amerika*, particularly on racial questions.[76] Semple keenly supplied him with statistical material, as well as sharing her views on the inner workings of segregation.[77] In 1893, for instance, she wrote to Ratzel to tell him of a "terrible illness" that had befallen "the race of the negroes and the Indians" and that was rooted in "indecency."[78] She was also keen to affirm Ratzel's belief that the "Teutonic" emigrants to North America were less prone to mix with "lower races" than other European migrant populations.[79] A decade later, Semple was still supplying her former professor with letters and newspaper clippings, which he would draw upon in his writing.[80] Ratzel, in turn, praised her work.[81] One of her papers was included in the 1904 *Festschrift* for Ratzel's sixtieth birthday.[82]

Ratzel's geography offered Semple a level of abstraction higher than that of the brand of teleological sociology she had encountered during her postgraduate studies. The Leipzig geographer allowed her to view American history, including slavery, as little more than the outcome of geographic forces, operating never far beneath the surface of the social fabric. As she wrote in the 1903 *American History and its Geographic Conditions*,[83]

> The question of slavery in the United States was primarily a question of climate and soil, a question of rich alluvial valley and fertile coastland plain, with a warm, moist, enervating climate, versus rough mountain upland and glaciated prairie or coast, with a colder, harsher, but more bracing climate. The morale of the institution, like the right of secession, was long a mooted question, until New England, having discovered the economic unfitness of slave industry for her boulder-strewn soil, took the lead in the crusade against it. The South, by the same token of geographical conditions, but conditions favorable to the plantation system which alone made slave labor profitable, upheld the institution both on economic and moral grounds.[84]

Semple felt so indebted to Ratzel for having opened up this perspective to her that she dedicated to him her 1911 magnum opus *Influences of Geographic Environment* and even gave it the subtitle *On the Basis of Ratzel's System of Anthropo-geography*.[85] She had in 1894 proposed a translation of *Anthropo-geographie* but had settled, with his consent, on a reinterpretation.[86] In 1903, he had written to her:

> Dear Miss Semple! I see not a single reason why I should not agree to your plan. We stand on the same ground, as I share your view that "Anthropogeographie" cannot be translated word by word; it has to be adapted to the Anglo-Keltic and specifically the Anglo-American spirit [*Geist*]. I know of nobody who is more cut out for the job than you are.[87]

Published seven years after Ratzel's death, Semple's book constructed humanity as the passive recipient of environmental influences, which "determined," "everywhere" and "inevitably," the course of history.[88] In a much-quoted opening passage from *Influences of Geographic Environment*, she argued

> Man is the product of the earth's surface. This means not merely that he is a child of the earth, dust of her dust; but that the earth has mothered him, fed him, set him tasks, directed his thoughts, confronted him with difficulties that have strengthened his body and sharpened his wits, given him his problems of navigation or irrigation, and at the same time whispered hints for their solution. She has entered into his bone and tissue, into his mind and soul.[89]

This was something of a radicalization of Ratzel's *Anthropogeographie*,[90] which had obscured its determinist streak behind unnecessary complexity and convoluted prose.

Semple never broke with Ratzel. Although he was to her "the great master who was my teacher and friend during his life, and after his death my inspiration," she did not read him entirely uncritically.[91] Grand as Ratzel's theoretical ambition was, it lacked rigor. Difficult reading "even for Germans," his writings were the outbursts of "a seer or poet" who tended to throw out claims and conclusions that, however "brilliant," were often "highly suggestive."[92] When thinking with Ratzel, she thus deemed it necessary to throw him "into the concrete form of expression demanded by the Anglo-Saxon mind," a tweak of Ratzel's own formulation.[93] Sharply argued and accessibly written, Semple's book received much praise in newspapers on both sides of the Atlantic.[94] Its academic reception was more mixed. Semple was aware of the accusations of environmental determinism against the Leipzig geographer's work and sought, unsuccessfully, to distance herself from this charge.

There were critics elsewhere too. Her reports to the Inquiry had been attacked in 1918 for being overly concerned with the interest of great powers rather than with the principle of self-determination.[95] There was, in other words, a geopolitical impulse in Semple's writing that was only confirmed when in 1932 she passionately endorsed the Japanese invasion of Manchuria.[96] Semple may have been a Wilhelmine in Wilsonian clothing, but this did not mean that she simply restated Ratzel's ideas. The Leipzig geographer, she argued in 1924, had bestowed space with agentic qualities and had thus failed to understand how humans gave space its significance. Semple had little time for *Lebensraum* and recognized that the idea was open to abuse.[97] Ultimately, there were important analytical differences between Ratzel's and Semple's anthropogeography. Not only was she more systematic in her quest for classes of geographic influence than he had been, but she rejected his organicism too. Some of her translations reveal that she was unhappy with Ratzel's use of the term *Geist* (spirit or mind) as something that could intervene in shaping the fortune of nations. Whenever she spoke of "spirit," "intellect," "mental life," and "mental constitution" in her own books, she treated the former as epiphenomena of spatial struggle or environmental influence, and not as the grand ideas of political leaders or entire *Völker*.[98] Given this disinterest in the agency of military and political leaders, her arguments were thus often cruder in their insistence on environmental control than Ratzel had been.

A second crucial interlocutor of Ratzel's in the United States was the German-born founder of modern anthropology and trained geographer Franz Boas (1858–1942). Like Ratzel, Boas had a background in the natural sciences and had engaged in travel writing as a young man.[99] Boas had not studied under Ratzel but had heard him speak in the early 1880s and corresponded with him after emigrating to the United States in 1887, particularly in relation to their common interest, the spread of bow and arrow.[100] The two men also shared intellectual sources of inspiration, such as von Humboldt and Kant, even though this did not result in an exact alignment of the two men's politics. Boas had experienced antisemitic attacks since his student days and had come to oppose nationalism in its various forms.[101] His view of race was more thoroughly cultural than Ratzel's, even though he too perpetuated a color line between white and nonwhite populations.[102]

Boas's earlier work was clearly under the influence of Ratzel, particularly the latter's *Anthropogeographie*.[103] He took Ratzel's ethnology seriously and had a particular sympathy for Ratzel's diffusionism, indebted of course to Wagner's migration thesis. Boas became increasingly skeptical of Ratzel's insistence that similar cultural traits in different regions must have resulted from transmission.[104] He was critical, moreover, of the role that the geographical *longue durée* could offer in explaining the belief systems of various human societies and began to set his anthropology up against a wave of anthropogeographical writing that dominated American geography of the twentieth century. And yet Boas never denounced the importance of environmental factors on cultural evolution. Even toward the end of his life, he drew on *Völkerkunde* and applauded Ratzel's broader views on the interdependence of cultures.[105] He wrote the following in his 1938 *General Anthropology*: "The very fact that cultures are interdependent permits us to reconstruct historical happenings with a fair degree of certainty. It is Ratzel's great merit to have emphasized this possibility."[106]

After Boas's death, the anthropologist Ruth Benedict, a student of Boas's and fellow diffusionist, would describe Ratzel, though not without emphasizing their disagreements, as Boas's "teacher."[107] And yet the exchange between the two men was more likely one of equals. Ratzel had reviewed a number of Boas's works in the 1890s, praising him and noting the "great overlap" between his own anthropogeography and the latter's anthropology.[108] Even if Ratzel was aware that Boas's anthropology was in fact a challenge to his own anthropogeography, he did not seem to have viewed it as a threat.

A less direct but nonetheless crucial connection from Ratzel back across the Atlantic can be found in the work of the sociologist, historian, and in later years socialist and civil rights activist W.E.B. Du Bois (1868–1963). From 1892 to 1894, Du Bois had studied history in Berlin, where he had been lectured by one of Ratzel's intellectual heroes, the nationalist and antisemite Heinrich von Treitschke.[109] Upon his return from Germany, Du Bois was appointed professor at Atlanta University. In 1906, a lecture by Boas played a crucial role in Du Bois abandoning a residual racism that had expressed itself in the view that Africans had "no history."[110] In 1909, he would cofound the National Association for the Advancement of Colored People. A tremendously prolific writer, Du Bois became a leading theorist of racial discrimination and a sharp critic of the racial order established after the American Civil War. Perhaps surprisingly, Du Bois remained positive about the German Empire, even at the end of his long life.[111] Before turning more systematically to Marxism, he had sought to model the agenda of black activism on the experience of German unification under Bismarck, even though the violence unleashed against the Herero would dampen his enthusiasm for the German Empire.[112]

Du Bois drew on a wide array of German thinkers, one of whom was

Ratzel. Gooding-Williams argues that Du Bois embraced Ratzel for his insistence that phenotype could not explain historical difference, while Harris contends that he saw in Ratzel an intellectual "opponent."[113] These opposing interpretations point to the fact that Du Bois was likely to have been ambivalent about the Leipzig geographer, whom he engaged with through the English translation of *Völkerkunde*. Du Bois used this book as an authority on African family life and the spatial layout of villages.[114] In his 1915 book *The Negro*, Du Bois concurred with Ratzel's insistence on the unity of mankind but held him up as one of the race theorists whose writings implied that blacks could, by definition, have no history.[115] He nonetheless recommended *Völkerkunde* as a starred reading, one of a number of "special works on which the author has relied for his statements or which amplify his point of view."[116] This may have been out of necessity rather than choice, for Du Bois recognized that there were as yet no general black histories he could build on. He was, in other words, forced to read colonial texts against the grain.[117] Had Du Bois engaged with Ratzel's travel writing on the United States, he would have found much to disagree with, such as his conviction that the black population had only itself to blame for the failures of Reconstruction, an interpretation that Du Bois's 1935 *Black Reconstruction* would precisely set out to disprove.[118] Du Bois was certainly not alone in ignoring these earlier writings. They would only receive a renewed interest among American scholars in the 1970s and 1980s.[119]

The final notable North American reader of Ratzel's was the geographer Isaiah Bowman (1878–1950). Although Bowman's works would ultimately prove to have a shorter shelf life than those of Boas and Du Bois, he was arguably one of the most influential American intellectuals of the first half of the twentieth century. One of the founding members of the Council on Foreign Relations, Bowman played a pivotal role in promoting the study of international relations, which would eventually supersede political geography as the dominant academic field that had interstate politics as its object. Bowman was no armchair intellectual. While he had served as an advisor to President Woodrow Wilson during the First World War, he counseled President Franklin D. Roosevelt in the Second World War and is sometimes credited as one of the architects of the United Nations and indeed of a century of American imperialism.[120] President of Johns Hopkins University from 1935 to 1948, he also became notorious for his attempts to limit what he deemed to be a disproportionate Jewish influence over the academy.[121]

Bowman's relationship with Ratzel was complex. He had first soaked up Ratzel's major works as a student at Harvard in the early 1900s and had used them as a basis for his research into the Central Andes region.[122] Much like

Semple, he had felt that Ratzel needed to be translated into the American context. After the war, Bowman became more critical of Semple's position. His most influential treatise on global politics, the 1921 *New World: Problems in Political Geography* placed an emphasis on the role that changing political attitudes played, what Ratzel would have referred to as *Raumauffassungen*.[123] Bowman thus emphasized that environmental conditions, particularly the distribution and availability of raw materials, only acted to condition, not determine, state action in the international system. The book was underwritten by a Darwinian tone, but his possibilism arguably brought his work closer to Ratzel's than to Semple's anthropogeography.[124]

Upon reading Ellen Semple's *Influences of Geographic Environment* in 1917, Bowman had been so impressed that he had invited her to join him in advising the US administration in drawing up plans for a peace agreement. Although her papers were read by Wilson in 1919, Semple was not permitted to attend the conference because of her gender.[125] W. E. B. Du Bois too was in Paris, having organized a Pan-African Congress to coincide with the Peace Conference. Although he certainly tried, he did not get through to Wilson, who was fearful that black veterans might demand equality upon their return across the Atlantic.[126] Whereas Du Bois was disappointed by the implications of the peace treaty for the African continent, Ratzel's German followers would become truly obsessed with Versailles. Among a new wave of Ratzelians, one stood out: the retired major general and national socialist sympathizer Karl Haushofer.

Ratzel and the Rise of National Socialism

Before publishing in 1913 an influential study of Japanese military power, the Munich geopolitician Karl Haushofer (1869–1946) had worked as a military observer in Japan. On the eve of the First World War, he had also completed a doctorate under Erich von Drygalski, whose polar expeditions Ratzel had supported before his death. Haushofer subsequently served on both the western and eastern fronts during the war. After retiring from the army, he took up an academic post in geography at the Ludwig Maximilian University in Munich in 1919.[127] Haushofer's geopolitics, a term he had taken from Kjellén, was fed by a personal frustration, which he shared with Hettner, Penck, and many others, with the terms of the Versailles settlement. Germany, he argued, had to secure economic self-sufficiency, break out of its unfavorable centric position in Europe, and establish *Lebensraum* for its population. Haushofer also promoted an alliance with Soviet Russia (*Kontinentalblock*) to counter what he believed to be the dangerous dominance of Anglo-American naval power.

Ratzel was not the only of Haushofer's geopolitical influences, which also included Kjellén and, though to a lesser degree, Halford Mackinder.[128] But Ratzel was by far the most important. Haushofer may have been more cartographically and less philosophically inclined than his master, but his project was in essence an operationalization and radicalization of the central tenets of Ratzel's political geography. Like his intellectual hero, whom he had met as a child,[129] Haushofer criticized political scientists and sociologists as oblivious to the soil (*bodenfremd*) and compared them to architects who were ignorant of foundations.[130] Geopolitics, on the other hand, was a necessary investigation of the way the cultural and political landscape had been shaped through geomorphology, geology, climate, and the biosphere. He also sought to forge a link to military strategy by examining the military significance of physical geography.[131] Drawing on Ratzel, who had held that the struggle for existence was desperate in narrow spaces, Haushofer highlighted the fundamental problem of Germany's "mutilated" *Lebensraum*. Germany's political situation had changed since Ratzel's death in 1904, but the lessons of Ratzel's geography were still valid. In 1941, Haushofer edited a reader of Ratzel's writings that included passages on death and ruin landscapes.[132] Clearly, his reading of Ratzel was a necropolitical one.

Although Haushofer never joined the National Socialist Workers' Party (NSDAP), it is beyond doubt that he sought to shape the Nazi ideological project. In 1924, he made an anthropogeographical plea for a form of national socialism (*nationaler Sozialismus*), which recognized its precursors on the Asian continent.[133] In 1933, he would publish a pamphlet under the title *Der Nationalsozialistische Gedanke in der Welt* (The national socialist idea in the world).[134] Through his student Rudolf Hess, Hitler's deputy in the NSDAP, Haushofer was in the early 1920s acquainted with Adolf Hitler and, a decade later, with the Nazi foreign minister Joachim von Ribbentrop. There were links too with the Reich leader of the *Schutzstaffel* Heinrich Himmler and propaganda minister Joseph Goebbels. In fact, Spang concludes, there was by the 1930s "almost nobody among the main representatives of the regime who was not at least superficially familiar with Haushofer and his ideas."[135] And yet by the early 1940s, Haushofer's Ratzelian brand of geopolitics had begun to fall out of favor with the Nazi regime, both because of Haushofer's pro-Russian leanings and his theory's ambivalence toward the Third Reich's biological theory of race.[136] Haushofer, who viewed the attack on the Soviet Union as a catastrophic mistake, was briefly imprisoned in 1944. In 1946, he and his wife Martha committed suicide only a few miles west of where Ratzel had died in 1904.[137]

Even more than Ratzel, Haushofer was keen to spread the lessons of political geography to the general public. Not only did a number of his books have an introductory character, but he was also frequently heard on the radio.[138] Founded in 1924, Haushofer's *Zeitschrift für Geopolitik* (Journal for geopolitics) would moreover serve as a crucial milieu for the diffusion of geopolitical ideas in Germany, ultimately leading to an explosive preoccupation with questions of national survival and space. Rather than an academic journal, *Zeitschrift für Geopolitik* was in essence a monthly current affairs magazine with a significant input from academic geographers who used a steady stream of often cartoon-like maps to illustrate their points. Overwhelmingly nationalist and territorially revisionist in content, the journal was a success and sold four thousand copies a month by the year 1928.[139] It also supported the transition from the colonial-era pursuit of *Lebensraum* in the Southern Hemisphere to one of territorial conquest in eastern Europe.[140]

From its inception in 1924, *Zeitschrift für Geopolitik* was Ratzelian in its orientation. The first issue proclaimed its allegiance to Ratzel's *Gesetz der wachsenden Räume*.[141] Haushofer's contribution to the same issue sought to tease out the implications of the monsoon region for the "tightest of European living spaces" (*allerengsten europäischen Lebensraum*) and ended with a call for *Mitteleuropa* to be "released from its artificial isolation and absorbed into the community of struggling *Großräume*."[142] While a further piece by Erich Obst opened with a Ratzel epigraph, Termer's article took a Ratzelian approach to American colonization, arguing for the correlation between successful state formation in North America and the absence of racial mixing.[143] The final contribution by Lautensach on the geopolitics of Mediterranean seas cited Ratzel too.[144] In the years that followed, the journal continued to publish on Ratzel and in a Ratzelian mode.[145] Indeed, its most frequent contributors elevated *Raum* "to the status of a fully-fledged actor, endowed with progenitive powers" in ways that clearly chimed with Ratzel's *Politische Geographie*.[146]

Ratzel would remain helpful to Haushofer in his attempt to fashion geopolitics as a national socialist science. In 1941, he would claim that Ratzel was "far more a contemporary of the Third Reich than of the Second Reich."[147] Adolf Hitler, who described geography and history as his favorite school subjects, used the term *Lebensraum* on a number of occasions in his two-volume manifesto *Mein Kampf*, from the first 1925/1927 edition onwards, to mark Germany's supposed need for territorial space. *Lebensraum*, Hitler explained, was the only way the German *Volk* could protect itself from its own demise or the enslavement by another *Volk*.[148] Indeed, the leader of the Nazi Party's unpublished 1928 second book (*Zweites Buch*) employed the term even more frequently than *Mein Kampf* had.[149] In the years that followed, Hitler would also invoke *Lebensraum* in his speeches.

Much ink has been spilled on the role that Ratzel's initial formulation of *Lebensraum* played in shaping the powerful ideological construct of Nazi geopolitics. A crucial focus within this debate has been on the influence that Haushofer exerted on Hitler during those crucial months of 1924, when the future Führer was writing *Mein Kampf*. Some have argued that Haushofer visited Hitler frequently in Landsberg, where the latter had been imprisoned after the failed beer hall putsch, and thus effectively "taught" the future Führer geopolitics.[150] But others have been more cautious. Plöckinger notes that the name Haushofer does not appear on the prison's visitor's register, thus raising the possibility that the two men did not meet at all in 1924.[151] Allied propaganda, a number of observers rightly point out, exaggerated Haushofer's influence on Hitler in the 1940s, giving birth to the myth of Haushofer as an éminence grise.[152] While Haushofer's impact on Hitler's right-hand man Hess is beyond doubt, his effect on Hitler, in this reading, was indirect at best.

But even if Haushofer did not systematically teach Hitler at Landsberg, the former's diaries and correspondence suggest that he did visit the two leading Nazis on a number of instances and that he met Hitler on later occasions too.[153] It is known, moreover, that when the inmates of Landsberg began to discuss the concept of *Lebensraum* in July 1924 and could not agree on its meaning, Hess felt the need to ask Haushofer for a firmer definition, which the latter gladly supplied in writing. Hitler used the term in those sections of *Mein Kampf* that were written *after* but not those written *before* July 1924.[154] There was a second way Hitler could have been exposed to Ratzel, for he had access to the 1924 issues of *Zeitschrift für Geopolitik* at Landsberg.[155] As we have already seen, these first issues were littered with references to Ratzel and his concepts. Third, as Haushofer stated in 1941, there had been a copy of *Politische Geographie* in Hitler's small library at Landsberg, implying that he himself had placed it there.[156] Haushofer may have been exaggerating his importance in the early 1940s precisely as his influence on the regime was slipping away. And yet he reiterated this occurrence both in private correspondence and during an Allied interrogation in October 1945, and thus at a time when he would have sought to disassociate himself from the Nazis.[157]

This of course still does not mean that Hitler read *Politische Geographie*, a book of considerable length and dense prose.[158] And even if he did, it would have likely been a selective reading, in line with his tendency to integrate only ideas that could be made to fit his worldview, of which geopolitics was but a part.[159] Any reader of *Mein Kampf* will note, moreover, that geopolitical arguments were less central to Hitler's ideological project than those that drew on racial theory.[160] And yet, even those who have sought to be balanced in their assessment of Haushofer's influence on *Mein Kampf* have conceded that there is a strikingly Ratzelian and Haushoferian tone in crucial passages of Hitler's manifesto.[161] One of the earliest examinations of Hitler's use of political geographical terminology in *Mein Kampf* was conducted by Karl Lange in 1965. Lange singles out the language of "ground and soil" (*Grund und Boden*), "ground area of a state" (*Grundfläche des Staates*), "settlement area" (Siedlungsgebiet) and "the necessary clod for our German people" (*notwendige Scholle für unser deutsches Volk*).[162] Jacobsen observes that Hitler also adopted Ratzel's understanding of colonization and of war as an expression of the growth of states.[163] As Spang notes, Hitler repeatedly spoke of the state as an organism.[164] The state was what made the nation's life on earth viable.[165] Schultz highlights the claustrophobic tone that is common to both Ratzel and Hitler, an idea captured too of course by the hugely popular 1926 novel *Volk ohne Raum* (People without space) by Hans Grimm.[166] We can add to this that much like Ratzel, Hitler compared the political organization of "negro tribes" to those of culturally higher developed *Völker*, arguing that each of these fulfilled a function within constraints and demands of their specific context.[167] Hitler also speculated about the kind of ruins that modern civilization would leave behind.

In recent years, the question of Hitler's American role model, his fascination with Fordism, the New Deal, Hollywood, and especially with American race law and second-class citizenship, has received renewed attention.[168] Hitler was, like Ratzel, concerned about German emigration to the United States, which he interpreted as a bloodletting on the German body politic.[169] Simms notes the degree to which Hitler himself was influenced by Madison Grant's 1916 book *The Passing of the Great Race*.[170] And yet Hitler had not read this book when he wrote *Mein Kampf*. Instead, it is more likely that Hitler took at least some of his views on America from the aforementioned piece in *Zeitschrift für Geopolitik* by the Americanist Franz Termer (1894–1968).[171] Indeed, the same distinction between an admirably segregationist "Germanic" North America, which had been spared racial mixing, and a rotten "Romanic" Central and South America, which was ethnically mixed, could be found in both Termer and Hitler.[172] Termer had been a student of the geographer and explorer Karl Sapper, who in turn had been mentored by Ratzel. A frequent contributor to the *Zeitschrift* in the 1920s, Termer remained a Ratzel enthusiast even after the war. He was working on a new edition of Ratzel's *Aus Mexico* in 1968 when he died. Published posthumously, Termer's foreword praised Ratzel as one of the greatest geographers since Alexander von Humboldt.[173] While there were thus, contra Schlögel, *more* than one route from Ratzel to Hitler,[174] it should be emphasized that Ratzel was only *one* of a number of influences on the formation of national socialism. He was, however, the most important reference point in the formation of *Geopolitik*.

The route from Ratzel to Hitler is thus also a route from Darwin to Hitler, albeit an indirect one. Contemporary Darwinists are keen to disassociate Darwin's theory from those who followed in his footsteps. The former, they argue, had only ever spoken about the struggle for existence among individual members of a species or race, while the Darwinists had operated with a problematic notion of group selection. Any "belief in a struggle between races" is thus declared "un-Darwinian."[175] This question has been hotly debated by proponents of creationism and intelligent design, who hope to have found in this connection a powerful moral argument against natural selection.[176] They would be well-advised to recall that Ratzel was for much of his life a believer in a Christian God. This ideological struggle over Darwin's legacy aside, Darwin's *The Descent of Man* did of course present history as an evolutionary struggle between "civilized" and "savage" races in which the former were replacing the latter. It is this trope that was also present in Ratzel's *Lebensraum* reasoning.

Lebensraum as Doctrine

By the time the Nazis took power in 1933, Ratzel had been dead for almost three decades. His former students, if still alive, were in the later stages of their careers. Some ended up becoming rank-and-file followers of the Nazi regime. Kurt Hassert was the most prominent of Ratzel's former students to sign the 1933 vow of allegiance of the professors of the German universities to Adolf Hitler.[177] Ratzel's keenest followers in the geographical community, men such as Karl Haushofer and Heinrich Schmitthenner, had however not been supervised by him.[178] Not all geographers, moreover, who sought to uphold Ratzel's memory were in the inner circle of geopoliticians, as the example of the regional geographers Friedrich Metz and Ewald Banse illustrates.[179] The 1930s geography also saw the rise of Walter Christaller's spatial science. Although spatial science would become instrumental to the practical organization of Nazi *Lebensraum* in the 1940s, its spatial abstractions otherwise signaled a break with the Ratzelian tradition.[180]

Adolf Hitler had not been the only Nazi leader to promote the idea of *Lebensraum*. In his book *Der Mythus des 20. Jahrhunderts* (The myth of the twentieth century), the party ideologue Alfred Rosenberg spoke of *Lebensraum* as agricultural farmland and territory that belonged to a race.[181] As a catch-phrase, *Lebensraum* soon served to legitimate a variety of policy aims, such as territorial revisionism, the establishment of a sphere of influence, and indeed conquest beyond Germany's borders of 1914. School children too had to learn about the problems supposedly associated with narrow living space. Franz Knieper's pedagogical book *Geopolitik für die Unterrichtspraxis* (Geopolitics for the classroom), which went through six editions from 1932 to 1938, explained that geopolitics "stood on the shoulders of the genius master Ratzel."[182] The term also found entry into broader public life, including artistic discourses.[183] Indeed, even factory designers used *Lebensraum* as a conceptual device to argue for ways of making workspaces more habitable.[184] Ratzel's political geography now spread to Hungary and Japan.[185]

By the 1940s, *Lebensraum* had become increasingly divorced from Ratzel's original formulation. Geographers were still debating his *Lebensraum* concept in their journals,[186] but Ratzel's version of *Lebensraum* had been superseded by other formulations. The NSDAP economist Werner Daitz, for instance, sought to bring geopolitics in line with Nazi racial doctrine.[187] *Lebensraum* now fulfilled the function of a bridge between the geo- and necropolitical dimensions of the new Nazi empire in eastern Europe, which entailed agricultural colonization through population transfer on the one hand, and the extermination of unwanted populations on the other. And yet these two dimensions, of settler colonization and extermination, were hardly incompatible with Ratzel's spatial theory of history. Newspapers throughout the new Nazi *Lebensraum* reported in 1944 on Ratzel's one hundredth birthday, including in the short-lived German newspapers for Norway and the *Reichskommissariat Ostland*.[188] Kurt Vowinckel, Karl Haushofer's coeditor of *Zeitschrift für Geopolitik*, described Ratzel as a "seer from the Upper Rhine."[189] Haushofer himself, now disconnected from the Nazi leadership and clearly under no illusion that the war was lost, claimed that it was Ratzel who had already predicted a struggle between the American way of life and its "Bolshevist" counterpart.[190]

The legal thinker, political theorist, and national socialist Carl Schmitt (1888–1985) had also studied Ratzel's political geography. Schmitt was neither a geographer nor a member of the Haushoferian school of geopolitics. But after spending the interwar years developing his critique of liberalism and concept of sovereignty, Schmitt had in the late 1930s turned to a theorization of the politics of global space. At the height of the Second World War, he argued that history was marked by spatial revolutions that not just transformed the measurement of space and extended the spatial horizons of whole societies, but that also reshaped hegemonic conceptions of space.[191] He would develop this idea further in his magnum opus *Nomos of the Earth*, published only during the denazification period.[192] It is worth discussing Schmitt in some detail because his works had a much longer half-life than many of the thinkers we have discussed thus far.

Although Schmitt is often seen as a political philosopher, it is important to recognize the geopolitical underpinnings of many of his writings, particularly in light of his foreign political advocacy in Nazi Germany.[193] A crucial concept in Schmitt's geopolitical writings was the prescriptive vision of a world order (*Nomos*) of greater spaces (*Großräume*). *Großraum* had in Schmitt the sense of a sphere of influence, a region that extended beyond the territorial boundaries of the state and that was embedded within a larger spatial order. Schmitt advocated for *Großräume* moreover as spaces of nonintervention that banned external powers (*raumfremde Mächte*) from intervention. This vision was modeled on the American Monroe Doctrine, whose principles, he hoped, could be applied to other parts of the world, thus forming an alternative to what he believed was the highly unstable liberal world order that had been established in the wake of the First World War. Schmitt was driven by the nostalgia for the *Nomos* of the *jus publicum Europaeum*, which had defined European history from the seventeenth to the nineteenth century. Although he had fallen out of favor with parts of the Nazi regime by the late 1930s, his writings in the 1940s continued to associate Jewish "spacelessness" with Anglo-American sea power, thus fusing geopolitics and antisemitism in ways that chimed with dominant ideology.[194]

Ulmen, following Schmitt himself, traces the origins of Schmitt's *Großraum* concept not to geography, but to the discourse of *Großraumwirtschaft* (economy of greater space), which had emerged after the First World War, particularly in relation to the governance of energy resources.[195] But the concept clearly had a Ratzelian ring to it, as a number of observers have noted.[196] Ratzel's influence on Schmitt was not insignificant. Not only did Schmitt hold a collection of Ratzel essays in his private library at the time of his death,[197] but he also used the notion of *Lebensraum* in some of his post-war writings as a reaffirmation of the *land* in an age defined by the nuclear and space race.[198] More crucially, Schmitt cited Ratzel's 1901 *Lebensraum* twice in an important 1941 essay that pioneered the ideas in *Nomos of the Earth*. Ratzel, in Schmitt's view, was the "founder of a new science of space" who had already recognized that the mastering of space (*Raumbewältigung*)[199] was "the defining trait of all life."[200] Schmitt used Ratzel to argue that *Großraum* was

> not a space that is only greater compared to a relatively smaller space; it is not an expanded minor space. The merely mathematical-physical-natural scientific neutrality of the hitherto existing concept of space should now be overturned. "There is," as Ratzel says, "already something greater—I would almost say creative and inspirational—in the wide space."[201]

Schmitt's aim was to translate Ratzel's point about the relationship between spatial consciousness and spatial ordering into the language of legal theory. In *Nomos of the Earth*, he invoked the geographer again to unpack the short-lived spatial divisions of the 1939 Panama Declaration, which had sought to establish a zone of neutrality in the Western Hemisphere by carving a line into the Atlantic Ocean three hundred miles from both American coasts. Schmitt argued that space itself had now "protruded into the expanse and evenness of the sea."[202] Ratzel of course had written in some detail about the Monroe Doctrine. He had also developed a critique of liberalism as a doctrine of British colonial policy. The Leipzig geographer shared with Schmitt a tendency for etymological speculation, including about the term *Raum*, as well as a critique of a mathematical, abstract, and empty conception of space. Like Ratzel, Schmitt made a connection between migration, settlement, and colonization and drew an opposition between settler and trade colonialism. Ratzel certainly would have found much to agree with in Schmitt's insistence that "history is a history of territorial conquest."[203]

It is crucial, however, to take note of the most important breaks between these two bodies of thought. Ratzel had only commented on international law (*Völkerecht*) in passing. Although his *Politische Geographie* recognized that international law regulated, to an extent at least, the conduct of war,[204] he was ultimately interested in very different laws, ones of a biological and geographical nature, such as his law of spatial growth.[205] Ratzel's and Schmitt's concepts of space, moreover, were hardly congruent. Whereas Ratzel had seen *großräumig* in opposition to *kleinräumig*, Schmitt treated *Großraum* as an antithesis to Anglo-American universalism and its discriminate "just war" concept of war.[206] Schmitt sought more consistently to historicize space. Ratzel's *Raumauffassungen* may have been historically specific, but they were all too often distinguished, counter to Schmitt's reading of Ratzel, simply in terms of scale.

Whereas Schmitt promoted a pluriverse of pan-regions as zones of nonintervention, Ratzel did not outline any such vision, although we have seen in chapter 5 that he was perhaps starting to move in that direction in the final year of his life. Schmitt neither committed to Ratzel's organismic concept of the state, nor did he stray into environmental determinism.[207] Ratzel, conversely, did not construct a Schmittian friend/enemy distinction between the supposedly spaceless Jews and the spatially rooted peoples.[208] Ratzel had, as we have seen in chapter three, read the "Jewish question" through the prism of racial struggle, but had not made a conceptual link between Jewishness and Anglo-American sea power that was so crucial to Schmitt's geopolitical theory.

After the war, Schmitt sought to reconnect with the remaining Ratzelians. Indeed, when the *Zeitschrift für Geopolitik* was revived in the early 1950s, Schmitt published an obituary in the journal.[209] With the Haushoferian school now discredited, Schmitt became something of an heir to the empty throne of German spatial thinking. His ideas would shape profoundly a new wave of political realist thinking in the Cold War era and would return again to the fore in the early twenty-first century.

Meanwhile in 1930s Baltimore, another antisemite had followed closely the formation of a Ratzelian brand of geopolitics in Germany. Although Isaiah Bowman had feared the damage that the association of political geography with national socialism might have on his own scholarly reputation, it was not until 1942, and thus after Washington's entry into the war against Nazi Germany, that he sought to distance himself from German geopolitics and from Ratzel. "In my view," he now claimed, the "'geographical-basis-of-power' idea of Ratzel, as set forth in the first edition of his political geography (1897), is completely unsound."[210] And yet Bowman's disavowal of German geopolitics was not absolute. There was "virtue in the argument of 'organic boundaries' and the philosophy of Lebensraum."[211] The problem, he would later add, was that German geography had expanded *Lebensraum* "from its earlier purely descriptive economic meaning to one that gives territorial expansion a pseudoscientific justification."[212] The issue with Ratzel, in other words, was not *Lebensraum*, which Bowman in fact endorsed, but the German geographer's territorial fetish. By distinguishing between geography and geopolitics, Bowman tried to formulate an alternative and more intellectually legitimate blueprint for American foreign policy, which valorized expansion and domination by means of free trade, migration and economic development.

By the 1940s, the tide had turned decisively against Karl Haushofer's Ratzelian geopolitics in North American political discourse. Haushofer, the supposed mastermind of German expansionism, was deemed to be "dominated" by the Ratzelian conviction "that every people must be educated up from smaller to larger space conceptions."[213] Indeed, Ratzel was increasingly seen to have "made the whole trend toward the building of new empires look scientific and irresistible."[214] And yet this wholesale rejection of Ratzel in 1942 did not quite herald the end of his geography.

Ratzel after the Fall

The fall of Ratzel and his *Lebensraum* idea could hardly have been more abrupt. In Germany, geopolitics was placed under something of a taboo after 1945. Politicians steered clear of its language. Those who did not were sometimes shouted down in parliament.[215] But even though politicians sought to avoid the terminology of geopolitics, tainted as it was by its association with national socialism, geopolitics experienced a stealthy renaissance. In 1951, *Zeitschrift für Geopolitik* was relaunched through the Vowinckel publishing house, which specialized in revisionist pamphlets on the most recent war. From the beginning, the journal sought to adapt the concepts and ideas of interwar *Geopolitik* to a bipolar and nuclear armed world.

Academic geographers too remained keen to revive *Geopolitik*. In 1951, a second edition of the Marburg geographer Heinrich Schmitthenner's *Lebensräume im Kampf der Kulturen* (Living spaces in the clash of civilizations) appeared, thirteen years after the first. Despite the Nazis' misuse of the term *Lebensraum*, Schmitthenner claimed, the latter had lost nothing of its purchase: "[s]truggle is a fact of all life and one that has no ethical content," he explained.[216] A wave of "national dying" (*Völkersterben*) would rip through the European continent if it did not seek new living space in Africa.[217] Franz Knieper's aforementioned textbook was reissued in 1957 under a new title, *Politische Geographie*, but still included the term *Geopolitik* (in brackets) as a subtitle. Despite this symbolic recognition of the new political climate, the book's content remained largely unchanged: Ratzel was still the "master," *Lebensraum* still the dominant conceptual frame. Knieper was nonetheless keen to use Haushofer's "Jewish wife" and his son's participation in the "open resistance" against Hitler as evidence that there had always been a fundamental rift between geopolitics and national socialism.[218] Anxieties about lost territories and colonies that might "feed" the German nation lived on in 1950s schoolbooks, which now sought to frame *Lebensraum* in a more "harmonic" fashion.[219] Even those who purported to reassess Ratzel in the light of national socialism endeavored to reclaim at least the biogeographical and ethnological dimensions of his oeuvre, concluding that the Leipzig geographer could "hardly be made responsible" for what others had done with his ideas.[220]

In East Germany, Marxist-Leninist critiques of German geopolitics soon appeared, in line with a rejection of Ratzel's political geography in the Soviet Union.[221] When Ratzel's old department was closed by the German Democratic Republic in the 1960s, the decision was taken in light of the reactionary nature of his thinking.[222] Critical voices emerged too among West German geographers,[223] but neither the proponents nor the critics of Ratzel were now at the forefront of public debate. It was in military circles that geopolitics found a more direct link to power. *Zeitschrift für Geopolitik* counted in the 1950s a number of former Wehrmacht generals among its authors, many of whom aired their frustrations with the postwar settlement, in which they saw a new "Versailles." Although the journal renewed its pledge of allegiance to interwar geopolitics and *Lebensraum* in 1954, it soon ran into financial difficulties and was in 1956 absorbed by a politics and sociology journal.[224] But by that time, geopolitical thought had yet again developed a life of its own. A number of geopoliticians were now actively involved in setting up what was then still the "new *Wehrmacht*," soon to be known as the *Bundeswehr*.[225]

The most prominent of these new German geopoliticians was Friedrich Ruge (1894–1985). Ruge had been a vice admiral during World War II and was in 1956 appointed the commander of the postwar German navy. Born into a family of geographers, Ruge had grown up discussing works of geography and history at the family dinner table.[226] Indeed, his father Walter Ruge had been a colleague and friend of Ratzel's at Leipzig.[227] Ruge, who had upon his retirement served in four German navies, had been close to Erwin Rommel during the war and had encountered Hitler and other leading Nazis in person. After the war, he was as comfortable advising the first federal German chancellor Konrad Adenauer, speaking at the US Naval War College and exchanging ideas with Henry Kissinger on a visit to the United States in 1960, as he had been in serving the Nazi regime.[228] An admirer of both Britain and the United States, Ruge became the proponent of a new Atlanticist geopolitics, staunchly survivalist and anti-Soviet, but concerned more than his interwar predecessors with the oceans rather than with the soil.

In the 1950s, when the West German army was yet to be revived by the Allied powers, Ruge published works of military history and strategy and would continue to do so after his retirement. In 1955, he outlined the contours of his new geopolitics in *Zeitschrift für Geopolitik*. Germany, he claimed, had finally joined the sea powers, a fact his compatriots had yet to fully comprehend.[229] In the same year, Ruge also published a pamphlet under the title *Seemacht und*

Sicherheit (Sea power and security), which explained that the new German army was "the expression of our *Volk*'s will to live."[230] With this continuation of geopolitical terminology, Ruge sought to unite the different parts of the old Wehrmacht elite, those who had in 1944 conspired to assassinate Hitler and those who had continued to fight. Both had been right, he argued.[231] While Hitler had been a "demon" who had betrayed his own *Volk*, the Soviet Union was a slave society that could only be deterred in the most aggressive way possible, including with the threat of nuclear weapons. Ultimately, Ruge's political versatility stood not just for the endurance of geopolitics, but also for the afterlife that Ratzel's esteem for the United States and his particular understanding of sea power would have in the second half of the twentieth century.

Civil defense planners too continued to read the state as an organism. The Federal Agency for Civil Defense's first president Erich Hampe spoke in his 1956 book *Im Spannungsfeld der Luftmächte* (In the field of tension between the air powers) of a struggle for existence in which West German's location as a "border people" (*Grenzvolk*) had placed it.[232] He decorated his discussion not just with geopolitical maps that showed territorial borders, urban centers, and bomber routes, but also with blueprints of underground civilian bunkers. Every individual body of the *Volk* was but a cell in a larger organism. Hampe fantasized about moving large parts of the population underground in an attempt to make this organism more resilient. His colleague Alexander Löfken, the head of the Federal Agency for Technical Relief, called for the subordination of civil defense to military strategy to secure subterranean "living space of the future."[233] These strategists were thus increasingly drawn to the *earth*, not just as the place where the nation's vital natural resources were held but as a possible retreat in an era of nuclear warfare. While *Lebensraum* had still been defined in horizontally territorial ways by Haushofer, it acquired a vertical dimension in postwar strategy. This was clearly a further development of the concept of living space for the nuclear age. Although Ratzel was no longer invoked by name, the basic framework of a mortal struggle of states for space that Ratzel's political geography had first established and that had reverberated through Germany in the 1920s was entirely compatible with an age of superpower competition.

In the United States, Ratzel's name was inseparably intertwined with the fate of Nazi geopolitics. But here too there was complexity. The influential political theorist Hans Morgenthau declared geopolitics a pseudo-science in his 1948 *Politics among Nations*. For Morgenthau, power was the fundamental currency of world politics, not space. But despite his misgivings about *Geopolitik*, in both its Mackinderian and Haushoferian variants, Morgenthau was happy to consult Ratzel as an authority on the question of national power.[234] His realism, which shared much ground with interwar geopolitics, would establish itself as a dominant paradigm in both the academic field of international relations and the wider sphere of American foreign policy for decades to come. Although Isaiah Bowman had died during the early stages of the Cold War, his brand of geopolitics too set the parameters of much American foreign political reasoning during the second half of the twentieth century. Despite his insistence to the contrary, Bowman was undoubtedly "the heir to a wider geographical tradition developed by Friedrich Ratzel."[235] Realism and geopolitics continued to intersect throughout and after the Cold War.

While many nuclear strategists continued to write in a geopolitical register, they tended to derive their claim to scientific authority less from the timeless truths of a Ratzelian geography than from the ahistorical models of game theory and an encyclopedic knowledge of nuclear weapons systems. When the famous American strategist Herman Kahn used the English term "living space" in his seminal 1960 *On Thermonuclear War* to refer to nuclear bunkers, he was thus not invoking Ratzel.[236] The latter's insistence that humanity was but an epiphenomenon of evolutionary processes would have perhaps seemed quaint in a world in which humans had acquired the technological means to make that world disappear at the push of a button. But there was nonetheless a basic ideological congruence here, particularly in discourses that treated nuclear war as no more than a further development of traditional war, and thus as winnable. Well into the later stages of the Cold War, geopoliticians like Saul Cohen and neoconservatives like Richard Pipes would cite Ratzel authoritatively.[237] Not only was the Chilean dictator Augusto Pinochet a keen follower of the Leipzig geographer, but the latter's ideas were popular throughout a number of Latin American military regimes during the entire Cold War period.[238]

Among geographers, there was much less enthusiasm for Ratzel. In the late 1980s, Anglophone geographers began to return to his ideas. The tone was critical but lacked the hyperbole with which Ratzel and Haushofer had been treated by the wartime propaganda.[239] Increasingly, Ratzel became a symbol of the discipline's past errors. Indeed, the school of "critical geopolitics" emerged in the 1990s precisely as an attempt to develop alternatives to a Ratzelian and Mackinderian imagination.[240] But even after the fall of the Berlin Wall and the dissolution of the Soviet empire, Ratzel still found followers. One of the intellectual figureheads of the German new right, Karlheinz Weißmann, invoked Ratzel's concept of *Mittelage* in 1992 to outline a neo-Bismarckian foreign policy for a reunified Germany.[241] The neofascist *National Democratic Party of Germany* continued to speak confidently of the need for Germany to survive in its "central European *Lebensraum*."[242] Admiration for the Leipzig geographer and his ideas was not confined to Germany. It could be found, for instance, in the pan-Islamic and neo-Ottoman writings of the one-time Turkish prime minister Ahmet Davutoğlu.[243] In Russia, the infamous Eurasianist and advocate of Russian territorial expansion Alexander Dugin preached about Ratzel and *zhiznennoe prostranstvo* (Russian for *Lebensraum*) in his popular pamphlets and podcasts.[244] Although Dugin's influence on the administration of President Vladimir Putin has often been overstated, he has remained a popular commentator in Russia and has even served as a spiritual guide for the North American "alt right."[245]

Ratzel's ideas featured too in discourses surrounding the Syrian refugee crisis. Weißmann, now allied with the 2013 right-wing populist party *Alternative für Deutschland*, revived the Leipzig geographer in his 2016 book *Rubikon*. Europe was subject, he claimed, to a new "great migration" (*Völkerwanderung*) from the Global South to the North that was but an epiphenomenon of a struggle for *Lebensraum*.[246] It was important, Weißmann claimed, to realize the military character of all migratory movements. He was tapping into an increasingly popular discourse of *Kulturkreise* ("cultural or civilizational circles"), invoked in conservative newspaper op-eds and in some ministerial press releases alike to highlight the supposed cultural incompatibility of Syrian refugees with "native" Germans. Weißmann's interpretation of Ratzel as a theorist of the conflict between "migrants" and "autochthones" certainly ignored the fact that Ratzel had in fact been critical of notions of indigeneity, but Ratzel was hardly as detached from *Kulturkreislehre* as some have suggested.[247] Although the notion was most frequently invoked in the version popularized by Oswald Spengler, it had been most thoroughly conceptualized by one of Ratzel's followers, Leo Frobenius.[248]

Conclusion

This chapter has sought to sketch the transnational, multiple, and sometimes opposing trajectories on which Friedrich Ratzel's ideas traveled after his death in 1904. And while we certainly encountered glowing admirers and staunch critics, the majority displayed some level of ambivalence toward Ratzel's writings, perhaps not unsurprising given how indecisive Ratzel had himself been in his writing. My account has covered the decades immediately following his death in some detail and has necessarily been more cursory for the period after 1945, during which his influence was less direct. My discussion is far from comprehensive and leaves sufficient room for further examinations, particularly of his reception beyond the German and Anglo-American contexts.

I have shown that Ratzel's ideas circulated widely but unevenly in the first half of the twentieth century. While North American thinkers were taken by his *Anthropogeographie* and the English translation of *Völkerkunde*, their European counterparts were particularly enamored by *Politische Geographie* and the *Lebensraum* essay. This had crucial implications. For while Ratzel bled into wider geographical and ethnological discourses in the Anglosphere, his necropolitical geography was instrumentalized in Germany to great effect. A number of observers have rightly highlighted the role the First World War played in giving an impetus to this operationalization.[249] In the years that followed the war, *Lebensraum* was transformed first from a vision of colonial expansion into one of territorial revisionism. One should be careful not to conclude from this that this new strain of Ratzelian thought was a "rediscovery" of a long-lost thinker. When Germany signed the armistice on 11 November 1918, Ratzel had been dead for a mere fourteen years.

In the 1920s and 1930s, *Lebensraum* proved a tremendously versatile idea. It was precisely its elasticity and lack of rigor that allowed it to be integrated into different ideological formulations and projects.[250] This flexibility allowed *Lebensraum* to be turned into a guiding principle for an eliminationist project of German colonization in eastern Europe, which was in effect a return to its original meaning. But it was its necropolitical underside too that allowed for such a transfer. Although the meaning of *Lebensraum* had increasingly drifted from Ratzel's biogeographical formulation, he was still recognized as its author, both in Nazi Germany and in the United States. It was thus that his ideas were discredited after 1945, both east and west of the Iron Curtain. This did not mean that his influence became entirely intangible. Not only were there many German geographers who sought to salvage what could be saved from the Ratzelian corpus, but generals and dictators too remained enthralled by the Leipzig geographer. Among West German military leaders and civil defense planners, a Ratzelian geopolitics survived tacitly into a new Atlanticist era. Even in North America, its presence could still be felt.

While Ratzel has retained his admirers on the political right into the early twenty-first century, he has primarily remained a figure of historical interest among scholars. Today he is frequently held up as an example of geography's past errors, chiefly its entanglements with colonialism and his environmental determinism. Only a few scholars have sought to find inspiration in Ratzel's theoretical body of thought. In what is left of our discussion I will turn to the question of what it might mean to think with and against Ratzel in the present.

Chapter 8

Conclusion

Friedrich Ratzel was both a theorist and something of a practitioner of empire. He developed his human geography in the 1880s both as a way to grapple with his personal experience of North American settler colonialism and as a theoretical edifice that fueled and legitimated his advocacy for German colonization. Well networked within Germany's colonial elite, the Leipzig geographer argued that colonization could be found as much in the animal kingdom as it could in the human realm, a view that imbued the phenomenon with a natural and transhistorical dimension. While he produced something of a typology of different forms of colonization, this conceptual matrix served primarily the purpose of allowing him to make a case for an overseas and agricultural form of settler colonization.

In the 1890s, Ratzel entered a geopolitical period. He did not abandon his scholarly endeavors, but many of his writings were now driven in the most explicit way by the fantasy of expanding Germany's empire and challenging British hegemony at sea. Ratzel's geopolitical writings shared common ground with the ideas of Alfred Thayer Mahan and Halford Mackinder, which were developed almost simultaneously. But the necropolitical impulse was specific to Ratzel. War, to him, was both an expression of evolutionary struggle and a medium through which nature imposed itself upon spatial consciousness. While Ratzel had more theoretical depth than Mahan and Mackinder, this did not prevent his interventions from descending into banal militarism or jingoism, particularly as he turned his attention to interimperial rivalry.

There was also a dreamier Ratzel who was concerned with life in all its dynamism and exuberance as well as with the material boundaries earthly foundations placed on it. This was a thinker who pursued earthworms and who was mesmerized by spiders wafting through space in their gondolas, a man who listened to farm animals at night and who pondered the meaning of decay he found all around him, be it in decomposing leaves or the abandoned settlements along the American frontier. Ratzel was an early proponent of the timespace continuum, a monist who questioned quite radically dichotomous understandings of nature and society, human and animal, organic and inorganic realms. He believed that death opened consciousness up to the universe and that nature enclosed art. His geography sought to restore humanity to what he believed was its rightful status: that of an earthworm that was always bound to return to the earth out of which it had crawled.

This pensive, Romantic, and often deliberately impractical streak in Ratzel, which dominates large parts of the oeuvre, chimes with recent developments in social theory. Geographers, in particular, are once again seeking to excavate the inorganic and inhuman forces that have subtended and provoked organic life and the political forms it takes throughout history. In the Anthropocene, we are told, the term geopolitics acts as a useful bridge between the earth system, biological life, and political organization. This resemblance between Ratzel's work and contemporary new materialisms has not gone unnoticed.[1] A temptation to collapse the question of meaning into matter, to sideline social structure, ideas, and ideology while opening up the political to nonhuman agencies, does chime with Ratzel's geography. And if Ratzel already stealthily features in contemporary debates, Verne rightly asks, then have we "been too radical in condemning his work"?[2]

Ratzel may not have identified as a vitalist, but he did integrate vitalist ideas into his worldview, alongside others that were more characteristic of mechanism, organicism, monism, and panpsychism. As I hope to have shown in this book, the analytical limitations and political problems in Ratzel's oeuvre considerably outweigh its promise. His explanatory model relied on a number of moves that limited his ability to truly comprehend the phenomena he set out to examine. The first was to assume a direct link between the soil and sociopolitical organization, as if there was no process of extraction and commodification of nature's resources. As noted by some of his contemporaries, this soil mysticism covered up a lack of conceptual precision. The second was to downplay the role that various ideologies, including Christianity, nationalism, colonialism, and liberalism, had played in fueling the very wars his writings naturalized as a biological urge. In the later Ratzel, a growing enchantment with the restless creativity of life may have diverted attention away from the immutable properties of the earth, but it did little to explain historical change.

Ratzel's political positions were not simply tagged on to his theoretical work but emerged organically from his larger theoretical edifice. His preoccupation with the power of geology drove him to explain even language and racial difference as geologic. He had no remorse for a species or race that was wiped off the earth's surface because there was nothing in his cosmos that allowed for regret, sadness, or anger, even though these emotions did occasionally make themselves felt in his writing. Attempts to save Ratzel's supposedly "original, complex, and multifaceted" political geography while disposing of his "pseudo-scientific" theory of the state rely on the tricky separation of science and pseudo-science.[3] Of course, much can be gained from an ontology that treats the earth's crust, hydrosphere, atmosphere, and biosphere, and therefore the humans who inhabit it, as one. But it is unclear why it is necessary to revive Ratzel in order to reach such an understanding. If the idea is to return to an original moment in Ratzel "before [his] agenda became toxified by imperialism,"[4] then the answer is that there was never such a moment.[5] As we have seen, it was precisely his theory of life and earth, established in the late 1860s from a reading of Darwin, Haeckel, and Spencer, that drove him to idealize the colony.

There are other lessons in Ratzel for contemporary debates. It has recently become popular among decolonially minded posthumanists to argue that colonial violence rests on and works through dualistic nature/society constructions and that monism and animism, inspired perhaps by various indigenous cosmologies, might be drawn upon as forms of intellectual resistance against colonial Eurocentrism.[6] And yet in Ratzel we find a thinker who was a monist, an animist, *and* an advocate for settler colonialism. Indeed, his entire intellectual project was one of imploding binaries. The misanthropy that sometimes underwrites contemporary materialist scholarship could be a starting point for further discussion. Like those who have in recent years found solace in the vision of a planet "that will still be here when we are not,"[7] Ratzel was mesmerized by a world of fossils and human ruins. He too was enthralled by the fantasy of a landscape whose beauty lay in the eerie absence of human but not all life.

Much like Ratzel and many other Darwinian thinkers, contemporary vitalists like Jane Bennett have made use of anthropomorphism to think about animal agencies and thus blur the boundary between nature and society.[8] And yet it has been less common to reverse this move and conceptualize phenomena such as war in the language of geology or microbiology. If the human subject is indeed caught up in assemblages of war, as Ratzel came to believe, which are defined and confined by powerful material agencies, then it is unclear how this subject would ever extricate itself from these mighty entanglements of nature, technology, and flesh.[9] It surely is difficult to comprehend Ratzel's own militarism as anything other than thickened ideology that had entered into the most capillary social practices, mobilizing bodies and materials, but not the other way around.

This links up with a second problem, which is the locus of responsibility. For if nature can wage war, as it certainly can in Ratzel, then this raises the question of culpability. Ratzel was incurious toward the very social structures that had made the war in which he fought possible. Captivated by the bodies and materials that had come together to produce the battlefield of the Franco-Prussian War, he would come in the 1890s to hail war as the creator of space itself. All this indicates that Bennett's distinction between a venomous politics of agonism and a virtuous politics of enchantment is difficult to sustain.[10] In the work of the late Ratzel, echoing back to us from the bellicose fin-de-siècle, we can find both a Romantic voice of wonder and an agonist understanding of politics in surprising harmony. The solution to the ethical problems raised by a Ratzelian geopolitics cannot simply be reduced to one of ontology or scale either. For while his political geography certainly performed a disembodied gaze, he did write about flesh and bone bodies, including his own. If the concept of geopolitics is indeed to be rehabilitated,[11] then it may be useful to bear in mind the complexity of the Ratzelian legacy.

Although Ratzel may not have been aware of it, much of his writing was, in some form or other, preoccupied with disappearance, decay, dying, and death, a fixation that shaped his human and political geography and that would live on in the work of some of his most important interlocuters and popularizers. Achille Mbembe develops necropolitics as a theory of the less-than-living, which approaches being itself as distributed along a spectrum from life to death. It is a concept that seeks to grapple with the rendering of colonized subjects as always already undead. In doing so, it unravels the conditions under which it becomes possible to inflict pain and indeed to exterminate populations. Schmitt's *jus publicum Europaeum*, drawn in part from the Ratzelian corpus, relies on the colony's "efficacy as a formation of terror," on a distinction between a world in which war is regulated and one in which it is not.[12] Necropolitics, in this reading, encompasses the racialized othering inherent in liberalism, its bordering practices, and its way of war. For Mbembe, the defining characteristics of our current epoch is a great escape from democracy, which is in fact the eternal return of "democracy's bitter sediment," first found in the colony and the plantation.[13]

As one of "discontinuous lines, of raw and rapid gestures," Mbembe's approach to necropolitics may not be naturally suited to the kind of work carried out in this book.[14] But here too, I hope to have shown, the term has value. My aim was not to humanize an individual colonizer by casting him as a traumatized subject, only to then explain away the violent geographies he gave birth to. The point was rather to show an instance of necropolitics in high resolution, to reveal its *where* and a *when*. Like Mbembe, I have approached necropolitics not just as a form of rule but as a set of intellectual discourses, which have their own sociomaterial and bodily context. The society of enmity Mbembe attributes to the present has its roots, he would be unlikely to disagree, in the numerous Ratzels of the eighteenth and nineteenth centuries, many of whom were, in some form or other, involved in the larger project of liberalism. To be clear, Ratzel was no unqualified enthusiast for liberal democracy. He wanted the state to be governed by a monarch in tandem with a bourgeois elite. But the young Ratzel had certainly been enticed by democracy. It was only when he witnessed its short-lived and half-hearted existence in post–Civil War America that he was repulsed by its very inclusiveness and called instead for a global form of apartheid. Ratzel was fascinated by the cheapening of life that the colonial encounter entailed. Bordering and war would become his fundamental preoccupations.

The figure of Ratzel then clearly stands at the intersection of a number of intertwined political projects, those of settler colonialism and liberalism but also of fascism and the particular form it would take in Germany. It is thus unsurprising perhaps that he has been used quite frequently to illustrate the so-called Windhoek to Auschwitz thesis. The problem with this continuity thesis, however, is not its focus on the endurance of colonial violence in fascist forms, but its tendency to present the issue as a specifically German one.[15] Ratzel's colonial geography may be branded as characteristically "German" due to its pan-German nationalism and soil mysticism. And yet he drew on British biologists, philosophers, and explorers. Many of the artifacts he wrote about had been sourced from the collections of the British Museum. His vision of a German settler colonialism was largely modeled on Anglo-Saxon colonization. US race law, which would later inspire the Nuremberg Laws, had impressed Ratzel too. The popularity of his ideas not just in western Europe and North America but in Latin America and eastern Europe attests moreover to the transnational appeal of Ratzel's necropolitical geography.

In this sense, Ratzel's *Lebensraum* idea can be read as part of a wider shift in the history of government. Its popular appeal in the early twentieth century was an expression of the need for some kind of meta-narrative that could support the growing desire to govern life by categorizing and sorting it, fostering and abandoning it. Unknowingly, Ratzel was providing a scientific, philosophical, and poetic language with which to think through such a politics of life and death. He used this intellectual platform to become a theorist of Zygmunt Bauman's "modern gardening state," which views society "as an object of designing, cultivating and weed-poisoning."[16] There was indeed something fundamentally mundane and domesticated about Friedrich Ratzel and his colonial fanaticism. He led a middle-class life, wrote sweet letters to his daughters, loved celebrating Christmas, and hid in the serenity of his garden. Nothing in his diaries, letters, or writings implies that he understood his role in the larger structures of colonial violence. Where he was able to see violence at all, he either naturalized it or denounced it as the violence of the "other." This was the case when he spoke of the outbreak of violence among *Naturvölker* or condemned the British Empire's crimes. Ratzel may neither have personally inflicted pain onto colonized bodies, nor was he quite an Arendtian *Schreibtischtäter* who organized violence remotely through bureaucratic means. Instead, the Leipzig geographer's power was exerted at a higher level of abstraction, by the incessant use of pen and paper. His personal and professional connections to explorers and administrators, many of whom reported back to him from the German colonies, suggest that he did not operate far removed from the praxis of colonization.

Ratzel's cultural concept of race was one of a number of mutations in late nineteenth-century racial discourse. Cosmopolitanism remained a crucial referent for him, but the reality of colonialism demanded racial hierarchy between the colonizers and the colonized. His diffusionist critique of the idea of indigeneity may have raised eyebrows among adherents of Aryanism, but it too was epistemologically violent. Ratzel did not construct the land that was to be colonized as "empty," but rather as subject to an endless succession of migratory movements in which the most recent was always the most evolutionary successful and thus legitimate. The consequence was to strip from the colonized any legitimate connection to their land, to treat the latter as if it had always only temporarily been theirs. His thinking had in 1904 culminated in the prediction of further racial struggle in the twentieth century and the call for global racial segregation. It was only "common sense" that there was a need to "keep clean and improve one's race."[17] There was thus in Ratzel an openness to eugenicist arguments, even though he did not pursue this line of reasoning. In the 1870s, he had also come into contact with a white supremacist narrative of the "lost cause," which continues to shape American politics a century and a half later.

Unlike many of his peers in the pan-German movement, Ratzel neither seems to have feared Judaism nor Jewishness. Nonetheless, Ratzel *was* preoccupied, even obsessed, with racial otherness, found both in the (formerly) enslaved and the supposedly self-annihilating *Naturvölker*. And although he did not, by and large, construct the latter as threatening to white settlers, he did cast them as a "chronic malady" on the settler colonial body politic. Not only was the extermination of uncivilized *Völker* inevitable, but it was driven by the very social practices that defined these peoples as uncivilized. This idea of self-extermination allowed Ratzel to obscure the violence of European colonialism by making "the victims responsible," to return to Mbembe.[18] When Germany's colonies were lost in the First World War, a similar logic of victim blaming re-emerged in political discourse. The dehumanization that had previously been reserved for the colonized was extended to parts of Europe's population in much the same way as *Lebensraum* itself became a project of territorial expansion in Europe.

Notes

CHAPTER 1

1. Friedrich Ratzel, "Diary entry 31 March 1897," *Tagebuch 4.10.1896–6.1.1898, Abschrift angefertigt von Günter Buttmann*, Bayerische Staatsbibliothek, Ratzeliana Supplement 17a VI 15a.

2. Friedrich Ratzel, "Über den Lebensraum," *Die Umschau* 1, no. 21 (1897): 366.

3. Hans Schrepfer, "Was heißt Lebensraum? Eine notwendige begriffliche Klärung," *Geographische Zeitschrift* 48, no. 11–12 (1942): 417.

4. Ratzel, "Über den Lebensraum," 366.

5. Niels Werber, *Geopolitik zur Einführung* (Hamburg: Junus, 2014), 77.

6. David Olusoga and Casper W. Erichsen, *The Kaiser's Holocaust: Germany's Forgotten Genocide* (London: Faber and Faber, 2010), 108.

7. Friedrich Ratzel, *Politische Geographie* (Munich: R. Oldenbourg, 1897); Adolf Hitler, *Mein Kampf: Zwei Bände in einem Band* (Munich: Eher, 1925/1927 [1936]).

8. Adolf Hitler, *Hitler's Second Book: The Unpublished Sequel to Mein Kampf* (New York: Enigma, 1928 [2006]).

9. Holger Herwig, "Geopolitik: Haushofer, Hitler and Lebensraum," *Journal of Strategic Studies* 22, no. 2–3 (1999): 218–41; Hans-Adolf Jacobsen, *Karl Haushofer: Leben und Werk* (Boppard: Boldt, 1979); Karl Lange, "Der Terminus 'Lebensraum' in Hitlers 'Mein Kampf,'" *Vierteljahrshefte für Zeitgeschichte* 13, no. 4 (1965): 426–37; Mark Mazower, *Hitler's Empire: Nazi Rule in Occupied Europe* (London: Penguin, 2008); Woodruff D. Smith, *The Ideological Origins of Nazi Imperialism* (Oxford: Oxford University Press, 1986); Geoffrey Stoakes, *Hitler and the Quest for World Domination: Nazi Ideology and Foreign Policy in the 1920s* (Lemington Spa: Berg, 1986).

10. Innes M. Keighren, "History and Philosophy of Geography III: The Haunted, the Reviled, and the Plural," *Progress in Human Geography* 44, no. 1 (2020): 161.

11. Ian Klinke, *Cryptic Concrete: A Subterranean Journey into Cold War Germany*. Oxford: Wiley-Blackwell, 2018.

12. Ernst Stein, "Die Zeit auf dem Buckel," *Die Zeit*, 25 November 1966, https://www.zeit.de/1966/48/die-zeit-auf-dem-buckel/komplettansicht.

13. Robert D. Kaplan, *The Revenge of Geography: What the Map Tells Us about Coming Conflicts and the Battle against Fate* (New York: Random House, 2012), 81; Lawrence Freedman, *Strategy: A History* (Oxford: Oxford University Press, 2013), 122.

14. Joshua Jelly-Schapiro, "What Are Borders For?" *The New Yorker*, 27 November

2019, https://www.newyorker.com/books/under-review/what-are-borders-for/amp; John Lewis-Stempel, "The Women Made to Boil Heads of Their Own People in Germany's First Holocaust," *Sunday Express*, 19 January 2014, https://www.express.co.uk/news/world-war-1/454733/The-women-made-to-boil-heads-of-their-own-people-in-Germany-s-first-Holocaust; George Monbiot, "Colonised and Coloniser, Empire's Poison Infects Us All," *The Guardian*, 8 October 2012, https://amp.theguardian.com/commentisfree/2012/oct/08/empire-torture-kenya-catastrophe-europe; Christoph Strauch, "Von Brest–Litowsk zum Generalplan Ost?" *Frankfurter Allgemeine Zeitung*, 13 March 13 2018, https://www.faz.net/aktuell/politik/der-erste-weltkrieg/erster-weltkrieg-von-brest-litowsk-zum-generalplan-ost-15471514.html; Enzo Traverso, "Warping History," *Jacobin*, 8 October 2016, https://www.jacobinmag.com/2016/08/timothy-snyder-black-earth-bloodlands-holocaust-soviets-nazis/; Michael Williams, "Hitler's Holocaust Blueprint." *Daily Mail*, 23 September 2010, https://www.dailymail.co.uk/news/article-1314399/amp/Hitlers-Holocaust-blueprint-Africa-concentration-camps-used-advance-racial-theories.html.

15. Peter Dittmar, "Die ewige Angst vor dem Chinesen," *Die Welt*, 26 January 2020, https://www.welt.de/geschichte/article205323439/Gelbe-Gefahr-Die-ewige-Angst-vor-dem-Chinesen.html; Dankwart Guratzsch, "Der Selbstmord der Städte." *Die Welt*, 2 October 2019, https://www.welt.de/kultur/plus201239634/Totale-Verkehrsblockade-Der-Selbstmord-der-Staedte.html; Fabian Somavilla, "Ohne Impfpass kein Grenzübertritt: Wie Corona Grenzen militarisiert." *Der Standard*, 3 May 2020. https://www.derstandard.at/story/2000117194363/ohne-impfpass-kein-grenzuebertritt-wie-corona-grenzen-mililtarisiert?amplified=True.

16. Karl Schlögel, *Im Raume lesen wir die Zeit: über Zivilisationsgeschichte und Geopolitik* (Munich: Hanser, 2003), 10, 55, 59.

17. Woodruff D. Smith, "Friedrich Ratzel and the Origins of Lebensraum," *German Studies Review* 3, no. 1 (1980): 51–68.

18. Jürgen Osterhammel, *Die Verwandlung der Welt: Eine Geschichte des 19. Jahrhunderts* (Munich: C. H. Beck, 2009), 142; Julia Hell, *The Conquest of Ruins: The Third Reich and the Fall of Rome* (Chicago: University of Chicago Press, 2019), 241.

19. Friedrich Ratzel, *Anthropogeographie Erster Teil: Grundzüge der Anwendung der Erkunde auf die Geschichte* (Stuttgart: J. Engelhorn, 1882); Friedrich Ratzel, *Anthropogeographie Zweiter Teil, Zweite Auflage: Die geographische Verbreitung des Menschen* (Stuttgart: J. Engelhorn, 1891 [1899]); Friedrich Ratzel, *The History of Mankind, Volume I* (London: MacMillan, 1885 [1896]); Friedrich Ratzel, *The History of Mankind, Volume II* (London: MacMillan, 1886 [1897]); Friedrich Ratzel, *The History of Mankind, Volume III* (London: MacMillan, 1888 [1898]).

20. Friedrich Ratzel, *Glücksinseln und Träume: Gesammelte Aufsätze aus dem Grenzboten* (Leipzig: Fr. Wilh. Grunow, 1905), 139.

21. Friedrich Ratzel, *Das Meer als Quelle der Volksgröße: Eine politisch–geographische Studie* (Munich: R. Oldenbourg, 1900 [1911]); Mark Bassin, "Imperialism and the Nation State in Friedrich Ratzel's Political Geography," *Progress in Human Geography* 11, no. 4 (1987): 473–95.

22. Friedrich Ratzel, "Nationalitäten und Rassen," in *Kleine Schriften von Friedrich Ratzel, herausgegeben von Hans Helmolt, 2. Band*, ed. Hans Helmolt (Munich: R. Old-

enbourg, 1904 [1906]), 462–87; see also Mark Bassin, "Race *contra* Space: The Conflict between German *Geopolitik* and National Socialism," *Political Geography* 6, no. 2 (1987): 115–34.

23. Günther Buttmann, *Friedrich Ratzel: Leben und Werk eines deutschen Geographen* (Stuttgart: Wissenschaftliche Verlagsgesellschaft MBH, 1977), 110; Friedrich Ratzel, "Lebensraum—A Biogeographical Study," trans. Tul'si [Tuesday] Bhambry," *Journal of Historical Geography* 61 (1901 [2018]): 59–80; Friedrich Ratzel, *Die Erde und das Leben, 1. Band* (Leipzig: Bibliographisches Institut, 1902); Friedrich Ratzel, *Die Erde und das Leben, 2. Band* (Leipzig: Bibliographisches Institut, 1902).

24. Achille Mbembe, "Necropolitics," *Public Culture* 15, no. 1 (2003): 11–40; Achille Mbembe, *Necropolitics* (Durham, NC: Duke University Press, 2019).

25. Neil Smith, *American Empire: Roosevelt's Geographer and the Prelude to Globalization* (Oakland: University of California Press, 2004).

26. Gerry Kearns, *Geopolitics and Empire: The Legacy of Halford Mackinder* (Oxford: Oxford University Press, 2009).

27. Innes M. Keighren, *Bringing Geography to Book: Ellen Semple and the Reception of Geographical Knowledge* (London: I. B. Tauris, 2010).

28. Gopal Balakrishnan, *The Enemy: An Intellectual Portrait of Carl Schmitt* (London: Verso, 2000); Reinhard Mehring, *Carl Schmitt: A Biography* (London: Polity, 2014); Claudio Minca and Rory Rowan, *On Schmitt and Space* (London: Routledge, 2015); Stephen Legg, *Spatiality, Sovereignty and Carl Schmitt: Geographies of the Nomos* (London: Routledge, 2011).

29. Holger Herwig, *The demon of geopolitics: How Karl Haushofer "educated" Hitler and Hess* (Lanham, MD: Rowman and Littlefield, 2016); Christian Spang, *Karl Haushofer und die OAG: Deutsch–japanische Netzwerke in der ersten Hälfte des 20. Jahrhunderts* (Munich: Iudicium, 2018).

30. Michael Cox, *E. H. Carr: A Critical Appraisal* (Basingstoke: Palgrave, 2000); Oliver Jütersonke, *Morgenthau, Law and Realism* (Cambridge: Cambridge University Press, 2010); William E. Scheuerman, *Hans Morgenthau: Realism and Beyond* (Cambridge: Polity, 2009); Michael C. Williams, *Realism Reconsidered: The Legacy of Hans Morgenthau in International Relations* (Oxford: Oxford University Press, 2007). There are, furthermore, bestselling books available on the former statesman and academic Henry Kissinger, which reach opposite verdicts: Christopher Hitchens, *The Trial of Henry Kissinger* (London: Verso, 2001); Niall Ferguson, *Kissinger: 1923–1968: The Idealist* (London: Penguin, 2015).

31. Richard C. Powell, "Echoes of the New Geography? History and Philosophy of Geography I," *Progress in Human Geography* 36, no. 4 (2009): 519.

32. Innes M. Keighren, Christian Abrahamsson, and Veronica della Dora, "On Canonical Geographies," *Dialogues in Human Geography* 2, no. 3 (2012): 296–312.

33. In its most literal sense, the canon is a "measuring rod," even though the term has "come to mean a body of work that inscribes the conventions of a bespoke community," as Withers notes. Following the theologian Delwin Brown, Powell characterizes the canon's defining features as its "boundedness, normativity, contestability, contemporaneity, curatorial character and existential nature." Charles W. J. Withers, "Geography's Evolving Traditions and Textual Critique," *Dialogues in Human Geogra-*

phy 2, no. 3 (2012): 317; Richard C. Powell, "Notes on a Geographical Canon? Measures, Models and Scholarly Enterprise," *Journal of Historical Geography* 49 (2015): 3.

34. Gillian Rose, "Tradition and Paternity: Same Difference?" *Transactions of the Institute of British Geographers* 20, no. 4 (1995): 414–16; Avril Maddrell, "To Read or Not to Read? The Politics of Overlooking Gender in the Geographical Canon," *Journal of Historical Geography* 49 (2015): 31–38; James D. Sidaway, "The (Re)making of the Western 'Geographical Tradition': Some Missing Links," *Area* 29, no. 1 (1997): 72–80; Federico Ferretti, "Rediscovering other Geographical Traditions," *Geography Compass* 13, no. 3 (2019): 1–15.

35. Felix Driver and Elizabeth Baigent, "Biography and the History of Geography: A Response to Ron Johnston," *Progress in Human Geography* 31, no. 1 (2007): 101–6.

36. Andrew Barry, "Geography and Other Disciplines: Genealogy, Anamnesis and the Canon," *Journal of Historical Geography* 49 (2015): 85–93.

37. John Agnew, "Of Canons and Fanons," *Dialogues in Human Geography* 2, no. 3 (2012): 322.

38. Hans Morgenthau, *Politics among Nations: The Struggle for Power and Peace* (New York: Knopf, 1948 [1993]), 174.

39. Klaus Dodds and David Atkinson, *Geopolitical Traditions: A Century of Geopolitical Thought* (London: Routledge, 2000); Gearóid Ó Tuathail and John Agnew. "Geopolitics and Discourse: Practical Geopolitical Reasoning in American Foreign Policy," *Political Geography* 11, no. 2 (1992): 190–204; Gearóid Ó Tuathail, *Critical Geopolitics: The Politics of Writing Global Space* (Minneapolis: University of Minnesota Press, 1996); Simon Dalby, *Creating the Second Cold War: The Discourse of Politics* (London: Bloomsbury, 1990 [2016]).

40. Lucian Ashworth, "Realism and the Spirit of 1919: Halford Mackinder, Geopolitics and the Reality of the League of Nations," *European Journal of International Relations* 17, no. 2 (2010): 279–301. See also John M. Hobson, *The Eurocentric Conception of World Politics: Western International Theory, 1760–2010* (Cambridge: Cambridge University Press, 2012).

41. Jean-François Drolet and Michael C. Williams, "The Radical Right, Realism, and the Politics of Conservatism in Postwar International Thought," *Review of International Studies* 47, no. 3 (2021): 273–93; Stefano Guzzini, *The Return of Geopolitics in Europe?* (Cambridge: Cambridge University Press, 2012); Matthew Specter, *The Atlantic Realists: Empire and International Political Thought between Germany and the United States* (Stanford, CA: Stanford University Press, 2021).

42. Mark Bassin, "Between Realism and the 'New Right': Geopolitics in Germany in the 1990s," *Transactions of the Institute of British Geographers* 28, no. 3 (2003): 350366; Ian Klinke, "Geopolitics and the Political Right: Lessons from Germany," *International Affairs* 94, no. 3 (2018): 495–514.

43. In addition, there is also the nonacademic hagiography by Haaß, a PhD thesis by Hunter, and an edited collection on the centenary of Ratzel's *Politische Geographie*, which includes a blend of laudatory and critical contributions: Wolfgang Haaß, *Vom Dorf in die Welt: Gedanken zu dem großen Geographen Friedrich Ratzel* (Kronau: Gemeinde Angelbachtal, 2013); James M. Hunter, *Perspective on Ratzel's Political*

Geography (Lanham, MD: University Press of America, 1983); Marco Antonsich, Vladimir Kolossov, and Maria Paola Pagnini, *Europe between Political Geography and Geopolitics: On the Centenary of Ratzel's Politische Geographie* (Rome: Società Geografica Italiana, 2001).

44. Johannes Steinmetzler, *Die Anthropogeographie Friedrich Ratzels und ihre ideengeschichtlichen Wurzeln* (Bonn: Im Selbstverlag des geographischen Instituts, 1956).

45. Harriet Wanklyn, *Friedrich Ratzel: A Biographical Memoir and Bibliography* (Cambridge: Cambridge University Press, 1961).

46. Buttmann, *Friedrich Ratzel.*

47. Gerhard H. Müller, *Friedrich Ratzel (1844–1904), Naturwissenschaftler, Geograph, Gelehrter* (Stuttgart: GNT, 1996).

48. Alexandros Stogiannos, *The Genesis of Geopolitics and Friedrich Ratzel: Dismissing the Myth of the Ratzelian Geodeterminism* (Cham: Springer, 2019).

49. Bassin, *Race contra Space*; Woodruff D. Smith, *Politics and the Sciences of Culture in Germany, 1840–1920* (Oxford: Oxford University Press, 1991); David T. Murphy, *Heroic Earth: Geopolitical Thought in Weimar Germany 1918–1933* (Kent, OH: Kent State University Press, 1997); Smith, "Friedrich Ratzel"; Ó Tuathail, *Critical Geopolitics.*

50. Christian Abrahamsson, "On the Genealogy of Lebensraum," *Geographica Helvetica* 68 (2013): 37–44; Matus Halas, "Searching for the Perfect Footnote: Friedrich Ratzel and the Others at the Roots of Lebensraum," *Geopolitics* 19, no. 1 (2014): 1–18; Ulrike Jureit, *Das Ordnen von Räumen: Territorium und Lebensraum im 19. und 20. Jahrhundert* (Hamburg: Hamburger Edition, 2012); Ulrike Jureit, "Mastering Space: Laws of Movement and the Grip on the Soil," *Journal of Historical Geography* 61 (2018): 81–85; Ian Klinke and Mark Bassin, "*Lebensraum* and its Discontents," *Journal of Historical Geography* 61 (2018): 53–58. Wolfgang Natter, "Friedrich Ratzel's Spatial Turn," in *B/ordering Space*, ed. Henk van Houtum, Oliver Kramsch, and Wolfgang Zierhofer (Aldershot: Ashgate, 2005), 171–87.

51. Dörte Lerp, "Farmers to the Frontier: Settler Colonialism in the Eastern Prussian Provinces and German Southwest Africa," *Journal of Imperial and Commonwealth History* 41, no. 4 (2013): 567–83; Benjamin Madley, "From Africa to Auschwitz: How German South West Africa Incubated Ideas and Methods Adopted and Developed by the Nazis in Eastern Europe," *European History Quarterly* 35, no. 3 (2005): 429–64; Olusoga and Erichsen, *The Kaiser's Holocaust*; Jürgen Zimmerer, "The Birth of the *Ostland* out of the Spirit of Colonialism: A Postcolonial Perspective on Nazi Policy of Conquest and Extermination," *Patterns of Prejudice* 39, no. 2 (2005): 197–219, but also Timothy Snyder, *Black Earth: The Holocaust as History and Warning* (London: Vintage, 2015). For earlier arguments about the continuities between imperial and Nazi German expansionism, see Fritz Fischer, *Germany's Aims in the First World War* (New York: W. W. Norton and Company, 1967) and Smith, *Ideological Origins.*

52. Hannah Arendt, *The Origins of Totalitarianism* (San Diego: Harvest, 1951 [1979]).

53. Jens-Uwe Guettel, *German Expansionism, Imperial Liberalism and the United States, 1776–1945* (Cambridge: Cambridge University Press, 2012); Birthe Kundrus,

"German Colonialism: Some Reflections on Reassessments, Specificities, and Constellations," in *German Colonialism: Race, the Holocaust, and Postwar Germany*, ed. Volker Langbehn and Mohammad Salama (New York: Colombia University Press. 2011), 29–47; Susanne Kuss, *German Colonial Wars and the Context of Military Violence* (Cambridge, MA: Harvard University Press, 2017); Woodruff D. Smith, "Review of Jürgen Zimmerer: Von Windhuk nach Auschwitz? Beiträge zum Verhältnis von Kolonialismus und Holocaust," *German History* 30, no. 2 (2012): 312–14.

54. Jureit, *Mastering Space*, 81; see also Jureit, *Das Ordnen*, 152.

55. Maan Barua, "Ratzel, Bio-geography and the More-than-Human," *Journal of Historical Geography* 61 (2018): 102–8; Patricia Chiantera-Stutte, "Ratzel's Stone Guest: The Art of Politics in the Work of Friedrich Ratzel," *Journal of Historical Geography* 61 (2018): 91–96; Ian Klinke, "Friedrich Ratzel, Lebensraum and the Death Motif," *Journal of Historical Geography* 61 (2018): 97–101; Ian Klinke, "Vitalist Temptations: Life, Earth and the Nature of War in Friedrich Ratzel," *Political Geography* 72 (2019): 1–9; Hans-Dietrich Schultz, "Friedrich Ratzel—Pionier moderner Sozialgeographie? Ein Missverständnis," *Kultursoziologie* 23, no. 3 (2014): 54–66; Mark Usher, "Territory *Incognita*," *Progress in Human Geography* 44, no. 6 (2020): 1019–46; Julia Verne, "The Neglected 'Gift' of Ratzel for/from the Indian Ocean: Thoughts on Mobilities, Materialities and Relational Spaces," *Geographica Helvetica* 72 (2017): 85–92.

56. Jane Bennett, *Vibrant Matter: A Political Ecology of Things* (Durham, NC: Duke University Press, 2010); Bruce Braun, "Environmental Issues: Inventive Life," *Progress in Human Geography* 32, no. 5 (2008): 667–79; Diana Coole, "Agentic Capacities and Capacious Historical Materialism: Thinking with New Materialisms in the Political Sciences," *Millennium: Journal of International Studies* 41, no. 3 (2013): 451–69; Beth Greenhough, "Vitalist Geographies," in *Taking Place: Non–representational Theories and Geography*, ed. Ben Anderson and Paul Harrison (Farnham, Ashgate, 2010), 37–54; Sue Ruddick, "Rethinking the Subject, Reimagining Worlds," *Dialogues in Human Geography* 7, no. 2 (2017): 119–39.

57. William Connolly, "The 'New Materialism' and the Fragility of Things," *Millennium: Journal of International Studies* 41, no. 3 (2013): 399–412; Jason Dittmer, "Geopolitical Assemblages and Complexity," *Progress in Human Geography* 38, no. 3 (2014): 385–401; Vicki Squire, "Reshaping Critical Geopolitics? The Materialist Challenge," *Review of International Studies* 41, no. 1 (2015): 139–59; Sarah Whatmore, "Materialist Returns: Practicing Cultural Geography in and for a More-than-Human World," *Cultural Geographies* 13, no. 4 (2006): 600–609.

58. Abrahamsson, *Genealogy of Lebensraum*, 38; see also Christian Abrahamsson, "Book Review Forum: Vibrant Matter: A Political Ecology of Things," *Dialogues in Human Geography* 1, no. 3 (2011): 399–402.

59. Edward Ross Dickinson, "Biopolitics, Fascism, Democracy: Some Reflections on Our Discourse about 'Modernity,'" *Central European History* 37, no. 1 (2004): 1–48; Michael Dillon and Andrew W. Neal, *Foucault on Politics, Security and War* (Basingstoke: Palgrave, 2008); Derek Gregory, "Vanishing Points," in *Violent Geographies: Fear Terror and Political Violence*, ed. Derek Gregory and Allan Pred (London: Routledge, 2007), 205–23; Mark Mazower, "Foucault, Agamben: Theory and the Nazis," *Boundary 2* 35, no. 1 (2008): 23–34; Claudio Minca, "Giorgio Agamben and

the New Biopolitical Nomos," *Geografiska Annaler: Series B* 88, no. 4 (2006): 387–403; Claudio Minca, "Geographies of the Camp," *Political Geography* 49 (2015): 74–83; Marcelo Svirsky and Simone Bignall, *Agamben and Colonialism* (Edinburgh: University of Edinburgh Press, 2012).

60. Giorgio Agamben, *Homo Sacer: Sovereign Power and Bare Life* (Stanford, CA: Stanford University Press, 1998); Michel Foucault, *Society Must Be Defended: Lectures at the Collège de France, 1975–1976* (New York: Picador, 1976 [2003]); see also Roberto Esposito, *Bios: Biopolitics and Philosophy* (Minneapolis: University of Minnesota Press, 2008).

61. Frantz Fanon, *The Wretched of the Earth* (London: Penguin, 1963 [2001]); Arendt, *Origins of Totalitarianism*. See also Aimé Césaire's reading of Nazism as colonial violence applied to European subjects. Aimé Césaire, *Discourse on Colonialism* (New York: Monthly Review Press, 1955 [1972]).

62. Mbembe, "Necropolitics," 21.

63. Mbembe, "Necropolitics," 12.

64. Joy James, "All Power to the People! Hannah Arendt's Theory of Communicative Power in a Racialized Democracy," in *Race and Racism in Continental Philosophy*, ed. Robert Bernasconi and Sybol Cook (Bloomington: Indiana University Press, 2003), 249–67; Patricia Owens, "Racism in the Theory Canon: Hannah Arendt and 'the one great crime in which America was never involved,'" *Millennium: Journal of International Studies* 45, no. 3 (2017): 403–24.

65. Michael Dillon and Luis Lobo-Guerrero, "Biopolitics of Security in the 21st Century: An Introduction," *Review of International Studies* 34, no. 2 (2008): 265–92; Stuart Elden, "Secure the Volume: Vertical Geopolitics and the Depth of Power," *Political Geography* 34 (2013): 35–51; Paolo Giaccaria and Claudio Minca, "Life in Space, Space in Life: Nazi Topographies, Geographical Imaginations, and *Lebensraum*," *Holocaust Studies: A Journal of Culture and History* 22, no. 2–3 (2016): 151–71.

66. Chiantera-Stutte, "Ratzel's Stone Guest."

67. Noam Leshem, "'Over our dead bodies': Placing Necropolitical Activism," *Political Geography* 45 (2015): 34–44; José Luis Romanillos, "Geography, Death, and Finitude," *Environment and Planning A* 43, no. 11 (2011): 2533–53; Avril Maddrell and James D. Sidaway, *Deathscapes: Spaces for Death, Dying, Mourning and Remembrance* (Farnham: Ashgate, 2010).

68. Wanklyn, *Friedrich Ratzel*, 4.

69. To be more precise, Ratzel published nineteen research monographs and popular science books during his lifetime, as well as two shorter pamphlets, a guide for hikers, and a collection of essays. He also updated his major works considerably in later editions, and a second edition of his *Völkerkunde* was translated into English during his lifetime. His autobiography and a collection of essays appeared shortly after his death. (Hans Helmolt, *Kleine Schriften von Friedrich Ratzel, 1. Band*. Munich: R. Oldenbourg, 1906). Second and third editions of his major books continued to be published throughout the twentieth century.

70. Helmolt, *Kleine Schriften von Friedrich Ratzel, 1. Band*; Hans Helmolt, *Kleine Schriften von Friedrich Ratzel, 2. Band* (Munich: R. Oldenbourg, 1906). The second volume includes a comprehensive bibliography.

71. Ratzel, *Glücksinseln und Träume*.

72. Ratzel, *History of Mankind, Vol. 1, 2, and 3*.

73. The 1896 essay "The Territorial Growth of States" appeared in English in the same year as the German original. Friedrich Ratzel, "The Territorial Growth of States," *Scottish Geographical Magazine* 12, no. 7 (1896): 351–61; Friedrich Ratzel, "Die Gesetze des räumlichen Wachstums der Staaten. Ein Beitrag zur wissenschaftlichen politischen Geographie," *Petermanns Mitteilungen* 42, no. 5 (1896): 97–107. A second English-language translation of this essay by Roland Bolin appeared in 1969. Friedrich Ratzel, "Laws of the Spatial Growth of States." In *The Structure of Political Geography*, edited by Roger E. Kasperson and Julian V. Minghi (London: Transactions, 1896 [1969]), 17–28. A further English-language essay is Friedrich Ratzel, "Man as a Life Phenomenon on the Earth's Surface," in *The History of the World*, vol. 1, ed. Hans F. Helmolt (New York: Dodd, Mead and Co., 1902), 61–106.

74. Friedrich Ratzel, "Studies in Political Areas: The Political Territory in Relation to Earth and Continent," *American Journal of Sociology* 3, no. 3 (1897): 449–63; Friedrich Ratzel, "Studies in Political Areas II: Intellectual, Political, and Economic Effects of Large Areas," *American Journal of Sociology* 3, no. 4 (1898): 449–63; Friedrich Ratzel, "Studies in Political Areas III: The Small Political Area," *American Journal of Sociology* 4, no. 3 (1898): 366–79.

75. Friedrich Ratzel, "Diary entry 13 July 1895," *Tagebuch 1895, Abschrift angefertigt von Günter Buttmann*, Bayerische Staatsbibliothek, Ratzeliana Supplement 17a VI 13. Note that Ellen Semple translated a collage of passages from *Politische Geographie* rather than full chapters, often giving new meaning to Ratzel's text by shortening sentences and employing unconventional translations of Ratzel's terms, an issue to which I will return in chapter 5. Semple's "Studies in Political Areas: The Political Territory in Relation to Earth and Continent" draws on chapter 13 of *Politische Geographie*, "Studies in Political Areas II" on chapter 14 and "Studies in political areas III: The small political area" on chapter 15 of the same book. A detailed English-language table of contents of *Politische Geographie* was moreover translated and published in *The Professional Geographer* in 1973: Dennis Rumley, Julian Minghi, and Frank M. Grimm, "The Content of Ratzel's Politische Geographie," *The Professional Geographer* 25, no. 3 (1973): 271–77. *Politische Geographie* has also been translated into French and partially into Italian. Friedrich Ratzel, *Géographie politique* (Paris: Éditions régionales européennes, 1897 [1988]); Or Rosenboim, "Geopolitics and Nationalism: Interpreting Friedrich Ratzel in Italian, 1898–1916," in *Radicals and Reactionaries in Twentieth-Century International Thought*, ed. Ian Hall (Basingstoke: Palgrave, 2015), 17–41.

76. Friedrich Ratzel, *Sketches of Urban and Cultural Life in North America* (New Brunswick, NJ: Rutgers University Press, 1876 [1988]).

77. Ratzel, *Lebensraum—A Biogeographical Study*.

78. I am of course responsible for any mistranslations.

79. Withers, "Geography's Evolving Traditions," 317.

80. Ken Lawrence, "The Ku Klux Klan and Fascism," in *The U.S. Antifascism Reader*, ed. Bill V. Mullen and Christopher Vials (London: Verso, 1982 [2020]), 289–99. Robert O. Paxton, *The Anatomy of Fascism* (London: Alfred A. Knopf, 2004). See also Richard Steigmann-Gall, "Star-Spangled Fascism: American Interwar Political Extremism in Comparative Perspective," *Social History* 42, no. 1 (2017): 94–119.

81. David. N. Livingstone, "The Geography of Darwinism," *Interdisciplinary Science Reviews* 31, no. 1 (2006): 34.

82. Ute Wardenga, "Writing the History of Geography: What We Have Learnt—and Where to Go Next," *Geographica Helvetica* 68 (2013): 31.

83. Livingstone, "Geography of Darwinism."

84. Gillian Rose, "Tradition and Paternity: Same Difference?" *Transactions of the Institute of British Geographers* 20, no. 4 (1995): 415.

85. Patricia Owens and Katharina Rietzler, *Women's International Thought: A New History* (Cambridge: Cambridge University Press, 2021).

86. Ratzel's Leipzig papers were first collected by the geographer and explorer Hans Meyer in the Institute for Colonial Geography. They survived the Second World War in a cellar beneath Leipzig University. The Soviet authorities later moved the files to the Institute of Regional Geography. See Ingrid Hönsch, "Friedrich Ratzel's Collected Papers in the Archive for Geography in Leipzig," in *Europe Between Political Geography and Geopolitics: On the Centenary of Ratzel's Politische Geographie*, ed. Marco Antonsich, Vladimir Kolossov, and Maria Paola Pagnini (Rome: Società Geografica Italiana, 2001), 97.

87. Felix Driver, "Hidden Histories Made Visible? Reflections on a Geographical Exhibition," *Transactions of the Institute of British Geographers* 38, no. 3 (2013): 420.

88. Sidaway, "The (Re)making," 77.

89. Joanne P. Sharp, "Publishing American Identity: Popular Geopolitics, Myth and the Reader's Digest," *Political Geography* 12, no. 6 (1993): 491–503; Klaus Dodds, "The 1982 Falklands War and a Critical Geopolitical Eye: Steve Bell and the If . . . Cartoons," *Political Geography* 15, no. 6–7 (1996): 571–92; Jason Dittmer and Nicholas Gray, "Popular Geopolitics 2.0: Towards New Methodologies of the Everyday," *Geography Compass* 4, no. 11 (2010): 1664–77.

90. John Agnew, "By Words Alone Shall We Know: Is the History of Ideas Enough to Understand the World to Which Our Concepts Refer?" *Dialogues in Human Geography* 4, no. 3 (2014): 318.

91. Stuart Elden, " Necessary, But Not Sufficient: Geography, Territory and the History of Ideas," *Dialogues in Human Geography* 4, no. 3 (2014): 320–23.

92. Isabel Hull, *Absolute Destruction: Military Culture and the Practices of War in Imperial Germany* (Ithaca, NY: Cornell University Press, 2005); Kuss, *German Colonial Wars.*

93. Jeff Bowersox, *Raising Germans in the Age of Empire: Youth and Colonial Culture* (Oxford: Oxford University Press, 2013).

94. Michelle R. Moyd, *Violent Intermediaries: African Soldiers, Conquest, and Everyday Colonialism in German East Africa* (Athens, OH: Ohio University Press, 2014).

95. David N. Livingstone, *The Geographical Tradition: Episodes in the History of a Contested Enterprise* (Oxford: Blackwell, 1992).

96. Rose, "Tradition and Paternity," 414.

97. Livingstone, *Geographical Tradition*, 4–5. See also David N. Livingstone, "*The Geographical Tradition* and the Challenges of Geography Geographized," *Transactions of the Institute of British Geographers* 44, no. 3 (2019): 458–62.

98. Ó Tuathail, *Critical Geopolitics*.

99. Ferretti, "Rediscovering other Geographic Traditions"; Kearns, *Geopolitics and Empire*, 78.

100. Barry, "Geography and Other Disciplines," 93.

CHAPTER 2

1. Ratzel, *Glücksinseln und Träume*, 121.

2. Michael Howard, *The Franco-Prussian War* (London: Routledge, 1961 [2000]).

3. Ratzel's only contribution to the topic in the two decades after the war was a very brief foreword to an 1886 book titled *Durch Krieg zum Frieden* (To peace through war). Friedrich Ratzel, "Vorwort," in Karl Stieler, *Durch Krieg zum Frieden: Stimmungsbilder aus den Jahren 1870–1871* (Stuttgart: Bonz und Comp., 1886 [1895]), i–ii.

4. Friedrich Ratzel, "Reste von Vorlesungsmanuskript: Allgemeine politische Geographie und politische Ethnographie," 1890–1894, Leibniz Institut für Länderkunde, Nachlass Friedrich Ratzel 155/317.

5. Hassert notes that Ratzel's account of the war includes descriptions of scenes that happened after his injury in mid-1870 and that he thus could not have personally experienced. Kurt Hassert, "Friedrich Ratzel. Sein Leben und Wirken," *Geographische Zeitschrift* 11, no. 6 (1905): 309.

6. Kurt Hassert, "Friedrich Ratzel. Sein Leben und Wirken," *Geographische Zeitschrift* 11, no. 7 (1905): 379.

7. Karl Lamprecht, "Friedrich Ratzel," *Berichte Über die Verhandlungen der Königlich Sächsischen Gesellschaft der Wissenschaften zu Leipzig*, Sitzung vom 10. Dezember 1904: 260.

8. Karl Haushofer, "Friedrich Ratzel im raumpolitischen Umbruch seiner und unserer Zeit," *Geographische Zeitschrift* 50, no. 3/4 (1944): 81.

9. Ratzel did nonetheless draw a distinction between the realms of war and science and shook his head at others who did not. When one of his colleagues appeared to an academic seminar in full military uniform in the autumn of 1876, he noted his disapproval. Friedrich Ratzel, "Diary entry 11 November 1876," Tagebuch begonnen 5.10.1875, *Abschrift angefertigt von Günter Buttmann*, Bayerische Staatsbibliothek, Ratzeliana Supplement 17a VI 3.

10. Brett Fairbairn, "Economic and Social Developments," in *Imperial Germany, 1871–1918*, edited by James Retallack (Oxford: Oxford University Press, 2008), 69.

11. Christopher Clark, *Kaiser Wilhelm II: A Life in Power* (London: Penguin, 2009).

12. Katherine A. Lerman, "Bismarckian Germany," in *Imperial Germany, 1871–1918*, ed. James Retallack (Oxford: Oxford University Press, 2008), 34.

13. Osterhammel, *Die Verwandlung der Welt*, 132–38.

14. Erik Grimmer-Solem, *Learning Empire: Globalization and the German Quest for World Status, 1875–1919* (Cambridge: Cambridge University Press, 2019).

15. Gerhard Sandner and Mechthild Rössler, "Geography and Empire in Germany, 1871–1945," in *Geography and Empire*, ed. Anne Godlewska and Neil Smith (Oxford: Blackwell, 1994), 115.

16. Sandner and Rössler, "Germany and Empire," 116. For a detailed breakdown, including unfilled and substitute professorships, see Franz-Josef Schulte-Althoff, *Studien zur politischen Wissenschaftsgeschichte der deutschen Geographie* (Paderborn: Schöningh, 1971), 236–40.

17. Ratzel would in later years claim that these interests had nurtured in him a certain shyness. Friedrich Ratzel and Hans Helmolt, "Friedrich Ratzel: Ein Lebensabriß von ihm selbst und vom Herausgeber," in *Kleine Schriften von Friedrich Ratzel, herausgegeben von Hans Helmolt, 1. Band*, ed. Hans Helmolt (Munich: R. Oldenbourg, 1906), xxii. His diaries nonetheless reveal a man with a large circle of friends and acquaintances whom he met regularly. Ratzel may have been keen to emphasize the Romantic in him.

18. Ratzel, *Glücksinseln und Träume*, 25.

19. Ratzel, *Glücksinseln und Träume*, 20.

20. In the 1930s and 1940s, a study conducted by the Kaiser Wilhelm Institute for Genealogy und Demography on the relationship between genius and madness among highly gifted Germans examined Ratzel as a proband. The researchers found a significant history of mental disorders among his extended family. Kaiser-Wilhelm-Institut für Genealogie und Demographie (n.d.) "Akte Friedrich Ratzel" Historisches Archiv des Max-Planck-Instituts für Psychiatrie, MPIP-HA, GDA 560. This chapter's discussion of Ratzel's siblings has been taken from Carl's medical file and information entered into a family tree. See also Florian Mildenberger, "Lebensraum oder Umwelt? Friedrich Ratzel (1844–1904) und Jakob von Uexküll (1864–1944)," in *Jahrbuch für europäische Wissenschaftskultur 5*, ed. Olaf Breidbach and Stefano Poggi (Stuttgart: Steiner, 2009), 124.

21. Ratzel, *Glücksinseln und Träume*, 98.

22. Ratzel, *Glücksinseln und Träume*, 8. See especially his essay "Mein Dorf" (My village) in the same volume.

23. Ratzel, *Glücksinseln und Träume*, 60.

24. Ratzel, *Glücksinseln und Träume*, 85.

25. Friedrich Ratzel, "Beiträge zur Anatomie von Enchytraeus vermicularis Henle," *Zeitschrift für Wissenschaftliche Zoologie* 18 (1867): 99–108; Friedrich Ratzel, "Zur Entwicklungsgeschichte des Regenwurms (Lumbricus Agricola Hoffm.)," *Zeitschrift für Wissenschaftliche Zoologie* 18 (1868): 547–62; Friedrich Ratzel, "Beiträge zur anatomischen und systematischen Kenntnis der Oligochaeten," *Zeitschrift für Wissenschaftliche Zoologie* 18 (1868): 563–91.

26. Friedrich Ratzel, *Sein und Werden: Eine populäre Schöpfungsgeschichte* (Leipzig: Gebhardt und Reisland, 1869).

27. Ratzel and Helmolt, "Friedrich Ratzel," xxiv; Friedrich Ratzel, "Geehrter Herr Professor," Letter to Ernst Haeckel, Carlsruhe, 25 September 1869, transliterated by Peter Zigman, *Würzburger medizinhistorische Mitteilungen* 28: 387–388.

28. The anecdote of the stolen microscope has been much cited as the motivation for Ratzel's move into journalism. This interpretation relies quite strongly on his own account. We may question the degree to which Ratzel was ever fully engaged in the practice of scientific observation. The content and style of his first book suggest that he saw himself primarily as a popularizer of Haeckel's ideas. Indeed, he had in the 1870s keenly reported back to his master about his achievements in spreading the latter's

ideas. The story of the lost microscope may well have served as an excuse to find an occupation that suited his eclecticism, panoramic worldview, and compulsion to write.

29. Ratzel, *Glücksinseln und Träume*, 157; Wilhelm Schilling von Cannstadt, *Das Großherzoglich Badische 5. Infranterie-Regiment, jetzt Königlich Preußische 5. Badische Infranterie-Regiment Nr.113 im Feldzuge 1870–71* (Berlin: Ernst Siegfried Mittler und Sohn, 1876), 30.

30. Schilling von Cannstadt, *Das Großherzoglich Badische*, 26.

31. Howard, *The Franco-Prussian War*, 274.

32. Karl Stieler, *Durch Krieg zum Frieden: Stimmungsbilder aus den Jahren 1870–1871* (Stuttgart: Bonz und Comp., 1895).

33. Ratzel, *Glücksinseln und Träume*, 170.

34. Friedrich Ratzel, "Eine vergessene Stadt in Südfrankreich," in *Kleine Schriften von Friedrich Ratzel, herausgegeben von Hans Helmolt, 1. Band*, ed. Hans Helmolt (Munich: R. Oldenbourg, 1869 [1906]), 10–15.

35. Ratzel, *Glücksinseln und Träume*, 130, 144.

36. Ratzel, *Glücksinseln und Träume*, 135, 150.

37. Ratzel, *Glücksinseln und Träume*, 120. Here and elsewhere, I have left the noun *Volk* in the original German, rather than translate it as "people." The German tends to carry with it the meanings of "nation" or "ethnic group," which the English "people" does not necessarily. Keeping it in the German thus serves as a reminder that Ratzel's usage of *Volk* was embedded within a larger, multifaceted, and internally inconsistent *völkish* discourse, both scholarly and political, which had come to dominate the German-speaking world by the end of the nineteenth century. For a recent discussion thereof, see J. Laurence Hare and Fabian Link, "The Idea of *Volk* and the Origins of *Völkisch* Research, 1800–1930s," *Journal of the History of Ideas* 80, no. 4 (2019): 575–96.

38. Schilling von Cannstadt, *Das Großherzoglich Badische*, 76.

39. Schilling von Cannstadt, *Das Großherzoglich Badische*, 200.

40. Ratzel, *Glücksinseln und Träume*, 192.

41. Buttmann, *Friedrich Ratzel*, 36–37; Cornelia Lüdecke, "Ein genußreiches Zusammenleben und–arbeiten' Friedrich Ratzels Zeit in München (1875–1886)," *Berichte zur Wissenschaftsgeschichte* 25, no. 1 (2002): 25–39.

42. Ratzel was also supported by the Munich palaeontologist Karl von Zittel (1839–1904).

43. Thomas H. Huxley, *Handbuch der Anatomie der Wirbeltiere* (Breslau: Kern, 1873).

44. Moritz Wagner, "Lieber Freund!", Letter to Friedrich Ratzel, Munich, 4 June 1882, Bayerische Staatsbibliothek, Ratzeliana, Supplement 13.

45. Ratzel, *Sketches of Urban and Cultural Life*, 62.

46. Carl O. Sauer, "The Formative Years of Ratzel in the United States," *Annals of the Association of American Geographers* 61 (1971): 245–54.

47. Ratzel, *Glücksinseln und Träume*, 476.

48. Marie was not the first woman Ratzel had sought to marry. He had in 1871 told his mentor Ernst Haeckel that he was engaged to an unnamed sister of his lieutenant. It is not known what happened to this earlier romantic engagement. Friedrich Ratzel,

"Verehrter Herr Professor," Letter to Ernst Haeckel, Petersdorf, 26 May 1871, transliterated by Peter Zigman, *Würzburger medizinhistorische Mitteilungen* 28: 395–97.

49. Buttmann, *Friedrich Ratzel*, 56.

50. Friedrich Ratzel, "Verehrte Freundin!" Letter to Marie Wingens, Boston, 2 November 1873, Bayerische Staatsbibliothek, Ratzeliana, Supplement 10; Friedrich Ratzel, "Diary entry 27 September 1873," *Tagebuch 24.7.1873–4.5.1875, Abschrift angefertigt von Günter Buttmann*, Bayerische Staatsbibliothek, Ratzeliana Supplement 17a VI 1.

51. Friedrich Ratzel, "Verehrte Freundin!" Letter to Marie Wingens, Yosemite, 21 August 1874, Bayerische Staatsbibliothek, Ratzeliana, Supplement 10.

52. Ratzel, *Glücksinseln und Träume*, 43.

53. Friedrich Ratzel, "Verehrte Freundin!" Letter to Marie Wingens, Charleston, 2 February 1874, Bayerische Staatsbibliothek, Ratzeliana Supplement 10. With Marie a devout Catholic and Ratzel at that time doubting the existence of god, their long-distance relationship was also marked by disagreements over questions of belief. These would resurface when the couple debated sending one of their daughters to a monastery, something Friedrich strongly opposed. Friedrich Ratzel, "Diary entry 9 October 1876," *Tagebuch, 5.10.1876–9.3.1878, Abschrift angefertigt von Günter Buttmann*, Bayerische Staatsbibliothek, Ratzeliana Supplement 17a VI 4.

54. Friedrich Ratzel, "Diary entry 23 January 1876," *Tagebuch begonnen 5.10.1875, Abschrift angefertigt von Günter Buttmann*, Bayerische Staatsbibliothek, Ratzeliana Supplement 17a VI 3.

55. Buttmann, *Friedrich Ratzel*.

56. Friedrich Ratzel, "Diary entry 9 March 1876," *Tagebuch begonnen 5.10.1875, Abschrift angefertigt von Günter Buttmann*, Bayerische Staatsbibliothek, Ratzeliana Supplement 17a VI 3.

57. An 1897 letter to Karl Lamprecht also mentions her "nervous cough." Friedrich Ratzel, "Lieber Freund!" Letter to Karl Lamprecht, Leipzig, 29 January 1897, Universitäts- und Landesbibliothek Bonn, Nachlass Lamprecht, S2714 Korr. 44.

58. Friedrich Ratzel, "Liebe Hedwig!" Letter to Hedwig Ratzel, Frankfurt, 14 October 1895, Bayerische Staatsbibliothek, Ratzeliana Supplement 10.

59. The Polytechnische Schule München was soon to be renamed the Technische Universität München.

60. Friedrich Ratzel, *Die chinesische Auswanderung: Ein Beitrag zur Cultur- und Handelsgeographie* (Breslau: J. U. Kern, 1876).

61. Ratzel and Helmolt, "Friedrich Ratzel," xxiv.

62. Hassert, "Friedrich Ratzel I," 312.

63. Smith, *Politics and the Sciences*, 206.

64. Friedrich Ratzel, "Lieber Kollege!" Letter to Kurt Hassert, Leipzig, 6 January 1902, Archiv für Geographie, Leibniz Institut für Länderkunde, Nachlass Friedrich Ratzel 163/657. For a discussion of the institute's history during the Ratzel years, see Buttmann, *Friedrich Ratzel*, 73–83.

65. The others were Ferdinand von Richthofen, Albrecht Penck, and Alfred Hettner. Ute Wardenga, "Friedrich Ratzel zum 100. Todestag am 9. August 2004," In *Jubiläen 2004. Personen-Ereignisse*, edited by Rektor der Universität Leipzig, 47–51. Leipzig: Universität Leipzig, 2004, 47.

66. Friedrich Ratzel, "Lieber Freund!" Letter to Rudolph Oldenbourg, Leipzig, 10 February 1900, Bayerische Staatsbibliothek, Ratzeliana Supplement 10. See also Müller, *Friedrich Ratzel*, 60.

67. Smith, *Politics and the Sciences*, 204.

68. Lesser known figures of this loose network include the economists Karl Bücher and Wilhelm Roscher, as well as the Anglicist Richard Wülker and the theologian Rudolf Kittel. Buttmann, *Friedrich Ratzel*, 74.

69. Smith, *Politics and the Sciences*, 204, 209.

70. Wardenga, "Friedrich Ratzel," 49. On Ratzel and Lamprecht see Roger Chickering, *Karl Lamprecht: A German Academic Life (1856–1915)* (Atlantic Highlands, NJ: Humanities Press, 1993). Ratzel treated Lamprecht as an ally in Leipzig university politics and spoke frankly about his colleagues' shortcoming with him. Friedrich Ratzel, "Verehrter Freund u. Kollege," Letter to Karl Lamprecht, Leipzig, 10 July 1893, Universitäts- und Landesbibliothek Bonn, Nachlass Lamprecht, S 2713: Korr. 44; Friedrich Ratzel, "Lieber Freund!" Letter to Karl Lamprecht, Leipzig, 18 July 1899, Universitäts- und Landesbibliothek Bonn, Nachlass Lamprecht, S 2713 UL 4: 3: 28.

71. Friedrich Ratzel, "Diary entry 29 April 1895," *Tagebuch 1895, Abschrift angefertigt von Günter Buttmann*, Bayerische Staatsbibliothek, Ratzeliana Supplement 17a VI 13. The young Ratzel had in 1867 attended lectures by Wundt. A letter from 1887 reveals Ratzel's eagerness to develop a friendship with the latter. Friedrich Ratzel, "Verehrter Herr Kollege!" Letter to Wilhelm Wundt, 16 August 1902, Ammerland, Universitätsarchiv Leipzig, Nachlass Wilhelm Wundt; NA Wundt/III/1301–1400/1375/339–342; Friedrich Ratzel, "Hochverehrter Herr Kollege!" Letter to Wilhelm Wundt, 6 February 1887, Leipzig, Universitätsarchiv Leipzig, Nachlass Wilhelm Wundt; NA Wundt/III/1601–1700/1601/187/489–490.

72. Friedrich Ratzel, "Diary entry 8 November 1875," *Tagebuch begonnen 5.10.1875, Abschrift angefertigt von Günter Buttmann*, Bayerische Staatsbibliothek, Ratzeliana Supplement 17a VI 3.

73. Max Eckert, "Friedrich Ratzel als politischer Lehrer," *Geographische Zeitschrift* 28, no. 7/8 (1922): 254.

74. Hassert, "Friedrich Ratzel II," 372; Keighren, *Bringing Geography to Book*, 16; J. Russell Smith, "Geography in Germany. II. The University," *Journal of Geography* 1, no. 10 (1902): 449. Ratzel also supervised a large number of dissertations, thirteen of which were published in 1899 alone. Eckert, "Friedrich Ratzel," 258–89.

75. Friedrich Ratzel, "Liebe Loll!" Letter to Lila Ratzel, Leipzig, 24 May 1902, Bayerische Staatsbibliothek, Ratzeliana Supplement 10.

76. Hassert, "Friedrich Ratzel II," 365. In his published work, Ratzel did not make great use of cartography either, although he did produce some new maps of Africa in 1885 that sought to establish the relationship between different types of civilization (nomadic, agricultural) and state development. Friedrich Ratzel, "Entwurf einer neuen politischen Karte von Afrika. Nebst einigen allgemeinen Bemerkungen über die Grundsätze der politischen Geographie," *Petermanns Mitteilungen* 31 (1885): 245–50. As correspondence with his publisher shows, he took a keen interest in the illustrations of his *Völkerkunde*. F. Heger, "Hochgeehrter Herr Professor," Letter to Friedrich Ratzel, Vienna, 2 May 1894, Archiv für Geographie, Leibniz Institut für Länderkunde, Nachlass Friedrich Ratzel 163/658.

77. Eckert, "Friedrich Ratzel," 258. Ratzel corresponded with a large number of scientists, explorers, politicians, diplomats, journalists, publishers, poets, and writers. The list includes Germanophone geographers like Alfred Hettner, Ernst Kapp, Eugen Oberhummer, Albrecht Penck, Ferdinand von Richthofen, Moritz Wagner, many of his former students, as well a number of German politicians, such as the conservative statesman Arthur von Posadowsky-Wehner. Ratzel also exchanged letters with international interlocutors, such as Simion Mehedinţi, a Romanian geographer with a degree from Leipzig and later supporter of the Romanian iron guard, the American geographer Ellen Churchill Semple, the French geographer Vidal de la Blache (correspondence with the latter is mentioned in Ratzel's diary, though is not included in the *Nachlaß*) and Emil Preetorius, a leader of the German-American community. I discuss his correspondence with German colonial administrators and explorers in chapter 5.

78. Friedrich Ratzel, "Diary entry 15 January 1876," *Tagebuch begonnen 5.10.1875, Abschrift angefertigt von Günter Buttmann*, Bayerische Staatsbibliothek, Ratzeliana Supplement 17a VI 3.

79. Friedrich Ratzel, "Diary entry 11 April 1895," *Tagebuch 1895, Abschrift angefertigt von Günter Buttmann*, Bayerische Staatsbibliothek, Ratzeliana Supplement 17a VI 13.

80. Friedrich Ratzel, "Lieber Kollege!" Letter to K. Hassert, Leipzig, 8 January 1903, Archiv für Geographie, Leibniz Institut für Länderkunde, Nachlass Friedrich Ratzel 163/657.

81. The cartographer Richard Lüdecke, for instance, reported back to Ratzel in 1892 that he had argued with Albrecht Penck on Ratzel's behalf about a map of Africa. R. Lüdecke, "Sehr geehrter Herr Professor!", Letter to Friedrich Ratzel, Gotha, 29 August 1892, Archiv für Geographie, Leibniz Institut für Länderkunde, Nachlass Friedrich Ratzel 155/319.

82. Friedrich Ratzel, "Lieber Freund!" Letter to Rudolph Oldenbourg, Leipzig, 10 February 1900, Bayerische Staatsbibliothek, Ratzeliana Supplement 10.

83. Friedrich Ratzel, "Lieber Kollege!" Letter to Kurt Hassert, Leipzig, 31 July 1899, Archiv für Geographie, Leibniz Institut für Länderkunde, Nachlass Friedrich Ratzel 163/657; Friedrich Ratzel "Diary entry 29 July 1902," *Tagebuch 4.7.1900–19.7.1903, Abschrift angefertigt von Günter Buttmann* Bayerische Staatsbibliothek, Ratzeliana Supplement 17a VI 20.

84. Karl Weule, "Friedrich Ratzel," *Mitteilungen des Vereins für Erdkunde zu Leipzig* (1904): 9.

85. Friedrich Ratzel, "Lieber Freund!" Letter to Rudolph Oldenbourg, Leipzig, 17 February 1901, Bayerische Staatsbibliothek, Ratzeliana Supplement 10.

86. Friedrich Ratzel, "Lieber Freund!" Letter to Rudolph Oldenbourg, Leipzig, 10 February 1900, Bayerische Staatsbibliothek, Ratzeliana Supplement 10. On Ratzel's auto-critique of some of his other works, see Müller, *Friedrich Ratzel*, 74–77.

87. Mildenberger, "Lebensraum oder Umwelt?"

88. Friedrich Ratzel "Diary entry 17 December 1876," *Tagebuch, 5.10.1876–9.3.1878, Abschrift angefertigt von Günter Buttmann*, Bayerische Staatsbibliothek, Ratzeliana Supplement 17a VI 4.

89. Friedrich Ratzel. (1874) "Verehrte Freundin!" Letter to Marie Wingens, Yosemite, 21 August 1874, Bayerische Staatsbibliothek, Ratzeliana, Supplement 10.

90. M. Ratzel, "Sehr geehrter Herr Professor!" Letter to unknown, Ammerland, 26 January 1905, Archiv für Geographie, Leibniz Institut für Länderkunde, Nachlass Friedrich Ratzel 163/657.

91. Weule notes that Ratzel was an incessant note taker who recorded not just his own thoughts but those of others. His manuscripts then emerged almost organically from a loose collection of ideas, only some of which were his own. Weule, "Friedrich Ratzel," 18.

92. "Gummiräder," *Die Grenzboten* 53 (1894): 110–111; "Unsere Volkstrachten." *Die Grenzboten* 56 (1896): 357–64; "Brutal." *Die Grenzboten* 56 (1897): 444.

93. "Unsere Bahnhofskneipen." *Die Grenzboten* 52 (1893): 190.

94. He mentioned his anger when in 1875 one of his fellow travelers in Mexico questioned whether his restlessness and ambition would ever allow him to realize his dream of settling down. Ratzel curiously responded that if his mother were to die, his drive would be extinguished. And while he did attempt to retreat to the countryside after his mother's death in 1888, he remained highly productive. Friedrich Ratzel, "Diary entry 17 February 1874," *Reisetagebuch 24.7.1873–4.5.1875, Abschrift angefertigt von Günter Buttmann*, Bayerische Staatsbibliothek, Ratzeliana Supplement 17a VI 2a.

95. Friedrich Ratzel, "Diary entry 9 August 1903," *Tagebuch 20.7.1903–9.8.1904, Abschrift angefertigt von Günter Buttmann*, Bayerische Staatsbibliothek, Ratzeliana Supplement 17a VI 21.

96. Friedrich Ratzel, "Diary entry 9 August 1904," *Tagebuch 20.7.1903–9.8.1904, Abschrift angefertigt von Günter Buttmann*, Bayerische Staatsbibliothek, Ratzeliana Supplement 17a VI 21.

97. Hassert, "Friedrich Ratzel I," 305.

98. Hans Helmolt, *Zu Friedrich Ratzels Gedächtnis. Geplant als Festschrift zum 60. Geburtstag, nun als Grabspende dargebracht von Fachgenossen und Schülern, Freunden und Verehrern* (Leipzig: Seele, 1904).

99. Lamprecht, "Friedrich Ratzel"; Steinmetzler, *Anthropogeographie Friedrich Ratzels*.

100. Steinmetzler, *Anthropogeographie Friedrich Ratzels*, 70.

101. Friedrich Ratzel, "Diary entry 5 February 1897," *Tagebuch 4.10.1896–6.1.1898, Abschrift angefertigt von Günter Buttmann*, Bayerische Staatsbibliothek, Ratzeliana Supplement 17a VI 15a.

102. Friedrich Ratzel, "Lieber Freund," Letter to Rudolph Oldenbourg, Leipzig, 22 February 1899, Bayerische Staatsbibliothek, Ratzeliana Supplement 10; also Müller, *Friedrich Ratzel*. Ratzel is sometimes credited with having coined the term "biogeography" itself, but see Barua, *Ratzel, Biogeography*.

103. Ratzel, *Anthropogeographie Erster Teil*; Ratzel, *Anthropogeographie Zweiter Teil*; Ratzel, *Politische Geographie*; Ratzel, *Sein und Werden*; Friedrich Ratzel, *Deutschland: Eine Einführung in die Heimatkunde* (Leipzig: Wilhelm Grunow, 1898); Ratzel, *Meer als Quelle*.

104. Friedrich Ratzel, *Wandertage eines Naturforschers. 1. Theil: Zoologische Briefe vom Mittelmeer. Briefe aus Süditalien* (Leipzig: Brockhaus, 1873); Friedrich Ratzel, *Wandertage eines Naturforschers. 2. Theil: Schilderungen aus Siebenbürgen und den Alpen* (Leipzig: Brockhaus, 1874); Ratzel, *Sketches of Urban and Cultural Life*; Ratzel, *Die chinesische Auswanderung*; Friedrich Ratzel, *Aus Mexico: Reiseskizzen aus den*

Jahren 1874 und 1875 (Stuttgart: Brockhaus, 1878 [1969]); Friedrich Ratzel, *Die Vereinigten Staaten von Nord-Amerika, Erster Band, Physikalische Geographie und Naturcharakter* (Munich: R. Oldenbourg, 1878); Friedrich Ratzel, *Die Vereinigten Staaten von Nord-Amerika, Zweiter Band: Culturgeographie unter besonderer Berücksichtigung der wirthschaftlichen Verhältnisse* (Munich: R. Oldenbourg, 1880); Friedrich Ratzel, *Die Vorgeschichte des europäischen Menschen* (Munich: R. Oldenbourg, 1874). Friedrich Ratzel, *Die Erde in vierundzwanzig gemeinverständlichen Vorträgen über Allgemeine Erdkunde* (Stuttgart: J. Engelhorn, 1881); Ratzel, *History of Mankind, I, II, III*; Ratzel, *Die Erde und das Leben I, II*. His ethnology, *Völkerkunde*, which will be examined in chapters 3 and 7, was influential in anthropology but drew extensively on the research of others. *Die Erde und das Leben*, which I discuss in chapter 4, was an ambitious attempt to fuse his various interests and write a panoramic work of all geography. It largely repeated, sometimes verbatim, ideas he had developed in *Anthropogeographie*.

105. Friedrich Ratzel, *Über Naturschilderung* (Munich: R. Oldenbourg, 1904 [1923]); Friedrich Ratzel, *Frommanns Taschenbuch für Fussreisende, herausgegeben und ergänzt von Dr. Friedrich Ratzel* (Stuttgart: Frommanns, 1880); Georg August Schweinfurth and Friedrich Ratzel, *Emin-Pascha: Eine Sammlung von Reisebriefen und Berichten Dr. Emin Paschas aus den ehemals ägyptischen Äquatorialprovinzen und deren Grenzländern* (Leipzig: F. A. Brockhaus, 1888).

106. Ratzel, "Territorial Growth"; Ratzel, "Lebensraum—A Biogeographical Study"; Ratzel, "Nationalitäten und Rassen."

107. E. G. R., "Obituary: Friedrich Ratzel," *Geographical Journal* 24, no. 4 (1904): 485–87.

108. Friedrich Ratzel, *Die Schneedecke, besonders in Gebirgen* (Stuttgart: Engelhorn, 1889); Friedrich Ratzel, "Die Kant-Lapclacesche Hypothese und die Geographie," In *Kleine Schriften von Friedrich Ratzel, herausgegeben von Hans Helmolt, 2. Band*, ed. Hans Helmolt (Munich: R. Oldenbourg, 1901 [1906]), 420–36. See also Hassert, "Friedrich Ratzel II," 361–362.

109. Friedrich Ratzel, *Anthropogeographie Zweiter Teil, Dritte Auflage: Die geographische Verbreitung des Menschen* (Stuttgart: J. Engelhorn, 1891 [1912]), xix.

110. Ratzel, *Anthropogeographie Erster Teil, Zweite Auflage*, 23.

111. Ratzel, *Sketches of Urban and Cultural life*, 72.

112. Ratzel, *Anthropogeographie Zweiter Teil, Zweite Auflage*, 76.

113. Natter, "Friedrich Ratzel's Spatial Turn," 177.

114. In 1899, Ratzel published a second edition of the first volume only. He did not find the need to update the second volume given that it had only appeared eight years earlier (the first volume had been published seventeen years earlier). Posthumous reprints of both volumes followed in 1909/12 and 1921/22. For a discussion of Ratzel's history of the concept of the *Umwelt* (environment), see Philip Conway, "The Historical Ontology of Environment: From the Unity of Nature to the Birth of Geopolitics" (PhD diss., University of Aberystwyth, 2019).

115. The term *Anthropogeographie* had already been in circulation since the early nineteenth century, which Ratzel and many of his contemporaries seem to have been unaware of. Müller, *Friedrich Ratzel*, 185.

116. "Geographische Handbücher," *Die Grenzboten* 48 (1889): 332. See also Ger-

hard H. Müller, "Zur Geschichte des Begriffs 'Anthropogeographie,'" *Geographische Zeitschrift* 80, no. 3 (1992): 185.

117. "Geographische Handbücher," 333.

118. Ratzel, *Anthropogeographie Erster Teil, Zweite Auflage*, 90.

119. Völker and States were "aggregate organisms" in *Anthropogeographie* and organisms whose levels of spiritual organization made them far more complex and even superior to other types of organisms. Ratzel, *Anthropogeographie Erster Teil, Zweite Auflage*, 2. Ratzel operated with different concepts of the *Volk*, one ethnic-cultural, the other statist, but he did not seem to mind the tensions between them. Hans-Dietrich Schultz, "'Eines schickt sich nicht für alle': Rassen, Völker und Nationen im länderkundlichen Denken Friedrich Ratzels," in *Prisma und Spiegel: Ethnologie zwischen postkolonialer Kritik und Deutung der eigenen Gesellschaft*, ed. Dorothea E. Schulz and Jochen Seebode (Hamburg: Argument, 2010), 121.

120. Ratzel, *Anthropogeographie Erster Teil, Zweite Auflage*, 65.

121. Ratzel, *Anthropogeographie Erster Teil, Zweite Auflage*, 41. He defined the environment (*Umwelt*) most succinctly in a lecture manuscript as "nature external to the living thing, which extends into the organism through light, air and sustenance, but is not itself part of this organism." Quoted in Steinmetzler, *Friedrich Ratzel*, 52.

122. Ratzel, *Anthropogeographie Erster Teil, Zweite Auflage*, 539.

123. Ratzel, *Anthropogeographie Zweiter Teil, Dritte Auflage*, 174. As Ratzel acknowledged with the phrase "old observation," this insistence on the influence of climate on national character was hardly his own invention. As Livingstone notes, "[t]he idea that climatic regions on both global and local scales found expression in what might be called an ethnic moral topography snakes its way throughout the corpus of nineteenth- and twentieth-century geographical writings," in Britain and North America as much as in Germany. David N. Livingstone, "Climate's Moral Economy: Science, Race and Place in Post-Darwinian British and American Geography," in *Geography and Empire*, ed. Anne Godlewska and Neil Smith (Oxford: Blackwell, 1994), 138.

124. Ratzel, *Anthropogeographie Erster Teil, Zweite Auflage*, 60.

125. Ratzel, *Anthropogeographie Erster Teil, Zweite Auflage*, 132.

126. Ratzel, *Anthropogeographie Zweiter Teil, Dritte Auflage*, 404–5.

127. Carl Ritter, *Ueber das historische Element in der geographischen Wissenschaft* (Berlin: Königliche Akademie der Wissenschaften, 1834), 155.

128. Friedrich Ratzel, *Politische Geographie: Geographie der Staaten, des Verkehrs und des Krieges* (Munich: R. Oldenbourg, 1897 [1903]), v. Unlike Weikart suggests, German colonial expansion was not a hidden aim in Ratzel's *Politische Geographie*, which he masked behind subtle formulations and only confessed to close friends. It was spelled out as clearly in this book as it was in many of his nonacademic writings. Richard Weikart, *From Darwin to Hitler: Evolutionary Ethics, Eugenics, and Racism in Germany* (Basingstoke: Palgrave, 2004), 193–94.

129. Ratzel, *Reste von Vorlesungsmanuskript.*

130. Heinrich Hertzberg, "Friedrich Ratzel's politische Geographie (Fortsetzung)," *Geographische Zeitschrift* 4, no. 8 (1898): 459, 463.

131. Ratzel belonged to the broad church of German Darwinism, which united a number of political streams and whose members often referred to themselves as *Dar-*

winisten, a term used even by those whose ideas deviated substantially from Darwin. Richard J. Evans, "In Search of German Social Darwinism: The History and Historiography of a Concept," in *Medicine and Modernity: Public Health and Medical Care in Nineteenth- and Twentieth-Century Germany*, ed. Manfred Berg and Geoffrey Cocks (Cambridge: Cambridge University Press, 1997), 55–89; Paul Weidling, "Ernst Haeckel, Darwinismus and the Secularization of Nature," in *History, Humanity and Evolution: Essays for John C. Greene*, ed. John R. Moore (Cambridge: Cambridge University Press, 1989), 311–53. As chapter 4 will show, the most significant influences on Ratzel's evolutionary worldview were Moritz Wagner, Ernst Haeckel, Charles Darwin, and Herbert Spencer (roughly in that order), although there were certainly Lamarckian logics in his writings too, imported in particular via his reading of Wagner and Spencer. I use the term "Darwinian" rather than "social Darwinian" throughout this book in light of the fact that both Darwin and Haeckel had themselves taken an interest in what were ostensibly social and historical processes and that "social Darwinism" was a term used by its opponents. None of this is to say that Ratzel did not perform a crucial transfer from biological to sociopolitical theory, but rather to refuse an overly easy distinction between "scientific" evolutionary theory (as applied to the natural world) on the one hand, and "pseudoscientific" evolutionary theory (as applied to human societies) on the other.

132. Ratzel, *Politische Geographie: Geographie der Staaten*, 4.

133. Ratzel, *Politische Geographie: Geographie der Staaten*, 13.

134. Ratzel, *Politische Geographie: Geographie der Staaten*, 13.

135. Ratzel, *Politische Geographie: Geographie der Staaten*, 8.

136. Andrew Reynolds, "Ernst Haeckel and the Theory of the Cell State: Remarks on the History of a Bio-political Metaphor," *History of Science* 46, no. 2 (2008): 123–52.

137. Ratzel made a number of subtle structural changes to the 1897 first edition, which had featured twenty-eight chapters organized into nine sections. More substantively, Ratzel added passages on war and transport. He acknowledged in the foreword to the 1903 edition that he had come to realize how fundamentally entwined the two were with the questions his political geography sought to address. Ratzel, *Politische Geographie: Geographie der Staaten*, vi.

138. Peter Schöller, "Wege und Irrwege der politischen Geographie und Geopolitik," *Erdkunde* 11, no. 1 (1957): 2.

139. Ratzel, *Politische Geographie: Geographie der Staaten*, 27.

140. Ratzel, *Politische Geographie: Geographie der Staaten*, 72.

141. Ratzel, *Politische Geographie: Geographie der Staaten*, 387.

142. Ratzel, *Politische Geographie: Geographie der Staaten*, 541.

143. Ratzel, *Politische Geographie: Geographie der Staaten*, 593.

144. Ratzel, *Politische Geographie: Geographie der Staaten*, 363.

145. Otto Kaemmel, "Friedrich Ratzel," *Die Grenzboten* 63 (1904): 420–21; Friedrich Ratzel, *Wider die Reichsnörgler: Ein Wort zur Kolonialfrage aus Wählerkreisen* (Munich: Oldenbourg, 1884), 3.

146. Kaemmel, "Friedrich Ratzel." There seems to have been some speculation among American exchange students that Kaiser Wilhelm II had visited one of Ratzel's lectures in 1902. Keighren, *Bringing Geography to Book*, 16. But while Ratzel found it

worthy of a note in his diary that he had seen the Kaiser ride past at a distance on one occasion in 1895, he made no note of a visit in 1902. Friedrich Ratzel, "Diary entry 26 October 1895," *Tagebuch 1895, Abschrift angefertigt von Günter Buttmann*, Bayerische Staatsbibliothek, Ratzeliana Supplement 17a VI 13.

147. Kaemmel, "Friedrich Ratzel," 421; see also Buttmann, *Friedrich Ratzel*, 54.

148. Buttmann, *Friedrich Ratzel*, 116.

149. Friedrich Ratzel, "Bismarck und die Sueddeutschen," unpublished manuscript, Archiv für Geographie, Leibniz Institut für Länderkunde, Nachlass Friedrich Ratzel 160/476.

150. "Europa," In *Brockhaus Konversationslexikon, 14. Auflage, Band VI*, no editor (Leipzig: Brockhaus, 1893), 434.

151. Friedrich Ratzel, "Lieber Freund," Letter to Rudolph Oldenbourg, S. Florent, 7 April 1899, Bayerische Staatsbibliothek, Ratzeliana Supplement 10.

152. "Zur Kenntnis der englischen Weltpolitik: 6. England in Südafrika," *Die Grenzboten* 54 (1895): 7–21.

153. "Die geographische Lage Deutschlands," *Die Grenzboten* 55 (1896): 390–97; 450–56.

154. Bascom Barry Hayes, *Bismarck and Mitteleuropa* (London: Associated University Presses, 1994), 107. See also Fritz Werner, "Die Grenzboten: Aus der Geschichte einer achtzigjährigen Zeitschrift nationaler Bedeutung," *Die Grenzboten* 81 (1922): 450.

155. Friedrich Ratzel, "Lieber Herr Doktor!" Letter to Kurt Hassert, Leipzig, 2 August 1898, Archiv für Geographie, Leibniz Institut für Länderkunde, Nachlass Friedrich Ratzel 163/657.

156. Eckert, "Friedrich Ratzel," 256.

157. "Verkehrter Bismarckkultus," *Die Grenzboten* 52 (1893): 141–42.

158. Ratzel, *Glücksinseln und Träume*, 119.

159. Friedrich Ratzel, "Diary entry 4 April 1895," *Tagebuch 1895, Abschrift angefertigt von Günter Buttmann*, Bayerische Staatsbibliothek, Ratzeliana Supplement 17a VI 13.

160. Ratzel, "Nationalitäten und Rassen," 465.

161. Ratzel, "Studies in Political Areas III," 6.

162. Schulte-Althoff, *Studien zur politischen Wissenschaftsgeschichte*, 195.

163. While Ratzel is often counted to the "core group" of the early pan-German movement, it should be noted that the executive board (*Gesamtvorstand*) had a membership of more than seventy members. The overall membership of the organization stood at twenty thousand in 1891. Arne Perras, *Carl Peters and German Colonialism 1856–1918: A Political Biography* (Oxford: Oxford University Press, 2004), 174; Lothar Werner, *Der Alldeutsche Verband, 1890–1918* (Berlin: E. Ebering, 1935), 32.

164. Alfred Kruck, *Geschichte des Alldeutschen Verbandes 1890–1939* (Wiesbaden: Franz Steiner, 1954), 19.

165. Kruck, *Geschichte des Alldeutschen Verbandes*.

166. "Deutschland, wach' auf!" *Deutsche Kolonialzeitung* 15 (1890): 184.

167. Roger Chickering, *We Men Who Feel Most German: A Cultural Study of the Pan-German League, 1886–1914* (Boston: Allen & Unwin, 1984), 101.

168. Chickering, *We Men Who Feel Most German*, 76; Hans-Dietrich Schultz, "Fantasies of *Mitte*: *Mittellage* and *Mitteleuropa* in German Geographical Discussion in the 19th and 20th century," *Political Geography Quarterly* 8, no. 4 (1989): 324.

169. "In welcher Richtung beeinflussen die afrikanischen Ereignisse die Tätigkeit des Kolonialvereins?" *Deutsche Kolonialzeitung* 2 (1885): 38–44; Friedrich Ratzel, "Verehrter Herr College!" Letter to Ernst Hasse, Leipzig, 6 August 1889, Archiv für Geographie, Leibniz Institut für Länderkunde, Nachlass Ernst Hasse 437/23.

170. Arendt, *The Origins of Totalitarianism*.

171. On the ideological connections and conflicts over political style between the Pan-German League and national socialist movement in the 1920s, see Chickering, *We Men Who Feel Most German*, 299–301.

172. Roger Griffin, *Fascism* (Cambridge: Polity, 2018).

173. Jureit, "Mastering Space," 85.

174. Friedrich Ratzel, "Weiße und Farbige im indischen Archipelagus," *Globus* 16 (1869): 377.

175. Smith, *Politics and the Sciences*, 220.

176. Bassin, "Imperialism and the Nation State," 483; see also Grimmer-Solem, *Learning Empire*, 59; Guettel, *German Expansionism*.

177. Moritz Wagner, "Lieber Freund!," Letter to Friedrich Ratzel, Munich, 4 June 1882, Bayerische Staatsbibliothek, Ratzeliana, Supplement 13.

178. Ratzel's position here was that particularly rigid patriarchal orders, such as those common in the Iberian Peninsula and in former Spanish colonies, effectively effeminized the male population. Ratzel was not in favor of an overturning of gender hierarchies per se, but of a mild form of patriarchy in which women were able to enter the workforce but in which gendered identities were not fundamentally disturbed.

179. Friedrich Ratzel, "Die Beurtheilung der Völker," *Nord und Süd* 6 (1878): 177–200.

180. Ratzel, *Anthropogeographie Erster Teil, Zweite Auflage*, 73.

181. Friedrich Ratzel, "Diary entry 14 September 1873," *Tagebuch 24.7.1873–4.5.1875, Abschrift angefertigt von Günter Buttmann*, Bayerische Staatsbibliothek, Ratzeliana Supplement 17a VI 1.

182. Ratzel, *Aus Mexico*, 178.

183. Ratzel, *Glücksinseln und Träume*, 200.

184. Ratzel, *Die Erde*, 338.

185. Friedrich Ratzel, "Emin Pascha: Eine Darstellung seines Lebens und Wirkens mit Benutzung seiner Tagebücher, Briefe und wissenschaftlichen Aufzeichungen von Georg Schweitzer," *Geographische Zeitschrift* 4, no. 5 (1898): 302.

186. Ratzel, *Vereinigten Staaten, Erster Band*, 186.

187. "In welcher Richtung," 39.

188. Founded in 1882, the Deutscher Kolonialverein merged in 1887 with the Gesellschaft für deutsche Kolonisation which had been founded by Carl Peters in 1884 to promote settler colonization more aggressively, becoming the influential Deutsche Kolonialgesellschaft. Smith, *The Ideological Origins*, 33, 38.

189. Weule, "Friedrich Ratzel," 24.

190. On von Richthofen, see Jürgen Zimmerer, "In the Service of Empire: Geogra-

phers at Berlin's University between Colonial Studies and Ostforschung (*Eastern Research*)," In *Hitler's Geographies: The Spatialities of the Third Reich*, ed. Paolo Giaccaria and Claudio Minca (Chicago: Chicago University Press, 2016), 67–92.

191. Sandner and Rössler, "Geography and empire," 120.

192. Kearns, *Geopolitics and Empire.*

193. Smith, *Politics and the Sciences*, 153.

194. "Eine 'Ehrung' für Virchow." *Die Grenzboten* 53 (1894): 335–36.

195. Smith, *Politics and the Sciences*, 11.

196. "Der Verschönerungsverein." *Die Grenzboten* 52 (1893): 574–75.

197. Ratzel, *Vereinigten Staaten, Erster Band*, 180.

198. Ratzel, *Glücksinseln und Träume*, 194; see also "Weltentwicklung und Weltschöpfung. Mit einem Anhang über Lyells und Darwins Gottesideen," *Die Grenzboten* 61 (1902): 669–84.

199. Smith, "Friedrich Ratzel," 57.

200. Ratzel, *Aus Mexico*, 268.

201. Ratzel, *Anthropogeographie Erster Teil, Zweite Auflage*, 2; Ernst Haeckel, *The history of creation Vol. 1* (London: Henry S. King, 1876), 6.

202. Ratzel, *Glücksinseln und Träume*, 201.

CHAPTER 3

1. Although the concept of settler colonialism was first used to denote forms of colonialism that depended on permanent settlements and an ideology of white supremacy in the Canadian and Australian context, it can usefully be applied to the United States. Anne Bonds and Joshua Inwood, "Beyond White Privilege: Geographies of White Supremacy and Settler Colonialism," *Progress in Human Geography* 40, no. 6 (2016): 715–33. On the particular iterations of settler colonialism in Latin America, see M. Bianet Castellanos, "Introduction: Settler Colonialism in Latin America," *American Quarterly* 69, no. 4 (2017): 777–81.

2. Bassin, "Race contra Space"; Buttmann, *Friedrich Ratzel*; Natter, "Friedrich Ratzel's Spatial Turn,"; Smith, "Friedrich Ratzel."

3. Hans-Dietrich Schultz, "Friedrich Ratzel: (k)ein Rassist?" *Geographische Hochschulmanuskripte* 2 (2006): 1–46.

4. Hans-Dietrich Schultz, "Herder und Ratzel: Zwei Extreme, ein Paradigma?" *Erdkunde* 52, no. 2 (1998): 127.

5. Smith, *The Ideological Origins.*

6. Arthur de Gobineau, *The Inequality of Human Races* (London: William Heinemann. 1852 [1915]).

7. Houston Stewart Chamberlain, *Die Grundlagen des neunzehnten Jahrhunderts* (Munich: Bruckmann, 1899 [1912]).

8. Chamberlain, *Die Grundlagen*, 453–54; see also Hans-Dietrich Schultz, "'Rasse' und Raum von Ritter zu Ratzel," in *'Rasse' und Raum: Topologien zwischen Kolonial-, Geo- und Biopolitik: Geschichte, Kunst, Erinnerung*, ed. Claudia Bruns (Wiesbaden: Reichert, 2017), 237.

9. Ratzel, "Nationalitäten und Rassen," 486; see also Friedrich Ratzel, "Geschichte,

Völkerkunde und historische Perspektive," in *Kleine Schriften von Friedrich Ratzel, herausgegeben von Hans Helmolt, 2. Band*, ed. Hans Helmolt (Munich: R. Oldenbourg, 1904 [1906]), 496.

10. Ratzel, "Nationalitäten und Rassen," 472.

11. Smith, "Friedrich Ratzel," 54.

12. Bassin, "Race contra Space," 119; see also Murphy, *Heroic Earth*, 11.

13. Natter, "Friedrich Ratzel's Spatial Turn," 184.

14. Gertjan Dijkink, "Ratzel's *Politische Geographie* and Nineteenth-Century German discourse," in *Europe Between Political Geography and Geopolitics: On the Centenary of Ratzel's Politische Geographie*, ed. Marco Antonsich, Vladimir Kolossov, and Maria Paola Pagnini (Rome: Società Geografica Italiana, 2001), 116.

15. Kaemmel, "Friedrich Ratzel," 421.

16. Schultz, "Friedrich Ratzel—Pionier moderner Sozialgeographie?" 61.

17. Bassin, "Race contra Space," 121.

18. Ratzel, "Nationalitäten und Rassen," 463.

19. Ratzel, "Nationalitäten und Rassen," 464.

20. Ratzel, "Nationalitäten und Rassen," 485.

21. Schultz, "Friedrich Ratzel: (k)ein Rassist?" 9, 24.

22. Robert Bernasconi, "Will the Real Kant Please Stand Up," *Radical Philosophy* 117 (2003): 13–23; Stuart Elden, "Reassessing Kant's Geography," *Journal of Historical Geography* 35, no. 1 (2009): 3–25; Saidiya Hartman, *Scenes of Subjection: Terror, Slavery, and Self-Making in Nineteenth-Century America* (New York: Oxford University Press, 1997); Mbembe, "Necropolitics"; Charles W. Mills, *Black Rights/White Wrongs: The Critique of Racial Liberalism* (Oxford: Oxford University Press, 2017).

23. Stewart Stehlin, "Introduction," in *Sketches of Urban and Cultural Life in North America*, ed. Stuart Stehlin (New Brunswick, NJ: Rutgers University Press, 1988), xiv.

24. Ratzel, *Wandertage eines Naturforschers 1*.

25. Ratzel, *Wandertage eines Naturforschers* 2, 4. Whereas Hermannstadt (Sibiu), the region's capital city, was to Ratzel an emblem "of a free German *Volk*," it was also populated by "brown people" (Walachians and Roma) "who cultivated a spirit of sweet idleness on streets and public squares." Ratzel, *Wandertage eines Naturforschers* 2, 54–55. He returned to the issue in his later writings. "Die Gegenwart und Zukunft der Siebenbürger Sachsen," *Die Grenzboten* 51 (1892): 449–57.

26. Ratzel, *Die chinesische Auswanderung*. Ratzel only rarely commented on Cuba in his papers and books. For an exception, see "Die Zeichen des cubanischen Aufstandes," *Die Grenzboten* 54 (1895): 479–81.

27. This book is sometimes counted among Ratzel's most notable works. Ratzel's former student Martha Krug-Genthe, for instance, described *Die Vereinigten Staaten von Nord-Amerika* as the best work on the United States written by a foreigner. Martha Krug-Genthe, "Tributes to Friedrich Ratzel," *Bulletin of the American Geographical Society* 36, no. 9 (1904): 551.

28. Hassert, "Friedrich Ratzel I," 311; Stogiannos, *The Genesis of Geopolitics*, 20.

29. Friedrich Ratzel, "Diary entry 15 November 1874" *Tagebuch 24.7.1873–4.5.1875, Abschrift angefertigt von Günter Buttmann* Bayerische Staatsbibliothek, Ratzeliana Supplement 17a VI 1.

30. Ratzel, *Sketches of Urban and Cultural Life*, 67.

31. Ratzel, *Sketches of Urban and Cultural Life*, 72.

32. E. G. R., "Obituary: Friedrich Ratzel," 486; see also Wanklyn, *Friedrich Ratzel*, 15.

33. Sauer, "The Formative Years of Ratzel," 252.

34. Friedrich Ratzel, "Verehrte Freundin!" Letter to Marie Wingens, New York, 14 August 1873, Bayerische Staatsbibliothek, Ratzeliana, Supplement 10.

35. Ratzel, *Sketches of Urban and Cultural Life*, 20.

36. Friedrich Ratzel, "Land und Landschaft in der nordamerikanischen Volksseele," in *Kleine Schriften von Friedrich Ratzel, herausgegeben von Hans Helmolt, 1. Band*, ed. Hans Helmolt (Munich: R. Oldenbourg, 1902 [1906]), 281.

37. Friedrich Ratzel, "Über Kalifornien," in *Kleine Schriften von Friedrich Ratzel, herausgegeben von Hans Helmolt, 2. Band*, ed. Hans Helmolt, 1–18 (Munich: R. Oldenbourg, 1877 [1906]), 4.

38. Friedrich Ratzel, "Ethnographie und Geschichtswissenschaft in Amerika," in *Kleine Schriften von Friedrich Ratzel, herausgegeben von Hans Helmolt, 1. Band*, ed. Hans Helmolt (Munich: R. Oldenbourg, 1897 [1906]), 350. By the time he reviewed in the most favorable terms both Frederick Turner's *The Significance of the Frontier in American History* and Theodore Roosevelt's *The Winning of the West*, his views on the United States and the frontier in particular had long been formed. Friedrich Ratzel, "F. Turner, The Significance of the Frontier in American History, Washington 1894," *Geographischer Literaturbericht* 574 (1895): 122; Friedrich Ratzel, "Th. Roosevelt, The Winning of the West, New York 1895," *Geographischer Literaturbericht* 816 (1895): 176.

39. Ratzel, "Über Kalifornien," 6.

40. Ratzel, "Land und Landschaft," 280,

41. Friedrich Ratzel, "Das Deutschtum in Nordamerika," In *Kleine Schriften von Friedrich Ratzel, herausgegeben von Hans Helmolt, 1. Band*, ed. Hans Helmolt (Munich: R. Oldenbourg, 1904 [1906]), 285.

42. Ratzel, *Sketches of Urban and Cultural Life*, 5; see also Sauer, "The Formative Years of Ratzel," 247.

43. Ratzel, *Sketches of Urban and Cultural Life*, 23.

44. Ratzel, *Sketches of Urban and Cultural Life*, 131.

45. Ratzel, *Sketches of Urban and Cultural Life*, 133.

46. Ratzel, *Sketches of Urban and Cultural Life*, 133.

47. Ratzel, *Sketches of Urban and Cultural Life*, 156.

48. Ratzel, *Sketches of Urban and Cultural Life*, 68.

49. Friedrich Ratzel, "Verehrte Freundin!" Letter to Marie Wingens, Boston, 2 November 1873, Bayerische Staatsbibliothek, Ratzeliana, Supplement 10.

50. Ratzel, *Vereinigten Staaten, Erster Band*, 43.

51. See also Guettel, *German Expansionism*, 87, 98.

52. Curtis Stokes, "Tocqueville and the Problem of Racial Inequality," *The Journal of Negro History* 75, no. 1/2 (1990): 6.

53. Other divergences between Ratzel and Tocqueville are highlighted by Stehlin, "Introduction," xxii.

54. Andrea Wulf, *The Invention of Nature: The Adventures of Alexander von Humboldt, the Lost Hero of Science* (London: Hodder and Stoughton, 2015), 98.

55. Jens-Uwe Guettel, "South-West Africa: German Colonialism, Indians, and American Westward Expansion," *Modern Intellectual History* 7 no. 3 (2010): 523–52.

56. S. W. Geiser, "Dr. Ernst Kapp, Early Geographer in Texas," *Field and Laboratory* 14, no. 1 (1946): 16–31.

57. Friedrich Kapp, *Geschichte der Sklaverei in den Vereinigten Staaten von Amerika* (Hamburg, Otto Meißner, 1854 [1861]).

58. "Bericht des Referenten Dr. Kapp," in *Im Bismarckschen Reich 1871–1890*, ed. Hans Fenske (Darmstadt: Wissenschaftliche Buchgesellschaft, 1978 [1880]): 258–59.

59. W. E. B. Du Bois, *Black Reconstruction in America 1860–1880* (New York: The Free Press, 1935 [1998]), 30.

60. Ratzel, *Sein und Werden*, 507; see also Ratzel, *Anthropogeographie Zweiter Teil, Dritte Auflage*, 380, and Schultz, "Friedrich Ratzel—Pionier moderner Sozialgeographie?" 61.

61. Ratzel, "Einige Aufgaben der politischen Ethnographie," in *Kleine Schriften von Friedrich Ratzel, herausgegeben von Hans Helmolt, 2. Band*, ed. Hans Helmolt, 409 (Munich: R. Oldenbourg, 1900 [1906]); Ratzel, *Die Erde und das Leben II*, 617; Ratzel, "Nationalitäten und Rassen," 482.

62. Friedrich Ratzel, "Über politische Verhältnisse in Innerafrika," in *Kleine Schriften von Friedrich Ratzel, herausgegeben von Hans Helmolt, 2. Band*, ed. Hans Helmolt, 172 (Munich: R. Oldenbourg, 1888 [1906]).

63. Friedrich Ratzel, *Die Vereinigten Staaten von Nord-Amerika, Zweiter Band, Zweite Auflage: Politische und Wirthschaftsgeographie* (Munich: R. Oldenbourg, 1880 [1893]), 293; Ratzel, "Das Deutschtum in Nordamerika," 359.

64. Ratzel, "Über politische Verhältnisse in Innerafrika," 169.

65. Ratzel, *Sketches of Urban and Cultural Life*, 67.

66. Ratzel, *Sketches of Urban and Cultural Life*, 147, 208, 211.

67. Ratzel, *Sketches of Urban and Cultural Life*, 6, 146.

68. See also Friedrich Ratzel, "Zukunft und Beurteilung der Neger," *Deutsche Revue* 4, no. 4 (1880): 97–111; Ratzel, *Die Vereinigten Staaten, Zweiter Band, Zweite Auflage*, 268.

69. Ratzel, *Politische Geographie, 1903*, 61–62.

70. "Aus den Baumwollenstaaten II," *Globus* 30 (1876): 344.

71. Ratzel, *Sketches of Urban and Cultural Life*, 146.

72. Hartman, *Scenes of Subjection*, 116.

73. "Aus den Baumwollenstaaten I," *Globus* 30 (1876): 317.

74. David N. Livingstone, *Adam's Ancestors: Race, Religion and the Politics of Human Origins* (Baltimore: Johns Hopkins University Press, 2008), 96.

75. James Moore, "Darwin's Progress and the Problem of Slavery," *Progress in Human Geography* 34, no. 5 (2010): 567.

76. Edward Lurie, "Louis Agassiz and the Races of Man," *History of Science Society* 45, no. 3 (1954): 234.

77. Moore, "Darwin's Progress," 567.

78. Agassiz's scientific racism would continue to inspire the far right throughout the twentieth century, from the fascist poet Ezra Pound (1885–1972) to the American far-right activist and Ku Klux Klan member John Kasper (1929–1998). Alex Marsh, *John Kasper and Ezra Pound: Saving the Republic* (London: Bloomsbury, 2015).

79. Ratzel, *Sein und Werden*, 442.

80. Ratzel, *Sketches of Urban and Cultural Life*, 105; Friedrich Ratzel, "Die Tagesansicht Gustav Theodor Fechners," *Die Grenzboten* 60 (1901): 169; Ratzel, *Glücksinseln und Träume*, 499; see also Steinmetzler, *Friedrich Ratzel*, 72.

81. Buttmann, *Friedrich Ratzel*, 44.

82. H. W. Brands, *The Man Who Saved the Union: Ulysses S. Grant in War and Peace* (New York: Doubleday, 2012).

83. Friedrich Ratzel, "Diary entry 25 January 1874," *Tagebuch 24.7.1873–4.5.1875, Abschrift angefertigt von Günter Buttmann* Bayerische Staatsbibliothek, Ratzeliana Supplement 17a VI, 1.

84. Ratzel, *Sketches of Urban and Cultural Life*, 148.

85. Ratzel, *Vereinigten Staaten, Zweiter Band*, 101.

86. Leeanna Keith, *The Colfax Massacre: The Untold Story of Black Power, White Terror and the Death of Reconstruction* (Oxford: Oxford University Press, 2008).

87. David W. Blight, *Race and Reunion: The Civil War in American Memory* (Cambridge, MA: Harvard University Press, 2001), 125–26.

88. Ratzel, *Sketches of Urban and Cultural Life*, 154.

89. Stehlin, "Introduction."

90. Ratzel, *Sketches of Urban and Cultural Life*, 152.

91. Ratzel, *Sketches of Urban and Cultural Life*, 152.

92. Ratzel, *Sketches of Urban and Cultural Life*, 170. Gottfried Leibniz, a source of intellectual inspirations for Ratzel, had spoken before him of nonwhites as animal-like who could be repatriated at will. Stuart Elden, "Leibniz and Geography: Geologist, Paleontologist, Biologist, Historian, Political Theorist and Geopolitician," *Geographica Helvetica* 68 (2013): 89. President Abraham Lincoln had supported a line similar to that of Ratzel's in which the racial problem was to be solved both by the abolition of slavery and the disappearance of the black Americans through recolonization. Richard Blackett, "Lincoln and Colonization," *OAH Magazine of History* 21 (2007), no. 4: 19–22.

93. "Aus den Baumwollenstaaten I," *Globus* 30 (1876): 317; see also Ratzel, "Zukunft und Beurteilung," 813.

94. Ratzel, *Vereinigten Staaten, Zweiter Band*, 99–101. He also used the term "scalawags" to denote white southerners who supported the Republican party.

95. Ratzel, "Zukunft und Beurteilung," 102. Ratzel also criticized the way in which the abolition of slavery in Cuba had been "thoughtlessly cheered by the masses." Ratzel, "Die Zeichen des cubanischen Aufstandes," 479.

96. Ratzel, "Zukunft und Beurteilung," 107. Ratzel did not entirely dismiss the craniology of his day. The second volume of his *Anthropogeographie* noted that it was "no coincidence" that prognathism, which he saw as being correlated with brain size, was more common in those races which stood "at the lowest level." Ratzel, *Anthropogeographie Zweiter Teil, Dritte Auflage*, 379–80.

97. Ratzel, *Die Vereinigten Staaten, Zweiter Band, Zweite Auflage*, 284.

98. Ratzel, *Die Vereinigten Staaten, Zweiter Band, Zweite Auflage*, 182. He believed that the climate of the American South was a more natural habitat for America's black population and he regretted that many of them were moving north following the failures of Reconstruction. It was no coincidence, he argued, that the terms "slave states" and

"cotton state" were used synonymously in the United States. Ratzel, "Zukunft und Beurteilung," 101.

99. "Beurteilung der Neger," *Die Grenzboten* 51 (1892): 20, 24.

100. Ratzel, *Die Vereinigten Staaten, Zweiter Band, Zweite Auflage*, 180.

101. Ratzel, *Die Vereinigten Staaten, Zweiter Band, Zweite Auflage*, 283; see also Guettel, *German Expansionism*, 137.

102. Friedrich Ratzel, "B. T. Washington: Vom Sklaven empor! Berlin 1902," *Geographischer Literaturbericht* 315 (1903): 97; Friedrich Ratzel, "America's Race Problems. Addresses at the Annual Meeting of the American Academy of Political and Social Science, Philadelphia, April Twelfth and Thirteenth, Philadelphia 1901," *Geographischer Literaturbericht* 521 (1902): 149–50. On the Atlanta Compromise of 1895, see Ibram X. Kendi, *Stamped from the Beginning: The Definitive History of Racist Ideas in America* (London: Bodley Head, 2016), 277–79.

103. Ratzel, "B. T. Washington," 97.

104. London Assurance Corporation, United States branch (1893) Legal advice, New York, 5 January 1893, Archiv für Geographie, Leibniz Institut für Länderkunde, Nachlass Friedrich Ratzel 163/646.

105. E. Semple, "Geehrter Herr Professor," Letter to Friedrich Ratzel, n.p., 2 January 1893, Archiv für Geographie, Leibniz Institut für Länderkunde, Nachlass Friedrich Ratzel 163/686; E. Semple, "Geehrter Herr Professor," Letter to Friedrich Ratzel, n.p., 30 January 1893, Archiv für Geographie, Leibniz Institut für Länderkunde, Nachlass Friedrich Ratzel 163/686.

106. Friedrich Ratzel, "Heinrich Driesmans: Rasse und Milieu, Berlin 1902," *Historische Vierteljahrschrift* 7 (1904): 394.

107. Die Redaktion. "Friedrich Ratzel," *Archiv für Rassen- und Gesellschaftsbiologie* 1 (1904): 736.

108. Ratzel, "Land und Landschaft," 280.

109. Ratzel, *Politische Geographie: Geographie der Staaten*, 438.

110. Friedrich Ratzel "Verehrte Freundin!" Letter to Marie Wingens, St. Louis, 29 May 1874, Bayerische Staatsbibliothek, Ratzeliana, Supplement 10.

111. Ratzel, *Aus Mexico*, iv, 13.

112. Friedrich Ratzel "Verehrte Freundin!" Letter to Marie Wingens, Yosemite, 21 August 1874, Bayerische Staatsbibliothek, Ratzeliana, Supplement 10. Ratzel had initially planned to visit the United States, Mexico, Nicaragua, Peru, and Argentina, as he reported in a letter to Ernst Haeckel in 1873. Friedrich Ratzel, "Verehrter Freund!" Letter to Ernst Haeckel, Berlin, 20 July 1873, transliterated by Peter Zigman, *Würzburger medizinhistorische Mitteilungen* 28: 401–402.

113. Ratzel, *Aus Mexico*, 174.

114. Ratzel, *Aus Mexico*, 280, 283. Note that Ratzel made similar comments about England and Wales in 1881. Friedrich Ratzel, *Tagebuch über Englandreise im Aug/Sep 1881, Abschrift angefertigt von Günter Buttmann*, Bayerische Staatsbibliothek, Ratzeliana Supplement 17a VI 9a.

115. Ratzel, *Aus Mexico*, 3.

116. Ratzel, *Aus Mexico*, 116.

117. Ratzel, *Aus Mexico*, 181.

118. Ratzel, *Aus Mexico*, 313.
119. Ratzel, *Aus Mexico*, 329.
120. Ratzel, *Aus Mexico*, 164.
121. Ratzel, *Aus Mexico*, 179.
122. Ratzel, *Aus Mexico*, 186.
123. Ratzel, *Aus Mexico*, 272.
124. Ratzel, *Aus Mexico*, 273.
125. Ann Laura Stoler, "Rethinking Colonial Categories: European Communities and the Boundaries of Rule," *Comparative Studies in Society and History* 31, no. 1 (1989): 139.
126. Ratzel, *Aus Mexico*, 388.
127. Ratzel, *Aus Mexico*, 315, original emphasis.
128. Ratzel, "Beurtheilung der Völker," 194.
129. Ratzel, "Beurtheilung der Völker," 196.
130. Ratzel, "Beurtheilung der Völker," 196.
131. n.a., (n.d.) "Deutsche Amerikaner," Bayerische Staatsbibliothek, Ratzeliana II Besprechungen von Ratzelschen Werken.
132. Ratzel, *Aus Mexico*, 371.
133. Ratzel, *Wider die Reichsnörgler*.
134. Ratzel used the term ethnography instead of ethnology, even though his was very much a synthetic approach.
135. Ratzel, *History of Mankind I*, 3.
136. Ratzel, *History of Mankind I*, 4, 9.
137. Ratzel, *Die Erde und das Leben, 1. Band*.
138. Ratzel, "Lebensraum—A Biogeographical Study," 73.
139. Ratzel, *Die Erde und das Leben, 2. Band*, 671; Schultz, "Ratzel und Herder." Indeed, as Schulz demonstrates, Ratzel used the term *Herrenvolk* (master *Volk*), which was closely related to the term *Herrenrasse* (master race) and which would play such a crucial role in the formation of national socialist ideology. Schultz, "Ratzel und Herder," 138. Ratzel employed the term to describe the position of whites in the United States, arguing that the only races that deserved this title were those which could grant subordinate races their own living conditions in ways that protected the interests of all. Ratzel, "Nationalitäten und Rassen," 485.
140. Johannes Fabian, *Time and the Other: How Anthropology Makes Its Object* (New York: Colombia University Press, 1983); Ratzel, *History of Mankind I*, 5.
141. Ratzel, *Sketches of Urban and Cultural Life*, 62.
142. Ratzel, *Die Vereinigten Staaten, Zweiter Band, Zweite Auflage*, 179.
143. Ratzel, *History of Mankind I*, 4.
144. Ratzel, *History of Mankind I*, 14.
145. Ratzel, *Aus Mexico*, 169.
146. Ratzel, *Aus Mexico*, 169, emphasis added.
147. Ratzel, *Aus Mexico*, 315.
148. Even here Ratzel made distinctions. He thus found indigenous peasants to have limp and mild facial features while the faces of nomads, regardless of whether they were indigenous or European, were hard, wild, and daring.

149. Benjamin Madley, "Understanding Genocide in California under United States Rule, 1846–1873," *Western Historical Quarterly* 47, no. 4 (2017): 449–61.

150. Ratzel, *Aus Mexico*, 250–51.

151. Ratzel, *Die Vereinigten Staaten, Zweiter Band, Zweite Auflage*, 188. While his earlier work consistently described the indigenous populations of North America as incapable of civilization (*kulturunfähig*), Ratzel later recanted, noting that the last word on the level of civilization that some American peoples had achieved before the arrival of Europeans had not been spoken. Ratzel, *Ethnographie und Geschichtswissenschaft in Amerika*, 349.

152. Ratzel, "Über Kalifornien," 12.

153. Ratzel, "Über Kalifornien," 12.

154. Ratzel, "Nationalitäten und Rassen," 483.

155. Ratzel, "Nationalitäten und Rassen," 487.

156. Ratzel, "Nationalitäten und Rassen," 485.

157. The passage is worth quoting in full: "Between a feeling of community with humanity, of which we are all members, and an equally justified racial sentiment (*Rassengefühl*), we are caught in an awkward conflict of inclinations and judgments. History teaches us that power belongs to those who are strong in spirit (*Geistesstarke*) and strong in body (*Körperkräftige*), and that every *Volk* that desires to survive requires at a minimum the power to maintain its soil and protect itself against harmful influences. And yet the height of culture we have attained instills in us a secret aversion to the open admission of the necessity of severe racial struggles. If only the Earth had more space! But the way in which we live crammed on this globe, whose 150 million square kilometers of land is already too narrow for today's humanity of 1.5 billion, it is impossible to evade one another. It is pointless to deny that the differences in the natural endowments of mankind's various races exclude the [possibility of] equality of achievement and aspiration. Here too, salvation thus lies only in the gradation and division of tasks, which should be combined with spatial segregation in order to counteract the risk of the higher race mixing [with others]." Ratzel, "Nationalitäten und Rassen," 483.

158. Klaus Kost, "Anti-semitism in German Geography 1900–1945," *GeoJournal* 46 (1998): 286; see also Mark Bassin, "Studying Ourselves: History and Philosophy of Geography," *Progress in Human Geography* 24, no. 3 (2000): 480.

159. Michael Fahlbusch, "Volk ohne Raum—Raum ohne Volk. Der lange Schatten der Deutsch-Völkischen in der Weimarer Republik," In *Demokratiegeschichte als Zäsurgeschichte: Diskurse der frühen Weimarer Republik*, ed. Heidrun Kämper, Peter Haslinger and Thomas Raithel (Berlin: Akademie, 2014), 268; Boris Michel, "Anti-semitism in Early Twentieth Century German Geography. From a 'Spaceless' People to the Root of the 'ills' of Urbanization," *Political Geography* 65 (2018): 1–7.

160. Schultz, "Rasse und Raum," 238.

161. Friedrich Ratzel, "Der Ursprung der Arier in geographischem Licht," In *Kleine Schriften von Friedrich Ratzel, herausgegeben von Hans Helmolt, 2. Band*, ed. Hans Helmolt (Munich: R. Oldenbourg, 1899 [1906]), 391; see also discussion in Ishan Ashutosh, "Mapping Race and Environment: Geography's Entanglements with Aryanism," *Journal of Historical Geography* 61 (2018): 15–23. See also his discussion on the origin of the *Indo-Germanen*. Friedrich Ratzel, "Die geographische Methode in der

Frage nach der Urheimat der Indogermanen," in *Kleine Schriften von Friedrich Ratzel, herausgegeben von Hans Helmolt, 2. Band*, ed. Hans Helmolt (Munich: R. Oldenbourg, 1904 [1906]), 534–35.

162. Ratzel, "Der Ursprung der Arier," 393.

163. Ratzel, *Die Erde und das Leben* 2, 626; Friedrich Ratzel, "Zur Frage der Indogermanen-Heimat," In *Kleine Schriften von Friedrich Ratzel, herausgegeben von Hans Helmolt, 2. Band*, ed. Hans Helmolt (Munich: R. Oldenbourg, 1904 [1906]), 541–42.

164. "1860er Antisemitismus," *Die Grenzboten* 53 (1894): 282.

165. Ratzel, "Nationalitäten und Rassen," 467.

166. Friedrich Ratzel, "Emin Pascha," *Deutsche Revue* 17 (1892): 218.

167. Ratzel, *Politische Geographie: Geographie der Staaten*, 39; Ratzel, "Nationalitäten und Rassen," 468.

168. Ratzel, *Aus Mexico*, 381.

169. Ratzel, *Politische Geographie: Geographie der Staaten*, 393.

170. Ratzel, *Aus Mexico*, 18; Ratzel, "Nationalitäten und Rassen," 468. Ratzel, for instance, did *not* include Jewish people in a discussion of non-German minorities on German soil. Ratzel, *Deutschland: Eine Einführung*, 286–97. On one occasion, however, Ratzel did make an antisemitic comment about an acquaintance, Bertha von Rothschild, noting approvingly in his diary that she "had no Jewish look about her apart from a slight squint." Friedrich Ratzel, "Diary entry 1 April 1880," *Tagebuch, 1879–1881, Abschrift angefertigt von Günter Buttmann* Bayerische Staatsbibliothek, Ratzeliana Supplement 17a VI 7.

171. Ratzel describes Zionism, which had gained traction during his lifetime, as partially motivated by antisemitism and "an experiment" that, if successful, would motivate other *Völker* without land to follow suit (Ratzel, "Nationalitäten und Rassen," 468; see also Schultz, "Rasse und Raum," 238. The founding father of modern Zionism, Theodor Herzl, took offense precisely with the belief, held by Ratzel, that Jews were somehow living in "host nations." Nonetheless, passages in his highly influential 1896 pamphlet *Der Judenstaat* seem to presuppose a Ratzelian political geography in which rootedness and soil (*Boden*) are crucial parameters of political legitimacy. Like Ratzel, broadly Darwinian in his worldview, Herzl argued that Jews would have to be "uprooted" and "placed in a better soil." Theodor Herzl, *Der Judenstaat: Staatsschrift* (n.p.: Gutenberg-DE, 1896 [2005]), 86.

172. Stoler, "Rethinking Colonial Categories," 137.

173. As I have noted, his views on the Sinti and Roma were, however, consistent with a more European form of racism.

174. Ratzel, *Die Erde und das Leben* 2, 618.

175. W. E. B. Du Bois, "The Study of the Negro Problems," *Annals of the American Academy of Political and Social Sciences* 11 (1898): 7.

176. Friedrich Ratzel, "F. Boas, Human Faculty as Determined by Race, New York 1894," *Geographischer Literaturbericht* 40 (1895): 12.

177. Charles Darwin, *The Voyage of the Beagle* (London: Penguin, 1839 [1989]), 62, emphasis added.

178. Ratzel, *History of Mankind I*, 131.

179. Ratzel, "Über politische Verhältnisse in Innerafrika," 172.

180. Philip S. Foner, "Alexander von Humboldt on Slavery in America," *Science & Society* 47, no. 3 (1983): 33–34; Moore, "Darwin's Progress."

181. Ratzel, *Die Vereinigten Staaten, Zweiter Band, Zweite Auflage*, 181.

182. Ratzel, *Die Erde und das Leben* 2, 543.

183. Kendi distinguishes between three traditions in American racial thought: segregationism, assimilationism, and antiracism. While segregationism has blamed the black population for racial disparities and antiracism has denounced racism itself, assimilationism sees both racial discrimination and the black population as part of the problem. Ratzel, according to this distinction, falls into the segregationist camp. This, however, does not mean that he did not take an interest in assimilationist arguments, perhaps because he realized that they were in fact more compatible with his diffusionist ontology. Kendi, *Stamped from the Beginning*, 2.

CHAPTER 4

1. Friedrich Ratzel, "Der Lebensraum: Eine biogeographische Studie," in *Festgaben für Albert Schäffle zur siebenzigsten Wiederkehr seines Geburtstages a, 24. Februar 1901*, ed. Karl Bücher et al. (Tübingen: Verlag der Laupp'schen Buchhandlung, 1901), 102–89. The longer version of the essay also appeared as a monograph with the same publisher. Friedrich Ratzel, *Der Lebensraum. Eine biogeographische Studie* (Tübingen: Verlag der Laupp'schen Buchhandlung, 1901).

2. Ratzel, "Lebensraum—A Biogeographical Study," 60.

3. Ratzel, "Lebensraum—A Biogeographical Study," 60.

4. Steinmetzler, *Anthropogeographie Friedrich Ratzels*, 12; Jureit, "Mastering Space."

5. Ratzel, "Lebensraum—A Biogeographical Study," 73.

6. Oscar Peschel, *Völkerkunde* (Leipzig: Duncker und Humblot, 1874), 21.

7. Johann Wolfgang v. Goethe, *Wahlverwandtschaften, 1. Band* (Tübingen: J. G. Cottaische Buchhandlung, 1809), 303.

8. Halas, "Searching for the Perfect Footnote"; Michael Heffernan, *The Meaning of Europe: Geography and Geopolitics* (London: Arnold, 2000).

9. Oscar Peschel, "Ursprung und Verschiedenheit der Menschenrassen," *Das Ausland* 33 (1860): 391–94.

10. Friedrich Ratzel, "Oskar Peschel," in *Kleine Schriften von Friedrich Ratzel, herausgegeben von Hans Helmolt, 1. Band*, ed. Hans Helmolt (Munich: R. Oldenbourg, 1887 [1906]), 441.

11. When Ratzel wrote a biographical entry on Peschel in 1887, he did not comment on Peschel's use of the term *Lebensraum*. Ratzel, "Oskar Peschel."

12. Jureit, "Mastering Space," 84.

13. We can find the *Kampf um Raum* section in the 1891 second volume of *Anthropogeographie*, the 1901 *Lebensraum* essay, and the 1902 *Die Erde und das Leben*; Ratzel, *Anthropogeographie Zweiter Teil, Dritte Auflage*, xxiv; Ratzel, "Lebensraum—A Biogeographical Study," 72; Ratzel, *Die Erde und das Leben*, 2, 593.

14. Friedrich Ratzel, "Diary entry 2 December 1896," *Tagebuch 4.10.1896-6.1.1898, Abschrift angefertigt von Günter Buttmann* Bayerische Staatsbibliothek, Ratzeliana Supplement 17a VI 15a.

15. Friedrich Ratzel, "Diary entry 9 March 1897," *Tagebuch 4.10.1896-6.1.1898, Abschrift angefertigt von Günter Buttmann* Bayerische Staatsbibliothek, Ratzeliana Supplement 17a VI 15a.

16. Friedrich Ratzel, "Diary entry 19 March 1897," *Tagebuch 4.10.1896-6.1.1898, Abschrift angefertigt von Günter Buttmann* Bayerische Staatsbibliothek, Ratzeliana Supplement 17a VI 15a.

17. Friedrich Ratzel, "Diary entry 21 March 1897," *Tagebuch 4.10.1896-6.1.1898, Abschrift angefertigt von Günter Buttmann* Bayerische Staatsbibliothek, Ratzeliana Supplement 17a VI 15a.

18. Friedrich Ratzel, "Diary entry 26 March 1897," *Tagebuch 4.10.1896-6.1.1898, Abschrift angefertigt von Günter Buttmann* Bayerische Staatsbibliothek, Ratzeliana Supplement 17a VI 15a. Three years later, he noted that he had discussed a longer version of the *Lebensraum* essay with colleagues at Leipzig, later referring to *Lebensraum* as one of two intellectual challenges for the second half of 1900 in his annual review entry, with which he liked to conclude his diary each year. Friedrich Ratzel, "Diary entry 9 December 1900," *Tagebuch 4.7.1900-19.7.1903, Abschrift angefertigt von Günter Buttmann* Bayerische Staatsbibliothek, Ratzeliana Supplement 17a VI 20; Friedrich Ratzel, "Diary entry 13 December 1900," *Tagebuch 4.7.1900-19.7.1903, Abschrift angefertigt von Günter Buttmann* Bayerische Staatsbibliothek, Ratzeliana Supplement 17a VI 20.

19. Ratzel reused significant passages from the 1901 essay in his 1902 *Die Erde und das Leben*. He also made use of the term *Lebensraum* in an essay on the origins of the Indo-Germans. Ratzel, *Die Erde und das Leben* 2, 590–606; Friedrich Ratzel, "Zur Frage der Indogermanen-Heimat," *Archiv für Rassen- und Gesellschafts-Biologie* 1, no. 4 (1904): 579. It found its way into the second and subsequent editions of his *Anthropogeographie*, too. Ratzel, *Anthropogeographie Erster Teil, Zweite Auflage*, 230–32; Ratzel, *Anthropogeographie Erster Teil, Dritte Auflage*, 149–50. Finally, he used *Lebensraum*, though only in passing, to refer to the bordered space of the peasant family, a dwelling, surrounded by a fence or hedge, Ratzel, *Glücksinseln und Träume*, 83.

20. Ratzel, *Politische Geographie: Geographie der Staaten*, 1.

21. Tul'si (Tuesday) Bhambry, "Translator's introduction," *Journal of Historical Geography* 61 (2018): 59.

22. Ratzel, "Lebensraum—A Biogeographical Study," 71; see also Ratzel, "Über den Lebensraum," 365.

23. Ratzel, "Lebensraum—A Biogeographical Study," 71. For an earlier version of his thinking in relation to the oecumene, see Friedrich Ratzel, "Über die Anwendung des Begriffs Oekumene auf geographische Probleme der Gegenwart (Mit einer Karte)," *Berichte über die Verhandlungen der Königlich Sächsischen Gesellschaft der Wissenschaften zu Leipzig, philologisch-historische Klasse* 40 (1888): 137–80.

24. Ratzel, "Lebensraum—a Biogeographical Study," 70.

25. Thomas Achelis, "Friedrich Ratzel," *Geographischer Anzeiger* 5 (1904): 244.

26. Ratzel thought highly of Peschel and applauded his scientificity and patriotism. Ratzel, "Oskar Peschel."

27. Ratzel also drew on the philosopher of technology Ernst Kapp and the transport geographer Johann Georg Kohl. Humboldt, Kapp, and Kohl had all either journeyed or emigrated to the Americas.

28. The list of philosophical influences here is not comprehensive. On Ratzel and Herder, for instance, see Schultz, *Herder und Ratzel*. On Ratzel and Hegel, see Carlotta Santini, "One Sea, One Humanity: Modelling the Man-Sea Relationship in Friedrich Ratzel's Anthropogeographical Project," *Journal of Interdisciplinary History of Ideas* 6, no. 12 (2017): 1–24.

29. Rudolf Kittel, "Zu Friedrich Ratzels Gedächtnis," *Die Grenzboten* 63 (1904): 516–24; K. Lamprecht, "Friedrich Ratzel," 259–69.

30. Ratzel, "Lebensraum—A Biogeographical Study," 64.

31. Ratzel, *Sketches of Urban and Cultural Life*, 253, 283.

32. Ratzel, *Sein und Werden*, 482.

33. Friedrich Ratzel, "Neue Untersuchungen über die Vogelnester," *Meyers Ergänzungsblätter zur Kenntnis der Gegenwart* 2 (1870): 496–97.

34. Ratzel, "Neue Untersuchungen über die Vogelnester," 498.

35. He argued in his later work that protoplasm was the basic vehicle of life in the organism, not the cell, as had previously been assumed. Ratzel, *Anthropogeographie Erster Teil, Zweite Auflage*, 4.

36. Ratzel, *Glücksinseln und Träume*, 506.

37. Ratzel, *Sein und Werden*, 468.

38. Friedrich Ratzel, "Diary entry 27 September 1873," *Tagebuch 24.7.1873-4.5.1875, Abschrift angefertigt von Günter Buttmann* Bayerische Staatsbibliothek, Ratzeliana Supplement 17a VI 1.

39. Ratzel, *Anthropogeographie, Erster Teil, Zweite Auflage*, 49; see also "Weltentwicklung und Weltschöpfung," 574.

40. Ratzel, *Die Vereinigten Staaten, Zweiter Band, Zweite Auflage*, 239.

41. D. R. Stoddart, "Darwin's Impact on Geography," *Annals of the Association of American Geographers* 56, no. 4 (1966): 695. Ratzel was certainly a believer in randomly occurring variants and was keen to apply this idea to the realm of culture (slight mutations of artifacts etc.). And yet it is certainly true that when discussing state behavior and historical events this approach was of little use and gave way to a more determinist reading. Darwin would himself de-emphasize chance in later editions of *On the Origin of Species*.

42. Robert J. Richards, *The Tragic Sense of Life: Ernst Haeckel and the Struggle over Evolutionary Thought* (Chicago: Chicago University Press, 2008), 2. One of Haeckel's later books sold 40,000 copies in its first year, while the *On the Origin of Species* only sold 39,000 copies over the course of its six English editions (in the period 1859–1890).

43. Ulrich Kutschera, Georgy S. Levit, and Uwe Hossfeld, "Ernst Haeckel (1834–1919): The German Darwin and his Impact on Modern Biology." *Theory in Biosciences* 138 (2019): 1–7.

44. Friedrich Ratzel, "Hochgeehrter Herr Professor," Letter to Ernst Haeckel, Munich, 24 March 1872, transliterated by Peter Zigman, *Würzburger medizinhistorische Mitteilungen* 28: 398–401. See Peter M. Zigman, "'. . . und meine Anschauungen sind natürlich selbstständiger geworden' – Friedrich Ratzels Briefe an Ernst Haeckel,

seinen Mentor und akademischen Kollegen" *Würzburger medizinhistorische Mitteilungen* 28 (2009): 383.

45. Bassin, "Imperialism and the Nation State," 487.

46. Ratzel, *Sein und Werden*, 476–77. Ratzel endorsed Haeckel again in 1872 in a bibliographical entry. Friedrich Ratzel, "Ernst Haeckel," In *Kleine Schriften von Friedrich Ratzel, herausgegeben von Hans Helmolt, 1. Band*, ed. Hans Helmolt (Munich: R. Oldenbourg, 1872 [1906]), 373–76.

47. Richards, *Tragic Sense of Life*, 125. Müller claims that Ratzel moved away from Haeckel's brand of monism in the 1870s. And yet a basic monist ontology can still be felt in his final writings. Müller, *Friedrich Ratzel*, 41.

48. Ratzel, *Sein und Werden*, 477.

49. Richards, *Tragic Sense of Life*, 125.

50. Ratzel, *Sein und Werden*, 19.

51. Diana Coole and Samantha Frost, "Introducing New Materialisms," in *New Materialisms: Ontology, Agency, and Politics*, ed. Diana Coole and Samantha Frost (Durham, NC: Duke University Press, 2010), 1–43. The vitalist Henri Bergson famously spoke of an *élan vital* to capture this exceptional quality, indeterminacy, excess, or spirit that caused things to happen in the world. In life adhered thus a creativity that subjects the world to constant evolution and innovation and which is difficult if not impossible to contain and discipline. As Bergson put it, "[i]n vain we force the living into this or that one of our moulds. All the moulds crack. They are too narrow, above all too rigid, for what we try to put into them." Henri Bergson, *Creative Evolution* (Lanham, MD: University Press of America, 1911 [1983]), x.

52. Richards, *Tragic Sense of Life*, 124

53. Ratzel, *Sein und Werden*, 10.

54. Ratzel, *Glücksinseln und Träume*, 79.

55. Ratzel, "Lebensraum—A Biogeographical Study," 63.

56. Ratzel, *Die Erde und das Leben* 2, 555.

57. As Abrahamsson points out, Ratzel has thus far been ignored in the philosophy and history of vitalism. Neither of the well-known twentieth-century vitalists, such as Hans Driesch or Henri Bergson, engaged with Ratzel directly, and Georges Canguilhem only mentioned him in passing. Abrahamsson, "Genealogy of Lebensraum"; Georges Canguilhem, "The Living and its Milieu," *Grey Room* 3 (2001): 15. And yet Ratzel's intellectual life was not detached from these thinkers. Driesch and Ratzel were both students of Ernst Haeckel. Indeed, their books would be published in the same book series with the publisher J. A. Barth. In general, the opposition between mechanism and vitalism was far less sharp in German evolutionary discourses than it was in the Anglophone world. Smith, *Politics and the Sciences*, 91.

58. Ratzel, "Lebensraum—A Biogeographical Study," 63.

59. Smith, *Politics and the Sciences*, 207.

60. Friedrich Ratzel, "Der Geist, der über den Wassern schwebt," in *Kleine Schriften von Friedrich Ratzel, herausgegeben von Hans Helmolt, 1. Band*, ed. Hans Helmolt (Munich: R. Oldenbourg, 1901 [1906]), 239.

61. Mildenberger, "Lebensraum oder Umwelt?"

62. "Weltentwicklung und Weltschöpfung," 574.

63. Bassin, "Imperialism and the Nation State," 487.

64. Zigman, ". . . und meine Anschauungen," 376; Steinmetzler, *Anthropogeographie Friedrich Ratzels*. The exchange is incomplete. Only one (impersonal) letter from Haeckel to Ratzel has survived.

65. Zigman, ". . . und meine Anschauungen," 383.

66. Friedrich Ratzel, "Mein verehrter Lehrer und lieber Freund!" Letter to Ernst Haeckel, Leipzig, 14 February 1904, transliterated by Peter Zigman, *Würzburger medizinhistorische Mitteilungen* 28: 408.

67. Haeckel was a proponent of a weak form of polygenesis, the belief that different human races had developed separately from one another. Haeckel's view of the origin of humanity emphasized the role of language. "As, according to the unanimous opinion of most eminent philologists, all human languages are not derived from a common primaeval language, we must assume a polyphyletic origin of language, and in accordance with this a polyphyletic transition from speechless Ape-like Men to genuine Men." Haeckel, *The History of Creation*, 294. Ratzel did not follow Haeckel here.

68. Weidling, "Ernst Haeckel, Darwinismus," 319.

69. Haeckel, *The History of Creation*, 256. On the heated debate over continuities and discontinuities between Haeckel's monism and fascism, see Daniel Gasman, *Haeckel's Monism and the Birth of Fascist Ideology* (New York: Peter Lang, 1998); Robert J. Richards, "Ernst Haeckel's Alleged Anti-semitism and Contributions to Nazi Biology," *Biological Theory* 2, no. 1 (2007): 97–103.

70. Haeckel, *The History of Creation*, 20. Haeckel uses the term *Vernichtung* twice, which translates as "extermination" or "annihilation." Ernst Haeckel, *Natürliche Schöpfungsgeschichte*. Berlin: Georg Reimer, 1868, 16.

71. Patrick Brantlinger, *Dark Vanishings: Discourse on the Extinction of Primitive Races, 1800–1930* (Ithaca, NY: Cornell University Press, 2003), 169.

72. Charles Darwin, *On the Origin of Species by Means of Natural Selection: Or, the Preservation of Favoured Races in the Struggle for Life* (London: Murray, 1859), 57, 490.

73. Ratzel, "Lebensraum—A Biogeographical Study"; Ratzel, 1902..

74. Charles Darwin, *The Descent of Man: Selection in Relation to Sex, Volume I*, (Cambridge: Cambridge University Press, 1871 [2009]).

75. Ratzel, *Glücksinseln und Träume*, 399. Ratzel also wrote in a letter to the Catholic priest and writer Heinrich Hansjakob that he was ashamed for his earlier staunch Darwinism and emphasized that by the mid-1890s he had recognized that science was only able to describe a thin layer of reality and that only religious belief could penetrate further. Haaß, *Vom Dorf in die Welt*, 59.

76. "Weltentwicklung und Weltschöpfung," 585.

77. Friedrich Ratzel, "H. Spencer, Prinzipien der Biologie, Bd. I, Stuttgart, 1876," *Archiv für Anthropologie* 10 (1878): 339–41; Friedrich Ratzel, "H. Spencer, Prinzipien der Biologie, Bd. II, Stuttgart, 1876," *Archiv für Anthropologie* 11 (1879): 181–82.

78. Herbert Spencer, *The Principles of Sociology, Vol. 1* (London: Williams and Norgate, 1877), 473.

79. On Spencer's views on imperialism and racial mixing, see Hobson, *Eurocentric Conception of World Politics*, 95–98.

80. Friedrich Ratzel, "Diary entry 12 November 1895," *Tagebuch 1895, Abschrift angefertigt von Günter Buttmann* Bayerische Staatsbibliothek, Ratzeliana Supplement 17a VI 13.

81. "Der Staat als Organismus." *Die Grenzboten* 55 (1896): 617. There were other roots of the state-as-organism analogy too, as Jureit discusses in some detail. Jureit, *Das Ordnen von Räumen*, 136–38.

82. Ratzel, *Politische Geographie: Geographie der Staaten*, 130.

83. Ratzel, *Politische Geographie: Geographie der Staaten*, 9.

84. Ziegler responded to Ratzel in a letter, noting that while "[o]ne can draw pleasure from a comparison of the human body's tissue with the strata of the social hierarchy, it is only ever a parable and nothing can be proven with it, other than perhaps the statement that the division of labor and differentiation is generally useful and becomes necessary on higher levels of development." H. E. Ziegler, H. E., "Geehrtester Herr Professor!" Letter to Friedrich Ratzel, 10 January 1896, Bayerische Staatsbibliothek, Ratzeliana I, 1–5. There were limits, moreover, to the work that a comparison of human societies with animal states (*Tierstaaten*) could do. Feeling the need to qualify his organicism, Ratzel included a caveat in *Politische Geographie* too. Ratzel, *Politische Geographie: Geographie der Staaten*, 15.

85. Herbert Spencer, *Social Statics, or the Conditions Essential to Human Happiness* (London: Williams and Norgate, 1851 [1868]), 415.

86. Spencer, *Principles of Sociology*, 43.

87. Schäffle would become known for his role in developing Bismarck's social laws (*Sozialgesetzgebung*) of the 1880s. Spencer's influence could already be felt in the first edition of Ratzel's *Anthropogeographie*, while Ratzel seems to have turned to Schäffle only in the 1890s.

88. Ratzel, *Wandertage eines Naturforschers. 1. Theil*, 227–29.

89. Ratzel, *Wandertage eines Naturforschers. 1. Theil*, 233.

90. Ratzel, *Anthropogeographie Erster Teil*, 399.

91. Ratzel, *Glücksinseln und Träume*, 195.

92. Fritz L. Kramer, "A Note on Carl Ritter, 1779–1859," *Geographical Review* 49, no. 3 (1959): 406–9.

93. Friedrich Ratzel, "Zu Karl Ritters hundertjährigem Geburtstage," in *Kleine Schriften von Friedrich Ratzel, herausgegeben von Hans Helmolt, 1. Band*, ed. Hans Helmolt (Munich: R. Oldenbourg, 1879 [1906]), 399.

94. Michael Heinrich, *Karl Marx and the Birth of Modern Society* (New York: Monthly Review Press, 2019), 215; Cornelia Lüdecke, *Militärische und zivile Geographie: Von Carl Ritters Geographie in Berlin bis zu Karl Haushofers Geopolitik in München* (Norderstedt: Die Bibliothek der Erde, 2019), 25, 44.

95. Ritter also had an influence on French geography. Both Vidal de la Blache and Élisée Reclus had heard his lectures. Lüdecke, *Militärische und zivile Geographie*, 133.

96. Ratzel, "Zu Karl Ritters hundertjährigem Geburtstage," 401.

97. Ratzel, *Anthropogeographie Erster Teil, Zweite Auflage*, 31.

98. Ritter, *Ueber das historische Element*, 158.

99. Carl Ritter, *Allgemeine Erdkunde: Vorlesungen an der Universität zu Berlin gehalten* (Berlin: Reimer, 1862), 12.

100. There was some overlap here with von Humboldt's *Cosmos* too, which had

argued that it was crucial "to consider each organism as part of the entire creation." Interestingly, von Humboldt acknowledged that a comparative geography that explained the phenomena on the earth's surface by reference to the earth system had been pioneered by Ritter. Alexander von Humboldt, *Cosmos: A Sketch of a Physical Description of the Universe, Vol. 1*. (Baltimore: Johns Hopkins University Press, 1845 [1997]), 42, 48.

101. Ritter, *Ueber das historische Element*, 1, original emphasis.

102. Ratzel, *Die Erde und das Leben* 2, 504.

103. von Humboldt, *Cosmos*, 24.

104. Ratzel, *Anthropogeographie Erster Teil, Zweite Auflage*, 33.

105. Ratzel, M. (1901) "Sehr geehrter Herr Professor!" Letter from Marie Ratzel to unknown, Leipzig, 18 November 1901, Archiv für Geographie, Leibniz Institut für Länderkunde, Nachlass Friedrich Ratzel 163/657. Wagner wrote to Ratzel in 1882: "I always thank fate that granted me in late age a friend who also fulfills the ideal of a son." Moritz Wagner, "Lieber Freund!", Letter to Friedrich Ratzel, n.p., 21 January 1882, Bayerische Staatsbibliothek, Ratzeliana, Supplement 13.

106. Ratzel, *Anthropogeographie Erster Teil*, 464; see also Schultz, "Eines schickt sich nicht für alle," 127–28.

107. Ratzel had encountered Wagner's work before the two men became close. Ratzel, *Sein und Werden*, 469.

108. Charlotte Weissman, "The Origins of Species: The Debate between August Weismann and Moritz Wagner," *Journal of the History of Biology* 43, no. 4 (2010): 729.

109. Hanno Beck, "Moritz Wagner als Geograph," *Erdkunde* 7, no. 2 (1953): 127.

110. Moritz Wagner, *The Darwinian Theory and the Law of the Migration of Organisms* (London: Edward Stanford, 1873), 75. Although now often considered of minor influence in the history of the emergence of evolutionary theory, Wagner's writings did have a notable afterlife in the mid-twentieth century. Ernst Mayr, "Isolation as an Evolutionary Factor," *Proceedings of the American Philosophical Society* 103, no. 2 (1943): 221–30.

111. Wagner, *The Darwinian Theory*, 75.

112. Frank J. Sulloway, "Geographic Isolation in Darwin's Thinking: The Vicissitudes of a Crucial Idea," *Studies in History of Biology* 3 (1979): 55.

113. Charles Darwin, *On the Origin of Species by Means of Natural Selection: Or, the Preservation of Favoured Races in the Struggle for Life, Sixth Edition, with Additions and Corrections to 1872* (London: Murray, 1859 [1876]), 81–82.

114. Peter J. Vorzimmer, *Charles Darwin: The Years of Controversy. The Origin of Species and its Critics 1859–1882* (Philadelphia: Temple University Press, 1970), 182.

115. John A. Campbell and David N. Livingstone, "Neo-Lamarckism and the Development of Geography in the United States and Great Britain," *Transactions of the Institute of British Geographers* 8, no. 3: 267–94; Halas, "Searching for the Perfect Footnote;" Ó Tuathail, *Critical Geopolitics*, 36. Wagner may have attacked Darwin from a Lamarckian position, but when debating critics of evolutionary theory, he self-evidentially spoke of the *Lamarck-Darwin'sche Descendenztheorie*. Moritz Wagner, *Die Entstehung der Arten durch räumliche Sonderung: Gesammelte Aufsätze von Moriz Wagner* (Basel: Benno Schwabe, 1889).

116. Friedrich Ratzel, *Raum und Zeit in Geographie und Geologie: Naturphiloso-*

phische Betrachtungen, herausgegeben von Dr Paul Barth (Leipzig: Johann Ambrosius Barth, 1907), 28.

117. Ratzel, *Sein und Werden*, 438.

118. "Weltentwicklung und Weltschöpfung," 582, 584. A social organism like a state may have experienced a life cycle for Ratzel, but it was difficult to determine such an organism's lifetime. There were no generations. On the question as to whether Haeckel too was ultimately a Lamarckian rather than a Darwinian, see Livingstone, *The Geographical Tradition*, 199; Weidling, "Ernst Haeckel, Darwinismus," 320, and rebuttal by Richards, *The Tragic Sense of Life*, 142–43.

119. Ratzel's letters to Wagner have not survived.

120. Moritz Wagner, "Lieber Freund," Letter to Friedrich Ratzel, Munich, 1 April 1877, Bayerische Staatsbibliothek, Ratzeliana, Supplement 13.

121. Moritz Wagner, "Mein herzlieber theurer Freund!" Letter to Friedrich Ratzel, Ammerland, 5 September 1881, Bayerische Staatsbibliothek, Ratzeliana, Supplement 13. See also Karl von Scherzer, "Biographische Skizze," In *Die Entstehung der Arten durch räumliche Sonderung: Gesammelte Aufsätze von Moriz Wagner*, ed. Moritz Wagner (Basel: Benno Schwabe, 1899), 9–32.

122. Moritz Wagner, "Mein herzlieber theurer Freund!" Letter to Friedrich Ratzel, Ammerland, 5 September 1881, Bayerische Staatsbibliothek, Ratzeliana, Supplement 13; Moritz Wagner, "Lieber Freund!" Letter to Friedrich Ratzel, n.p., 1 July 1885, Bayerische Staatsbibliothek, Ratzeliana, Supplement 13.

123. Friedrich Ratzel, "Moritz Wagner," in *Kleine Schriften von Friedrich Ratzel, herausgegeben von Hans Helmolt, 1. Band*, ed. Hans Helmolt, 461–75 (Munich: R. Oldenbourg, 1895 [1906]), 470.

124. Beck, "Moritz Wagner als Geograph," 126.

125. Moritz Wagner (1882) "Lieber Freund!", Letter to Friedrich Ratzel, Munich, 4 June 1882, Bayerische Staatsbibliothek, Ratzeliana, Supplement 13; Moritz Wagner, "Lieber Freund," Letter to Friedrich Ratzel, n.p., 24 December 1884, Bayerische Staatsbibliothek, Ratzeliana, Supplement 13; Moritz Wagner, "Die Kulturzüchtung des Menschen gegenüber der Naturzüchtung im Tierreich," in *Die Entstehung der Arten durch räumliche Sonderung: Gesammelte Aufsätze von Moriz Wagner*, Moritz Wagner (Basel: Benno Schwabe, 1886 [1889]), 519–39. On the way in which Ratzel appropriated Wagner's theory, see Jureit, *Das Ordnen von Räumen*, 142–46.

126. Sulloway, "Geographic Isolation in Darwin's Thinking," 58.

127. Steinmetzler, *Anthropogeographie Friedrich Ratzels*, 87. That said, Ratzel may have entertained doubts about Wagner's thesis. Steinmetzler has found a qualification of Wagner's separation thesis in the handwritten manuscript of *Die Erde und das Leben* that was not included in the book. Steinmetzler, *Anthropogeographie Friedrich Ratzels*, 95. Nonetheless, Ratzel stuck with Wagner until the end of his life and promised to make his ideas the basis of a new general biography and which he would sketch in the 1901 *Lebensraum* essay. Ratzel, *Anthropogeographie Erster Teil, Zweite Auflage*, xi.

128. Ratzel, "Moritz Wagner," 462, 473–74.

129. Moritz Wagner, *Darwin'sche Theorie und das Migrationsgesetz der Organismen* (Munich: Akademische Buchdruckerei von F. Straub, 1868), 28.

130. Ratzel, *Anthropogeographie Zweiter Teil, Dritte Auflage*, 348.

131. Ratzel, *Anthropogeographie Erster Teil*, 123.

132. Ratzel, *Anthropogeographie Zweiter Teil, Dritte Auflage*, 424.

133. Kathryn Yusoff, *A Billion Black Anthropocenes or None* (Minneapolis: University of Minnesota Press, 2018), 77.

134. Ratzel, *Anthropogeographie Zweiter Teil, Dritte Auflage*, 424.

135. Ratzel, *Wider die Reichsnörgler*, 20.

136. Ratzel, *Das Meer als Quelle*.

137. Friedrich Ratzel, "Die deutsche Tiefsee-Expedition," in *Kleine Schriften von Friedrich Ratzel, herausgegeben von Hans Helmolt, 2. Band*, ed. Hans Helmolt (Munich: R. Oldenbourg, 1898 [1906]), 382–90.

138. Ratzel, "Lebensraum—A Biogeographical Study," 73.

139. Carlotta Santini, "At the Origins of Modern Geography. The Oecumene: An Anthropogeographical Pattern," *History of European Ideas* 43, no. 6 (2017): 560–69.

140. One book that did use maps and drawings quite extensively was his 1902 *Die Erde und das Leben*.

141. Elden, "Leibniz and Geography."

142. Halford Mackinder, *Britain and the British Seas* (London: William Heinemann, 1902), 357.

143. Ratzel, *Anthropogeographie Zweiter Teil, Dritte Auflage*, V, 7. It is instructive to compare Ratzel's work to that of Halford Mackinder in this regard. For a longer discussion see Ian Klinke, "On the History of a Subterranean Geopolitics," *Geoforum* 127 (2021): 356–63. Mackinder very much *was* a flat earth thinker, despite an occasional insistence on the vertical. Halford Mackinder, "Modern Geography, German and English," *The Geographical Journal* 6, no. 4 (1895): 369. Although similarly concerned with climatic conditions and the distribution of natural resources, Mackinder's materialism was primarily of a horizontal kind. If we get a sense of the vertical in Mackinder, it is only in his elevation above geographic space in a God-like position, either as a political geographer who observed an objectified map, or as an explorer of Mount Kenya. Halford Mackinder, "A Journey to the Summit of Mount Kenya, British East Africa," *The Geographical Journal* 15, no. 5 (1900): 453–76. See also Kearns, *Geopolitics and Empire*, Ó Tuathail, *Critical Geopolitics*. He even actively resisted the vertical turn in military strategy that had made itself felt in the interwar period. Halford Mackinder, "The Round World and the Winning of Peace," *Foreign Affairs* 21, no. 4 (1943): 602.

144. Natter, "Friedrich Ratzel's Spatial Turn." Much the same could be said about Ratzel's political geography, which he hoped would compete with political science and sociology for the interpretive supremacy over *the state*.

145. Ratzel, *Anthropogeographie Erster Teil, Zweite Auflage*, 76.

146. Ratzel, *Anthropogeographie Zweiter Teil, Dritte Auflage*, xxiii.

147. Friedrich Ratzel, "Die Zeitforderung in den Entwicklungswissenschaften," *Annalen der Naturphilosophie* 1/2 (1902): 40–98; Ratzel, *Raum und Zeit*. While the 1907 book covered similar ground to that in his 1881 *Die Erde*, the latter did not have any of the theoretical depth of his later work.

148. Ratzel, *Raum und Zeit*, 1.

149. Ratzel, *Raum und Zeit*, 13.

150. von Humboldt, *Cosmos*, 65. Unlike von Humboldt, Ratzel had come under the influence of thermodynamics, particularly via the influence of Julius von Mayer's energeticism, a theory the 1899 first volume of *Anthropogeographie* flagged up in its very first pages. Thomas M. Turnbull, "From Incommensurability to Ubiquity: An Energy History of Geographic Thought," *Journal of Historical Geography* 73 (2021): 13.

151. Friedrich Ratzel, "Geschichte, Völkerkunde und historische Perspektive," *Historische Zeitschrift* 93, no. 1 (1904): 28. See also Schultz, "Friedrich Ratzel—Pionier moderner Sozialgeographie?" 64.

152. Stephan Günzel, *Raum: Eine kulturwissenschaftliche Einführung* (Bielefeld: Transcript, 2017), 21. See also May, Jon and Thrift, Nigel. "Introduction," In *Timespace: Geographies of temporality*, ed. Jon May and Nigel Thrift (London: Routledge, 2001), 11.

153. "Weltentwicklung und Weltschöpfung," 576–77.

154. Ratzel, *Raum und Zeit*, 2.

155. Note that Ratzel gave a slightly different account in *Die Erde und das Leben*, emphasizing here the difficulty of accounting for "the cosmic proportions of space and time" through the spatiotemporal measurements on our planet. Ratzel, *Die Erde und das Leben* 2, 85.

156. Ratzel, *Die Erde und das Leben* 2, 36.

157. Elden, "Reassessing Kant's Geography," 21.

158. Ratzel, "Lebensraum—A Biogeographical Study," 60. Schultz notes that this emphasis of the earth's finitude can already be found in Kant. Hans-Dietrich Schultz, "Pax geographica: Räumliche Konzepte für Krieg und Frieden in der geographischen Tradition," *Geographische Zeitschrift* 75, no. 1 (1987): 7. The conflict between an expansive population and a finite earth is also contained in Thomas Robert Malthus, *An Essay on the Principle of Population* (Oxford: Oxford University Press, 1798 [2004]), 14.

159. Ratzel, "Lebensraum—A Biogeographical Study," 71 original emphasis; see also Barua, "Ratzel, Bio-geography."

160. Buttmann, *Friedrich Ratzel*, 107.

161. Ratzel, *Über Naturschilderung*, 249.

162. Ratzel, *Über Naturschilderung*, 250.

163. Ratzel, *Über Naturschilderung*, 137, 249.

164. Smith, "Friedrich Ratzel," 52.

165. Ratzel, *Über Naturschilderung*, 8.

166. Wanklyn, *Friedrich Ratzel*, 38.

167. Ratzel, *Über Naturschilderung*, 33. Ratzel had no sense of historical change in aesthetic regimes, he simply wanted to determine the laws of aesthetics.

168. Ratzel, *Über Naturschilderung*, 34.

169. Ratzel, *Über Naturschilderung*, 64.

170. Ratzel, *Über Naturschilderung*, 16.

171. Ratzel, *Glücksinseln und Träume*, 13.

172. Lüdecke, "Ein genußreiches Zusammenleben," 27.

173. Karl A. Wittfogel, "Geopolitics, Geographical Materialism and Marxism," *Antipode* 17, no. 1 (1929 [1985]): 25.

174. Ratzel believed that the agricultural revolution had been stimulated by nature itself. Ratzel, *Anthropogeographie Erster Teil, Zweite Auflage*, 492.

175. Although Ratzel did, on occasion, mention a socioeconomic sphere that interfered in between the soil and the *Volk*, he did not pursue this line of thought. Hans-Dietrich Schultz, "'Hätte doch die Erde mehr Raum!' Friedrich Ratzel und sein (politisch-) geographisches Weltbild," *Mitteilungen der Geographischen Gesellschaft in München* 89 (2007): 22.

176. August Bebel, *Woman under Socialism* (New York: New York Labor News Company, 1879 [1904]), 323. Müller, *Friedrich Ratzel*, 41.

177. Ratzel, *History of Mankind I*, 125.

178. Mark Bassin, "Nature, Geopolitics and Marxism: Ecological Contestations in Weimar Germany," *Transactions of the Institute of British Geographers* 21, no. 2 (1996): 322.

179. Bassin, "Race contra Space;" Chiantera-Stutte, "Ratzel's Stone Guest," 96; Tom Johnston, "Environmental Determinism," in *AAG International Encyclopaedia of Geography*, ed. Douglas Richardson et al. (Oxford: Wiley-Blackwell, 2017), 1–3; David N. Livingstone, "Environmental Determinism," in *The Sage Handbook of Geographical Knowledge*, ed. John Agnew and David N. Livingstone (London: Sage, 2011), 374; Schultz, *Herder und Ratzel*, 140.

180. Natter, "Friedrich Ratzel's Spatial Turn;" Stogiannos, *The Genesis of Geopolitics*.

181. Ratzel, *Anthropogeographie Zweiter Teil, Dritte Auflage*, 389.

182. David T. Murphy, "'Retroactive Effects': Ratzel's Spatial Dynamics and the Expansionist Imperative in Interwar Germany," *Journal of Historical Geography* 61 (2018): 90.

183. Ratzel, "Studies in Political Areas II," 457, emphasis added.

184. Smith, "Friedrich Ratzel," 52.

185. Schultz, *Herder und Ratzel*, 140; Rainer Sprengel, *Kritik der Geopolitik. Ein deutscher Diskurs, 1914–1944* (Berlin: Akademie, 1996), 192.

186. Ratzel, *Anthropogeographie Erster Teil, Zweite Auflage*, 5. Steinmetzler has argued that Ratzel abandoned his radical evolutionary materialism in his later work and began to distinguish more systematically between body and mind from his 1881 *Die Erde* onwards. Steinmetzler, *Anthropogeographie Friedrich Ratzels*, 50. But as the discussion in this chapter shows, there was considerably more ambiguity in Ratzel than Steinmetzler acknowledges.

187. Ratzel, *Anthropogeographie Erster Teil, Zweite Auflage*, 6.

188. Ratzel, *Anthropogeographie Erster Teil, Zweite Auflage*, 1.

189. Ratzel, *Anthropogeographie Erster Teil*, 76–77.

190. Ratzel, "Über Kalifornien," 4.

191. Ratzel, *Die Erde und das Leben* 2, 296.

192. Ratzel, *Die Erde und das Leben* 2, 651.

193. Ratzel, *Anthropogeographie Zweiter Teil, Dritte Auflage*, xxiv.

194. Ratzel, "Lebensraum—A Biogeographical Study," 75.

195. Friedrich Ratzel, "Über die anthropogeographischen Begriffe Geschichtliche Tiefe und Tiefe der Menschheit," in *Kleine Schriften von Friedrich Ratzel, herausgege-*

ben von Hans Helmolt, 2. Band, ed. Hans Helmolt (Munich: R. Oldenbourg, 1890 [1906]), 288.

196. Ratzel, *Anthropogeographie Erster Teil, Zweite Auflage*, 49.
197. Ratzel, *Politische Geographie: Geographie der Staaten*, 759.
198. Ratzel, "Lebensraum—A Biogeographical Study," 70.
199. Sauer, "The Formative Years of Ratzel."
200. Mildenberger, "Lebensraum oder Umwelt?" 119.

CHAPTER 5

1. Otto von Bismarck, "Rede vor dem Reichstag, 26. Juni 1884," in *Otto von Bismarck, Werke in Auswahl: Reichsgestaltung und Europäische Friedenswahrung, 1883–1890*, ed. Gustav Adolf Rein et al. (Darmstadt: Wissenschaftliche Buchgesellschaft, 1884 [2001]), 162–80. The fact that German colonies would be known as "protectorates" (*Schutzgebiete*) was a consequence of this earlier hesitation toward the idea of territorial annexation. Sebastian Conrad, *German Colonialism: A Short History* (Cambridge: Cambridge University Press, 2012), 23.

2. Daniel Frymann, *Wenn ich der Kaiser wär': Politische Wahrheiten und Notwendigkeiten* (Leipzig: Dieterich, 1912), 200; see also Arendt, *Origins of Totalitarianism*, 151.

3. Ratzel, *Wider die Reichsnörgler*, 8.

4. Ratzel, *Wider die Reichsnörgler*, 23.

5. Ratzel's speech was published a few months later as a pamphlet under the title *Wider die Reichnörgler*, loosely translatable as "Against those who whine about the empire."

6. Ratzel, *Wider die Reichsnörgler*, 8. Ratzel believed that although Germanness (*Deutschtum*) in North America was being replenished via emigration, there was not enough German offspring. Even in states in which Germans had managed to establish their own newspapers and schools, only very few of them made it into the political elite, he bemoaned. "Aus Deutschamerika," *Die Grenzboten* 54 (1895): 403–4; "Zur Geschichte des Deutschtums in Nordamerika," *Die Grenzboten* 56 (1897): 519–522.

7. Ratzel, *Wider die Reichsnörgler*, 21.

8. Ratzel, *Wider die Reichsnörgler*, 11.

9. Schulte-Althoff, *Studien zur politischen Wissenschaftsgeschichte*, 93–94.

10. Ratzel, *Politische Geographie.*

11. Ratzel, *Politische Geographie: Geographie der Staaten*, 382.

12. Ratzel, *Wandertage eines Naturforschers. 2. Theil.*

13. F. R., "Ein gutes Ziel für deutsche Auswanderung I," *Allgemeine Zeitung* 118 (1880): 1713–1714; R. F., "Ein gutes Ziel für deutsche Auswanderung II," *Allgemeine Zeitung* 119 (1880): 1729–30. Even in later years, Ratzel did not entirely give up the hope that Germans would continue to colonize Argentina. Ratzel, *Glücksinseln und Träume*, 471.

14. "Dr Karl Peters," *Die Grenzboten* 56 (1897): 252–56.

15. Friedrich Fabri, *Bedarf Deutschland der Kolonien? Eine politisch-ökonomische*

Betrachtung (Gotha: Perthes, 1879); Bartholomäus Grill, *Wir Herrenmenschen: Unser rassistisches Erbe: Eine Reise in die deutsche Kolonialgeschichte* (Munich: Siedler, 2019), 24. Ratzel later praised Fabri's pamphlet as instrumental in having bundled early colonial energies into a coherent movement. F. R., "5 Jahre deutscher Kolonialpolitik," *Allgemeine Zeitung* 157 (1889): 2377–78.

16. Hell, *The Conquest of Ruins*, 279.

17. F. Fabri, "Wertester Herr Professor!" Letter to Friedrich Ratzel, Berlin, 15 November 1885, Bayerische Staatsbibliothek, Ratzeliana Supplement 11. In a later letter, Fabri shared with Ratzel his fears about how his criticism of Bismarck's colonial policy would be reviewed in the German press. It is clear from these surviving letters that the two men shared a personal bond and were politically aligned. F. Fabri, "Wertester Herr und Freund!", Letter to Friedrich Ratzel, Godesberg, 9 May 1889, Bayerische Staatsbibliothek, Ratzeliana Supplement 11.

18. Karl Weule, "Friedrich Ratzel," *Deutsche Erde* 3 (1904): 151. Ratzel was a signatory of an early appeal for colonization that was reprinted in most German newspapers. In Ratzel's Bavarian domicile, the new pan-German colonialism was less popular. The appeal was rejected by one Bavarian newspaper editor with the words, "I and most of my readers have no interest in colonies, except if you find me one into which one could export a few thousand Prussians." Deutsche Kolonialgesellschaft, *Die Deutsche Kolonialgesellschaft: 1882–1907* (Berlin: Dietrich Heimer, 1908).

19. "Die Aussichten unsers südwestafrikanischen Schutzgebietes," *Die Grenzboten* 51 (1892): 171–75; "Deutschands Stellung und Rechte am Niger," *Die Grenzboten* 56 (1897): 545–51; Ratzel, "Über politische Verhältnisse in Innerafrika." Ratzel was commenting here at a distance on these endeavors, which had both a scientific and a military function. German geographers played a crucial role in the reconnaissance of the German colonial hinterland. Schulte-Althoff, *Studien zur politischen Wissenschaftsgeschichte*, 123.

20. Georg Hartmann, *Einfluss des Treibeises auf die Bodengestalt der Polargebiete: Inaugural-Dissertation zur Erlangung der Doktorwürde* (Leipzig: Verein für Erdkunde, 1891).

21. Ratzel, "Über politische Verhältnisse in Innerafrika," 159; see also Matthew Unagst, "Men of Science and Action: The Celebrity of Explorers and German National Identity, 1870–1895," *Central European History* 50, no. 3 (2017): 315.

22. "Koloniale Biergespräche," *Die Grenzboten* 54 (1895): 246–47.

23. Arendt, *Origins of Totalitarianism*, 189.

24. F. Ratzel, "Verehrter Herr College!" Letter to Ernst Hasse, Leipzig, 6 August 1889, Archiv für Geographie, Leibniz Institut für Länderkunde, Nachlass Ernst Hasse 437/23.

25. F. Ratzel, "Sehr geehrter Herr!" Letter to Carl Peters, Ammerland am Würmsee, 28 August 1890, Bundesarchiv, Nachlass Carl Peters N 2223/32.

26. F. Ratzel, "Notitzen zu zwei Reden (eine davon an Dr. Peters gerichtet)" Bayerische Staatsbibliothek, Ratzeliana Supplement 4.

27. Friedrich Ratzel, "Das Buch des Dr. Karl Peters," *Die Grenzboten* 50 (1891): 601.

28. In the period 1894–1897 there was a growing concern, particular in the Social

Democrat and Zentrum parties, with the violent manner in which a number of colonial administrators, including Heinrich Leist in Cameroon and Carl Peters in East Africa, ruled over their subjects. Indeed, Peters was soon known as *Hängepeters* because of his habit of hanging for personal reasons. He would be stylized into a hero of the Third Reich in the 1941 film *Carl Peters*, which starred the famous Hans Albers in the lead role. Winfried Speitkamp, *Deutsche Kolonialgeschichte* (Stuttgart: Reclam, 2008), 138–39.

29. "Dr Karl Peters," 254.

30. W. Hübbe-Schleiden, "Verehrter Herr Professor!" Letter to Friedrich Ratzel, Barmen, 18 July 1881, Bayerische Staatsbibliothek, Ratzeliana Supplement 12.

31. Letters from Drygalski to Ratzel reveal that the former kept the latter updated about Germany's first Antarctic expedition and detailed his difficulties in getting the imperial bureaucracy to sign off on the budget. E. von Drygalski, "Sehr geehrter Herr Doktor!" Letter to unknown, Königsberg, 5 January 1899, Bayerische Staatsbibliothek, Ratzeliana Supplement 11; E. von Drygalski, "Hochverehrter Herr Geheimrat!" Letter to Friedrich Ratzel, Potsdam, 5 January 1899, Bayerische Staatsbibliothek, Ratzeliana Supplement 11. On Hans Meyer see Boris Michel, "Making Mount Kilimanjaro German: Nation Building and Heroic Masculinity in the Colonial Geographies of Hans Meyer," *Transactions of the Institute of British Geographers* 44, no. 3 (2019): 493–508.

32. O. Baumann, "Hochgeehrter Herr Professor," Letter to Friedrich Ratzel, Vienna, 11 May 1887, Bayerische Staatsbibliothek, Ratzeliana Supplement 11.

33. H. Grothe, "Hochverehrter Herr Professor!" Letter to Friedrich Ratzel, Munich, 30 August 1903, Bayerische Staatsbibliothek, Ratzeliana Supplement 12.

34. H. Klutschak, "Sehr geehrter Herr!" Letter to Friedrich Ratzel, Bremen, 29 November 1881, Bayerische Staatsbibliothek, Ratzeliana Supplement 12.

35. O. Baumann, "Sehr geehrter Herr Professor," Letter to Friedrich Ratzel, Berlin, 8 December 1889, Bayerische Staatsbibliothek, Ratzeliana Supplement 11; Grill, *Wir Herrenmenschen*, 36.

36. Nachtigal, G. (1880) "Hochgeehrter Herr Professor," Letter to Friedrich Ratzel, Berlin, 28 June 1880, Bayerische Staatsbibliothek, Ratzeliana Supplement 12.

37. Friedrich Ratzel, "Gustav Nachtigal," in *Allgemeine Deutsche Biographie XXIII*, ed. Historische Commission bei der Königl. Akademie der Wissenschaften, 193–99. (Leipzig: Duncker und Humblot, 1886), 197.

38. G. Hartmann, "Hochgeehrter Herr Professor!" Letter to Friedrich Ratzel, Chorab, 6 October 1900, Bayerische Staatsbibliothek, Ratzeliana Supplement 12.

39. F. Ratzel, "Verehrter Herr Kollege!" Letter to Ernst Hasse, Leipzig, 18 October 1888, Archiv für Geographie, Leibniz Institut für Länderkunde, Nachlass Ernst Hasse 437/5.

40. Friedrich Ratzel, "H. Wißmann, L. Wolf, C. v. Francois und H. Müller, Im Innern Afrikas, Leipzig 1888," *Geographischer Literaturbericht* 359 (1888): 80; Friedrich Ratzel, "H. Wißmann, Unter deutscher Flagge quer durch Afrika von West nach Ost, Berlin 1889," *Geographischer Literaturbericht* 135 (1890): 19.

41. F. Rl., "H. F. v. Behr, Kriegsbilder aus dem Araberaufstand in Deutsch-Ostafrika, Leipzig 1891," *Literarisches Zentralblatt* 14 (1891): 453.

42. Friedrich Ratzel, "Stabsarzt Dr. Ludwig Wolf," *Beilage zur Allgemeinen Zeitung* 71 (1890): 1. See also Friedrich Ratzel, "Stabsarzt Dr. Ludwig Wolf," *Mitteilungen des Vereins für Erdkunde zu Leipzig* 1889 (1890d): 104–14. Wolf had helped Ratzel edit some of his essays. Ratzel, "Über politische Verhältnisse in Innerafrika," 160.

43. Matthew Unagst, "Men of Science and Action: The Celebrity of Explorers and German National Identity, 1870–1895," *Central European History* 50, no. 3 (2017): 305–27.

44. F. R., "H. M. Stanley, Durch den dunkeln Weltteil, Bd. I, Leipzig 1878," *Literarisches Zentralblatt* (1878): 1570–71. In 1890, he noted that he felt both "admiration" and "disgust" for Stanley. Friedrich Ratzel, "Stanleys neues Buch." *Beilage zur Allgemeinen Zeitung* 168 (1890): 1.

45. "Briefe und Berichte Emin Paschas," *Beilage zur Allgemeinen Zeitung* 81, 1888: 1185–1187; Ratzel, *Anthropogeographie Zweiter Teil, Dritte Auflage*, 246; Ratzel, "Emin Pascha," 211–23; Friedrich Ratzel, "Emin Pascha," In *Kleine Schriften von Friedrich Ratzel, herausgegeben von Hans Helmolt, 1. Band*, ed. Hans Helmolt (Munich: R. Oldenbourg, 1904 [1906]), 511–18; Friedrich Ratzel et al., "Eingabe an das Reichskanzleramt in Sachen Emin Pascha," In *Kleine Schriften von Friedrich Ratzel, herausgegeben von Hans Helmolt, 1. Band*, ed. Hans Helmolt (Munich: R. Oldenbourg, 1891 [1906]), 519–21; Schweinfurth and Ratzel, *Emin-Pascha*.

46. Felix Driver, *Geography Militant: Cultures of Exploration and Empire* (Oxford: Blackwell, 2001), 139–43; Jeffrey Meyers, "Henry Morton Stanley and Emin Pasha: New Historical Sources for *Heart of Darkness*," *Conradiana* 48, no. 1 (2016): 77–82.

47. Despite his service to the Ottoman Empire, the signatories of the letter described Emin Pascha as an embodiment of Germany's status as a "scientific great power." Ratzel, "Über politische Verhältnisse in Innerafrika."

48. F. Ratzel, "Verehrter Herr College!" Letter to Ernst Hasse, Leipzig, 15 December 1890, Archiv für Geographie, Leibniz Institut für Länderkunde, Nachlass Ernst Hasse 437/24.

49. "Zur Kenntnis der englischen Weltpolitik," 13.

50. Ratzel, "Zukunft und Beurteilung," 100.

51. Ratzel, *Die Vereinigten Staaten, Zweiter Band, Zweite Auflage*, 95.

52. Ratzel, *Politische Geographie: Geographie der Staaten*, 155.

53. Ratzel, *Die Vereinigten Staaten, Zweiter Band, Zweite Auflage*, 96.

54. Ratzel, *Politische Geographie: Geographie der Staaten*, 158; Ratzel, "Geschichte, Völkerkunde," 487.

55. Ratzel, *Politische Geographie: Geographie der Staaten*, 151.

56. Ratzel, *Politische Geographie: Geographie der Staaten*, 152.

57. Ratzel, "Der Ursprung der Arier," 393.

58. Ratzel, *Politische Geographie: Geographie der Staaten*, 136.

59. Ratzel, "Lebensraum—A Biogeographical Study," 66.

60. Hell argues that ancient Rome was the model both for Ratzel's imperial fantasy and for his ruin gaze, an issue to which we will turn in the following chapter. But it is crucial to note that more often it was in fact the United States, read too as a colonial state, that served as the prototype for Ratzel. Hell, *The Conquest of Ruins*.

61. Felicity Rash, *The Discourse Strategies of Imperialist Writing: The German*

Colonial Idea and Africa, 1848–1945 (London: Routledge, 2017), 89. See also Zygmunt Bauman, *Modernity and the Holocaust* (Cambridge: Polity, 1989).

62. "Deutscher Kolonien- und Zeitungsklatsch," *Die Grenzboten* 55 (1896): 527.

63. "Beurteilung der Neger," 20.

64. Ó Tuathail, *Critical Geopolitics*; Werber, *Geopolitik zur Einführung*; Murphy, *Heroic Earth*; George Steinmetz, "Geopolitics," In *The Wiley-Blackwell Encyclopedia of Globalization*, ed. George Ritzer (Hoboken, NJ: Wiley-Blackwell, 2012), 1–23.

65. On a French variant of geopolitics and how it relates to its British and German counterparts, see Herman van der Wusten and Gertjan Dijkink, "German, British and French Geopolitics: The Enduring Differences," *Geopolitics* 7, no. 3 (2002): 19–38.

66. Guzzini, *The Return of Geopolitics*.

67. Even more than political realism, geopolitics has operated as a policy-oriented and low-brow political theory. While realism tends to express a status quo doctrine, geopolitics often promotes territorial expansionism; whereas realism moreover emerged in the fields of diplomatic history, political theory, and international relations, geopolitics was institutionalized in the discipline of geography. For a detailed discussion see Guzzini, *The Return of Geopolitics*.

68. Alfred Thayer Mahan, *The Influence of Sea Power upon History 1660–1805* (London: Bison, 1890 [1980]), 33.

69. Murphy, *Heroic Earth*, 3.

70. Kearns, *Geopolitics and Empire*; Halford Mackinder, *The First Ascent to Mount Kenya* (Athens: Ohio University Press, 1991).

71. Mackinder, "The Geographical Pivot of History."

72. Mackinder, "The Geographical Pivot of History"; Halford Mackinder, *Democratic Ideals and Reality: A Study in the Politics of Reconstruction* (London: Faber and Faber, 1919 [1942, reprinted 2009].

73. Mackinder, *Britain and the British Seas*.

74. At the heart of this German "school" of geopolitics, founded by Karl Haushofer in Friedrich Ratzel's name after the First World War (more on which in chapter 7), was the belief in war as the natural condition of the interstate system, a concern with Germany's jammed centric position (*Mittellage*) in Europe, and, ultimately, the aim of constructing a version of the Monroe doctrine: a greater continental space under German control. Bassin, "Between Realism and the 'New Right.'"

75. Ratzel, *Anthropogeographie Erster Teil, Zweite Auflage*, 579. Mackinder, on the other hand, was a close reader of Ratzel's *Anthropogeographie*. Kearns, *Geopolitics and Empire*, 4.

76. Sprengel, *Kritik der Geopolitik*, 72. Mackinder, too, spoke of the world as a "political and economic organism" in his 1904 piece. Mackinder, "The Geographical Pivot," 422.

77. Stogiannos, *The Genesis of Geopolitics*, 26.

78. Ratzel, *Anthropogeographie Erster Teil, Zweite Auflage*, 64. Neither Mahan, Mackinder, nor Ratzel used the term "geopolitics." Ratzel spoke of *Weltpolitik* (world politics), which complicates Smith's distinction between *Weltpolitik* and *Lebensraum* traditions in German political discourse. Woodruff D. Smith, "'Weltpolitik' und 'Lebensraum,'" in *Das Kaiserreich transnational: Deutschland in der Welt 1871–1914*, ed.

Sebastian Conrad and Jürgen Osterhammel (Göttingen: Vandenhoeck & Ruprecht, 2004), 29–48.

79. Ratzel, *Anthropogeographie Zweiter Teil, Dritte Auflage*, 247.

80. Ratzel, *Politische Geographie: Geographie der Staaten*, 248–49.

81. Ratzel was not making an argument here for the effects of the absence of overarching authority above the sovereign state (anarchy) on state behavior. As Hobson argues, Eurocentric and racist theories of international relations operate with an understanding of "formal hierarchy," where sovereignty is unequally distributed. Hobson, *Eurocentric Conception*, 19.

82. Ratzel, *Die Erde und das Leben* 2, 672.

83. "Deutschlands Lage," *Die Grenzboten* 55 (1896): 105–8.

84. "Nicht schielen!" *Die Grenzboten* 51 (1892c): 412. Ratzel did not always distinguish rigidly between states, *Völker*, or races as the decisive agents of history. In the late Ratzel, states were treated as the key actors.

85. Karl Haushofer, *Geopolitik der Pan-Ideen* (Berlin: Zentral, 1931); Carl Schmitt, *Völkerrechtliche Großraumordnung mit Interventionsverbot für raumfremde Mächte* (Berlin: Duncker and Humblot, 1941 [1991]).

86. Ratzel, *Politische Geographie*: *Geographie der Staaten*, 329, 342, 343, 504. More frequently, however, he spoke of *weite Räume* (wide spaces).

87. Ratzel, *Politische Geographie: Geographie der Staaten*, 402.

88. Ratzel, *Politische Geographie: Geographie der Staaten*, 380.

89. Ratzel, *Die Vereinigten Staaten, Zweiter Band, Zweite Auflage*, 86.

90. Ratzel, *Politische Geographie*, 189; Friedrich Ratzel, "Der Nicaraguakanal und die Monroe-Doktrin," *Beilage zur Allgemeinen Zeitung* 227 (1897): 1–7; "Zur Kenntnis der englischen Weltpolitik: 1," *Die Grenzboten* 54 (1895): 52; Friedrich Ratzel, "Die nordatlantischen Mächte. Eine politisch-geographische Studie [II]," *Marine-Rundschau* 7–12 (1903): 1055.

91. Ratzel, *Die Vereinigten Staaten, Zweiter Band, Zweite Auflage*, 86.

92. Friedrich Ratzel, "Der mitteleuropäische Wirtschaftsverein," *Die Grenzboten* 63 (1904): 253–54.

93. Ratzel, "Der mitteleuropäische Wirtschaftsverein," 255. Stogiannos certainly goes too far in classifying Ratzel as an early proponent of European integration. Stogiannos, *The Genesis of Geopolitics*, 164. Ratzel clearly limits his vision for Europe to that of a customs union. Bassin, "Imperialism and the Nation State," 489. And yet Ratzel's 1904 article was certainly uncharacteristic of the broader thrust of his political interventions, both because it was willing to think Germany's national interest in a more transnational frame and because it saw the United States as an opponent.

94. Ratzel also published in 1903 two pieces on the "North Atlantic powers." Friedrich Ratzel, "Die nordatlantischen Mächte. Eine politisch-geographische Studie [I]," *Marine-Rundschau* 7–12 (1903): 911–39; Ratzel, "Die nordatlantischen Mächte [II]." These essays emphasized the practical aim of his political geography, its objective of informing policy. The Atlantic, he claimed was dominated by the "culturally able white race." Ratzel, "Die nordatlantischen Mächte [I]," 920. Germany was not an Atlantic power in terms of its geographic position but its rising profile as a colonial and trading power was drawing into the formation of a greater political and economic space, "the

North Atlantic family, as we should prefer to call it." Ratzel, "Die nordatlantischen Mächte [II]," 1062.

95. Ratzel, *Politische Geographie: Geographie der Staaten*, 259.

96. "Die geographische Lage Deutschlands." *Die Grenzboten* 55 (1896): 390.

97. Ratzel, *Deutschland: Eine Einführung*, 16.

98. "Die geographische Lage Deutschlands," 393.

99. Ratzel, *Politische Geographie: Geographie der Staaten*, 310; *Deutschland: Eine Einführung*, 18.

100. "Die geographische Lage Deutschlands," 456.

101. Ratzel, *Deutschland: Eine Einführung*, 20.

102. Schulte-Althoff, *Studien zur politischen Wissenschaftsgeschichte*, 165. The naval debate had gripped the German public in full force shortly after the publication of *Politische Geographie* in 1897.

103. Ratzel stood out from a vast array of German naval officers and enthusiasts who wrote vigorously about the need for naval armament at the turn of the century by framing the issue as one of *geographical* necessity. Eckart Kehr, *Schlachtflottenbau und Parteipolitik 1894–1901* (Berlin: Emil Ebering, 1930), 389.

104. Santini "One Sea, One Humanity," 22.

105. Hassert, "Friedrich Ratzel. Sein Leben und Wirken I," 323; Rolf Hobson, *Imperialism at Sea: Naval Strategic Thought, the Ideology of Sea Power, and the Tirpitz Plan, 1875–1914* (Boston and Leiden: Brill, 2002), 290–91; Curt von Maltzahn, "Friedrich Ratzel. Ein Gedenkwort," *Marine-Rundschau* 2 (1905): 217–20.

106. Ratzel, *Das Meer als Quelle*, v; see also Ratzel *Politische Geographie: Geographie der Staaten*, 723.

107. Ratzel, *Das Meer als Quelle*, 5.

108. Ratzel, *Das Meer als Quelle*, 54.

109. Ratzel, *Das Meer als Quelle*, 58.

110. Ratzel, *Das Meer als Quelle*, 43.

111. Ratzel, "Weiße und Farbige," 377.

112. Eduardo Mendieta, "War as the School of Space: The Space of War and the War of Space," *Ethics, Place and Environment* 9, no. 2 (2006), 217.

113. Ashworth, "Realism and the Spirit of 1919," 283. Although Mahan was a military strategist and not a biologist, he too occasionally resorted to Darwinian language in making sense of state competition. Kearns, *Geopolitics and Empire*, 611.

114. Ratzel, *Das Meer als Quelle*, 73, 85.

115. Ratzel, *Das Meer als Quelle*, 85.

116. Ratzel, *Politische Geographie*, 20.

117. Ratzel, *Wider die Reichsnörgler*, 28.

118. F. Ratzel, "Diary entry 13 September 1881," *Tagebuch über Englandreise im Aug/Sep 1881, Abschrift angefertigt von Günter Buttmann* Bayerische Staatsbibliothek, Ratzeliana Supplement 17a VI 9a. In August and September of 1881, Ratzel visited England and Wales with stops in London, Salisbury, Bath, Gloucester, Aberystwyth, Bangor, and Chester. His diary entries are filled with landscape description and architectural and anthropological observations. Aside from a conversation with a fellow traveler about the slave trade, they convey little about England and its politics. He seems not

to have met any geographers on his journey, but then geography was yet to develop as a discipline within the British academy. F. Ratzel, *Tagebuch über Englandreise im Aug/ Sep 1881, Abschrift angefertigt von Günter Buttmann* Bayerische Staatsbibliothek, Ratzeliana Supplement 17a VI 9a.

119. His student Hettner thought it was his reading of *The Times* which made him such a versatile political commentator. Alfred Hettner, "Bücherbesprechungen: Friedrich Ratzel. Glücksinseln und Träume; Kleine Schriften." *Geographische Zeitschrift* 13 (1907): 528.

120. See "Zur Kenntnis der englischen Weltpolitik 1" for his opening statement.

121. "Die Maske ab!" *Die Grenzboten* 53 (1894): 338; "Zur Kenntnis der englischen Weltpolitik 6," 9. Ratzel's position on the Boers was actually more ambiguous than this statement suggests, and he was criticized by some of his contemporaries for his lack of enthusiasm for the Boers. F. Ratzel, "Lieber Kollege!" Letter to K. Hassert, Leipzig, 8 January 1900, Archiv für Geographie, Leibniz Institut für Länderkunde, Nachlass Friedrich Ratzel 163/657.

122. "Nochmals die englische Heuchelei." *Die Grenzboten* 53 (1894): 473.

123. Ashworth, "Realism and the Spirit of 1919," 282.

124. "Die Maske ab!" 340.

125. "Zur Kenntnis der englischen Weltpolitik," 6.

126. Ratzel, *Wider die Reichsnörgler*, 28–29.

127. Friedrich Ratzel, "Flottenfrage und Weltlage," in *Kleine Schriften von Friedrich Ratzel, herausgegeben von Hans Helmolt, 2. Band*, ed. Hans Helmolt (Munich: R. Oldenbourg, 1898 [1906]), 377.

128. F. Ratzel, "Lieber Kollege!" Letter to K. Hassert, Leipzig, 8 January 1900, Archiv für Geographie, Leibniz Institut für Länderkunde, Nachlass Friedrich Ratzel 163/657.

129. F. Ratzel, "Lieber Kollege!", Letter to K. Hassert, Leipzig, 8 January 1900, Archiv für Geographie, Leibniz Institut für Länderkunde, Nachlass Friedrich Ratzel 163/657.

130. "Zur Kenntnis der englischen Weltpolitik," 21.

131. Wanklyn, *Friedrich Ratzel*, 10. Unwilling to acknowledge the high human cost of the war, he claimed that France and Germany had easily coped well with their losses. Ratzel, *Anthropogeographie Erster Teil, Zweite Auflage*, 125.

132. Ratzel, *Glücksinseln und Träume*, 139.

133. A number of Haushofer's works covered questions of military strategy, chiefly his *Wehrgeopolitik*. Karl Haushofer, *Wehrgeopolitik: Geographische Grundlagen einer Wehrkunde* (Berlin: Junker und Dünnhaupt), 1941.

134. For exceptions, see Beatrice Heuser, *The Evolution of Strategy: Thinking War from Antiquity to the Present* (Cambridge: Cambridge University Press, 2010); Virginie Mamadouh, "Geographers and war, geographers and peace," in *The Geography of War and Peace: From Death Camps to Diplomats*, ed. Colin Flint, 26–60 (Oxford: Oxford University Press, 2005); Mendieta, "War as the School of Space."

135. Ratzel, *The History of Mankind, Volume I*, 130.

136. Ratzel, *The History of Mankind, Volume I*, 140. Ratzel developed this line of reasoning further elsewhere. "In peaceful competition, as in armed contests, the rule is

that the aggressor must advance on to the ground occupied by his opponent, and in overwhelming him he assimilates himself to him. States bordering on steppes in their struggle with the inhabitants of the steppes must acquire so much of the same character as will enable them to avail themselves of the advantages the steppe affords." Ratzel, "The Territorial Growth," 361. By the mid-1890s Ratzel not just blurred the distinction between war and migration, he also held that war itself necessitated some form of cultural assimilation.

137. Ratzel, *Anthropogeographie Zweiter Teil, Dritte Auflage*, 246, 99.

138. Ratzel, *Politische Geographie: Geographie der Staaten*.

139. Ratzel, *Politische Geographie: Geographie der Staaten*, 451.

140. Niels Werber, *Die Geopolitik der Literatur: Eine Vermessung der medialen Weltraumordnung* (Munich: Hanser, 2007), 40.

141. Ratzel, *Politische Geographie: Geographie der Staaten*, 371. This is an idea that still finds followers among some philosophers. Though not without nuance, Mendieta argues that "[w]hat makes Ratzel's work singularly important for an understanding of the ways in which space has been determined by war is that for him life is a struggle for space, and war is the school of space." Mendieta, "War as the School of Space," 213.

142. Ratzel, *Politische Geographie: Geographie der Staaten*, 94; see also Heuser, *Evolution of Strategy*, 126.

143. Ratzel, *Politische Geographie: Geographie der Staaten*, 93.

144. Ratzel *Politische Geographie: Geographie der Staaten*, 93.

145. Ratzel, *Glücksinseln und Träume*, 120.

146. Schultz, "Pax Geographica," 8.

147. Ratzel, "Territorial Growth," 360.

148. Carl von Clausewitz, *On War* (Princeton, NJ: Princeton University Press, 1832 [1989]), 75.

149. Ratzel, *Politische Geographie: Geographie der Staaten*, 93.

150. Ratzel, *Politische Geographie: Geographie der Staaten*, 38.

151. Charles Darwin and Alfred Wallace, "On the Tendency of Species to Form Varieties; and on the Perpetuation of Varieties and Species by Natural Means of Selection," *Journal of the Proceeding of the Linnean Society of London: Zoology* 3, no. 9 (1858): 46–47.

152. Ratzel, *Anthropogeographie Erster Teil, Zweite Auflage*, 556–57.

153. For an illustration of how Ratzel continued to think in environmentally determinist ways, observe the following excerpt: "The administration of European colonies in tropical regions must always proceed from the point of view that Europeans do not in the long run acclimate themselves to the tropics. Their upkeep creates unusual difficulties and costs. European civil servants, officers, and soldiers in the tropical colonies have to be paid, dressed, and fed well and surrounded by more comfort. They often have to be moved to cooler climates, be it back home or into high-altitude health stations." Ratzel, *Politische Geographie: Geographie der Staaten*, 287.

154. Ratzel, *Politische Geographie: Geographie der Staaten*, 381.

155. Ratzel, "Flottenfrage und Weltlage," 376.

156. For a reading of Ratzel as an idealist, see "Ratzel's *Politische Geographie*," 118.

157. Ratzel, *Politische Geographie: Geographie der Staaten*, 370.

158. Ratzel, *Politische Geographie: Geographie der Staaten*, 374.

159. Ratzel, *Politische Geographie: Geographie der Staaten*, 373.

160. Ratzel, "Lebensraum—A Biogeographical Study," 76.

161. Races, he believed, were governed by the same laws: "Races go into decline in narrow spaces; only in wide spaces do they develop twigs and branches and thus develop a mighty trunk like that of the Aryans." Ratzel, "Der Ursprung der Arier," 392.

162. Ratzel, "Der Ursprung der Arier," 397.

163. Hell, *The Conquest of Ruins*, 36.

164. Smith, "Friedrich Ratzel," 54.

165. A. C. Grayling, *War: An Enquiry* (New Haven, CT: Yale University Press, 2017); Mary Kaldor, *New and Old Wars: Organised Violence in a Global Era* (Cambridge: Polity, 2012).

166. Kaplan, *The Revenge of Geography*; Tim Marshall, *Prisoners of Geography: Ten Maps that Tell You Everything You Need to Know about Global Politics* (London: Elliot and Thompson, 2015).

167. Ratzel, *Politische Geographie: Geographie der Staaten*, 92–93.

CHAPTER 6

1. Ratzel, *Glücksinseln und Träume*, 13.

2. Ratzel, *Glücksinseln und Träume*, 46.

3. F. Ratzel, "Verehrte Freundin!" Letter to Marie Wingens, St. Louis, 29 May 1874, Bayerische Staatsbibliothek, Ratzeliana Supplement 10.

4. Ratzel, *Glücksinseln und Träume*, 46.

5. Ratzel, *Glücksinseln und Träume*, 383.

6. Agamben, *Homo sacer*; Foucault, *Society Must Be Defended*; Michel Foucault, *The History of Sexuality Part I: The Will to Knowledge* (London: Penguin, 1978).

7. Esposito, *Bios*, 111.

8. Mbembe, *Necropolitics*, 92, original emphasis.

9. Arendt, *Origins of Totalitarianism*.

10. Mbembe, *Necropolitics*, 122.

11. A. Bettelheim, "Friedrich Ratzel und der deutsche Nekrolog," *Münchner Allgemeine Zeitung* 197 (1904): 399.

12. Ratzel preferred the German term *Nekrolog* (necrology) to the perhaps more common, though perhaps more euphemistic, word *Nachruf*, which literally translates as "after-cry" in English.

13. "Der Verfall der Nekrologie," *Die Grenzboten* 53 (1894): 239; see also Wanklyn, *Friedrich Ratzel*, 55.

14. Ratzel, *Sketches of Urban and Cultural Life*, 206.

15. Ratzel, *Sketches of Urban and Cultural Life*, 182.

16. Ratzel, *Sketches of Urban and Cultural Life*, 182.

17. Ratzel, *Die Erde*, 326.

18. Ratzel, *Anthropogeographie Zweiter Teil, Dritte Auflage*, 333–34.

19. F. Ratzel, "Diary entry 11 September 1881," *Tagebuch über Englandreise im*

Aug/Sep 1881, Abschrift angefertigt von Günter Buttmann Bayerische Staatsbibliothek, Ratzeliana Supplement 17a VI 9a.

20. Ratzel, *Sketches of Urban and Cultural Life*, 295.

21. Ratzel, *Sketches of Urban and Cultural Life*, 293.

22. Ratzel, *Sketches of Urban and Cultural Life*, 294.

23. Ratzel, *Sketches of Urban and Cultural Life*, 219.

24. Ratzel, *Anthropogeographie Zweiter Teil, Dritte Auflage*, 335.

25. Ratzel, *Anthropogeographie Zweiter Teil, Dritte Auflage*, 332.

26. "Neger," in *Brockhaus Konversationslexikon, 14. Auflage, Bd XII*, n.e. (Leipzig: Brockhaus, 1884), 228.

27. F. Ratzel, "Diary entry 11 October 1894" *Tagebuch 1894, Abschrift angefertigt von Günter Buttmann* Bayerische Staatsbibliothek, Ratzeliana Supplement 17a VI 12.

28. Steinmetzler, *Anthropogeographie Friedrich Ratzels*, 137, 140; see also Hans-Dietrich Schultz, "Friedrich Ratzel: Bellizistischer Raumtheoretiker mit Naturgefühl oder Vorläufer der NS Lebensraumpolitik?" in *Auf der Suche nach Vielfalt: Ethnologie und Geographie in Leipzig*, ed. Claus Deimel, Sebastian Lentz, and Bernhard Streck (Leipzig: Leibniz-Institut für Länderkunde, 2009), 138.

29. In the twentieth century, Fechner would become one of the influences upon the notion of a "differential unconscious" developed by the French philosopher Gilles Deleuze. Christian Kerslake, *Deleuze and the Unconscious* (London: Continuum, 2007), 138.

30. William R. Woodward, "Fechner's Panpsychism: A Scientific Solution to the Mind-Body Problem," *Journal of the History of the Behavioural Sciences* 8 (1972): 370.

31. Woodward, "Fechner's Panpsychism," 370.

32. "Weltentwicklung und Weltschöpfung."

33. Weidling, "Ernst Haeckel, Darwinismus," 311.

34. Gustav Theodor Fechner, *Einige Ideen zur Schöpfungs- und Entwicklungsgeschichte* (Leipzig: Breitkopf und Härtel, 1873).

35. William James, "Introduction," in *The Little Book of Life after Death*, ed. Gustav Theodor Fechner (Boston: Little Brown and Company, 1904), x.

36. While this view did entail a rejection of a Judeo-Christian vision of heaven and hell, Fechner claimed that the basic idea of death as an awakening of the soul had indeed been demonstrated by the figure of Jesus Christ. Gustav Theodor Fechner, ed., *The Little Book of Life after Death* (Boston: Little Brown and Company, 1836 [1904]), 16–17.

37. Fechner, *The Little Book*, 4.

38. Fechner, *The Little Book*, 13.

39. Woodward, "Fechner's Panpsychism," 386.

40. "Staat als Organismus," 618.

41. Ratzel, "Die Tagesansicht," 169–70.

42. Steinmetzler even locates the source of Ratzel's *hologäische Erdauffassung* and organicism in Fechner. But Ratzel had already taken these on board by the time he developed an interest in Fechner. They are thus more likely to have originated in his reading of Ritter and Spencer. Steinmetzler, *Anthropogeographie Friedrich Ratzels*, 142.

43. Ratzel, *Über Naturschilderung*, 125.
44. Ratzel, *Über Naturschilderung*, 206.
45. Stehlin, "Introduction," xxvi.
46. Ratzel, *Anthropogeographie Zweiter Teil, Dritte Auflage*, 340.
47. Ratzel, *Die Erde*, 429.
48. Ratzel, *Sein und Werden*.
49. Ratzel, "Beurtheilung der Völker," 199.
50. Ratzel, *History of Mankind I*, 3.
51. Ratzel, *History of Mankind I*, 328.
52. When discussing "the Europeans" on the final pages of his third and last volume, he did not focus on burial rituals. Ratzel, *History of Mankind I*, 329, 731.
53. Ratzel, *History of Mankind I*, 365.
54. Brantlinger, *Dark Vanishings*.
55. Patrick Wolfe, "Settler Colonialism and the Elimination of the Native," *Journal of Genocide Research* 8, no. 4 (2006): 287–409.
56. Ratzel, "Nationalitäten und Rassen," 465.
57. Achelis, "Friedrich Ratzel," 246.
58. Brantlinger, *Dark Vanishings*, 1. Alfred Russel Wallace, for instance, fused evolutionary biology with anti-imperialist and anticapitalist thinking in ways that are fundamentally alien to Ratzel's geography. Brantlinger, *Dark Vanishings*, 186.
59. Darwin, *The Descent of Man*, 238.
60. Darwin, *Voyage of the Beagle*, 112.
61. Darwin, *Voyage of the Beagle*, 322.
62. Darwin, *Origin of Species*, 36.
63. Darwin, *The Descent of Man*, 238.
64. Darwin, *The Descent of Man*, 201.
65. Ratzel, *History of Mankind I*, 254.
66. Darwin *The Descent of Man*, 94, 238; Georg Gerland, *Ueber das Aussterben der Naturvölker* (Leipzig: Fleischer, 1868); Ratzel, *Anthropogeographie Zweiter Teil, Dritte Auflage*, 217. Gerland's 1868 essay had sought to develop a typology of reasons for the decline of *Naturvölker*. The short book was largely synoptic and relied on eighteenth- and early nineteenth-century travel writers, including on Darwin himself.
67. Ratzel, *History of Mankind I*, 129.
68. Ratzel, *History of Mankind I*, 129.
69. Patrick Brantlinger, *Taming Cannibals: Race and the Victorians* (Ithaca, NY: Cornell University Press, 2011).
70. On Truganini and the continuing political disputes surrounding her fate, see Suvendrini Perera, "Claiming Truganini: Australian National Narratives in the Year of Indigenous Peoples," *Cultural Studies* 10, no. 3 (1996): 393–412.
71. Ratzel, "Geschichte, Völkerkunde," 491.
72. Ratzel, *History of Mankind I*, 129.
73. Ratzel, *Anthropogeographie Zweiter Teil, Dritte Auflage*, 224.
74. Ratzel, *Anthropogeographie Zweiter Teil, Dritte Auflage*, 256.
75. Ratzel, *Anthropogeographie Zweiter Teil, Dritte Auflage*, 233.
76. Ratzel, *Anthropogeographie Zweiter Teil, Dritte Auflage*, 226; see also Ratzel, *History of Mankind I*, 11.

77. Ratzel, *Anthropogeographie Zweiter Teil, Dritte Auflage*, 228.
78. Ratzel, *Anthropogeographie Zweiter Teil, Dritte Auflage*, 240.
79. Murphy, *Heroic Earth*, 10.
80. Ratzel, *History of Mankind I*, 11.
81. Ratzel, *Anthropogeographie Zweiter Teil, Dritte Auflage*, 236.
82. Ratzel, *Anthropogeographie Zweiter Teil, Dritte Auflage*, 245, 252.
83. Thomas Achelis, "Friedrich Ratzel," *Nord und Süd* 86 (1898): 273.
84. Ratzel, *Anthropogeographie Zweiter Teil, Dritte Auflage*, 231.
85. Ratzel, *Anthropogeographie Zweiter Teil, Dritte Auflage*, 227.
86. Jureit, *Das Ordnen von Räumen*, 150.
87. Ratzel, *Politische Geographie: Geographie der Staaten*, 204.
88. Ratzel, *Politische Geographie: Geographie der Staaten*, 19.
89. Ratzel, *Politische Geographie: Geographie der Staaten*, 20.
90. Ratzel, *Politische Geographie: Geographie der Staaten*, 22.
91. F. Ratzel, "Diary entry 2 February 1894," *Tagebuch 1894, Abschrift angefertigt von Günter Buttmann* Bayerische Staatsbibliothek, Ratzeliana Supplement 17a VI 12; F. Ratzel, "Diary entry 29 May 1894," *Tagebuch 1894, Abschrift angefertigt von Günter Buttmann* Bayerische Staatsbibliothek, Ratzeliana Supplement 17a VI 12.
92. Ratzel, *Politische Geographie: Geographie der Staaten*, 69–70. Ratzel had made this argument in his regional geography too. North America, in his view, was the realm of racially "pure" "Germanic," particularly Anglo-Saxon, colonization. Ratzel, *Die Vereinigten Staaten, Zweiter Band, Zweite Auflage*, 7, 216–19.
93. Ratzel, *Politische Geographie: Geographie der Staaten*, 50.
94. Ratzel, *Politische Geographie: Geographie der Staaten*, 429–30.
95. Ratzel, *Politische Geographie: Geographie der Staaten*, 204.
96. Ratzel, *Politische Geographie: Geographie der Staaten*, 204–5.
97. Ratzel, *Deutschland: Eine Einführung*, 158.
98. Ratzel, "Lebensraum—A Biogeographical Study," 74.
99. Ratzel, "Lebensraum—A Biogeographical Study," 74.
100. Ratzel, "Lebensraum—A Biogeographical Study," 76.
101. Ratzel, "Lebensraum—A Biogeographical Study," 72. Unlike the term *Aussterben* (extinction), *Vernichtung* (extermination, annihilation, or total destruction) assumes agency. Whereas extinction takes a passive construction, extermination is an active process. Ratzel described the colonization of the Americas by European settlers as a "struggle for extermination" (*Vernichtungskampf*).
102. Ratzel, "Lebensraum—A Biogeographical Study," 73–74.
103. Ratzel, *Die Vereinigten Staaten, Zweiter Band, Zweite Auflage*, 224.
104. Ratzel, "Lebensraum—A Biogeographical Study ," 62.
105. Ratzel, *Anthropogeographie Erster Teil, Zweite Auflage*, 520.
106. Ratzel, *Glücksinseln und Träume*, 128.
107. Ratzel, *Glücksinseln und Träume*, 202, 231, 259, 145, 147, 232.
108. Ratzel, *Glücksinseln und Träume*, 232.
109. Ernst Jünger, *Storm of Steel* (London: Penguin, 1920 [2004]); Andreas Huyssen, "Fortifying the Heart—Totally Ernst Jünger's Armoured Texts," *German Critique* 59 (1993): 15. Given the semblance between Ratzel and Jünger's texts, it is perhaps unsur-

prising to learn that much like Ratzel, Jünger was a keen biologist, indeed an entomologist. Jünger was also a student of the very vitalist who continues to inspire a new generation of vital materialists, the aforementioned Hans Driesch. Bennett, *Vibrant Matter*.

110. See also Cynthia Enloe, "Wounds: Militarised nursing, feminist curiosity, and unending war," *International Relations* 33, no. 3 (2019): 393–412; Derek Gregory, "Trauma Geographies: Broken Bodies and Lethal Landscapes." The 2018 Antipode RGS-IBG lecture, 29 August 2018, Cardiff.

111. Ratzel, *Glücksinseln und Träume*, 196.

112. Ratzel, *Glücksinseln und Träume*, 197.

113. Buttmann, *Friedrich Ratzel*, 35.

114. Ratzel, *Glücksinseln und Träume*, 207.

115. Ratzel, *Glücksinseln und Träume*, 201.

116. Ratzel, *Glücksinseln und Träume*, 200, 230.

117. Mbembe, "Necropolitics," 36.

118. Mbembe, "Necropolitics," 36.

119. Foucault, *Society Must Be Defended*, 247.

120. Ratzel, *Glücksinseln und Träume*, 201.

121. Minca, "Giorgio Agamben"; Thomas Lemke, *Biopolitics: An Advanced Introduction* (New York: New York University Press, 2011), 13.

122. Ratzel, *Die Erde und das Leben*, 530.

123. Ratzel, *Die Vereinigten Staaten, Zweiter Band, Zweite Auflage*, 189, emphasis added.

124. Ratzel, *Sketches of Urban and Cultural Life*, 33.

125. Ratzel, *Glücksinseln und Träume*, 133.

126. Conrad, *German Colonialism*, 92.

127. Madley, "From Africa to Auschwitz"; Zimmerer, *Birth of Ostland*.

128. Hull, *Absolute Destruction*, 119.

129. Guettel, *German Expansionism*, 80.

130. Susanne Kuss, *German Colonial Wars and the Context of Military Violence* (Cambridge, MA: Harvard University Press, 2017), 176.

131. Eduard von Liebert, "Die politische, militärische und volkswirtschaftliche Bedeutung einer starken Seemacht," in *Verhandlungen des Deutschen Kolonialkongresses 1905* (Berlin: Georg Reimer, 1906), 929.

132. Geoff Eley, *Reshaping the German Right: Radical Nationalism and Political Change under Bismarck* (New Haven, NC: Yale University Press, 1980), 109–10; Zimmerer, *Birth of Ostland*, 208.

133. Georg Hartmann, "Die Mischrassen in unseren Kolonien, besonders in Südwestafrika," in *Verhandlungen des Deutschen Kolonialkongresses 1905* (Berlin: Georg Reimer, 1906), 909.

134. Hartmann, "Die Mischrassen," 910–16.

135. Quoted in Horst Drechsler, *Südwestafrika unter deutscher Kolonialherrschaft: Der Kampf der Herero und Nama gegen den deutschen Imperialismus (1884–1915)* (Berlin: Akademie, 1966), 106.

136. Drechsler, *Südwestafrika*, 119.

137. "Die Maske ab!" Although Ratzel had previously commented on Chinese-German relations in the late 1890s in *Die Grenzboten*, he did not do so in relation to the Boxer War and Wilhelm II's infamous "hun speech" of 1900 that told German soldiers to take no prisoners. Previously, he had eyed with great skepticism efforts at closer relations between the two nations for potentially contaminating Germany with "a drop of Mongol blood," a "patriotic anxiety" he could only disperse, he noted in a jingoistic tone, with a "good glass of German beer, not Chinese tea." "Deutsch-Chinesisch." *Die Grenzboten* 55 (1896): 622–26.

138. "Die Aussichten," 175.

139. "C. Pearson, The Chances of Death and Other Studies in Evolution," *Literarisches Zentralblatt* 8 (1898): 254–55; Karl Pearson, *The Chances of Death and Other Studies in Evolution* (London: Arnold, 1897).

140. Todd May, *Death* (London: Routledge, 2009), 26, 38. In this then there is a crucial difference between a Ratzelian and a Heideggerian concept of death.

141. Ellen Churchill Semple, *Influences of Geographic Environment on the Basis of Ratzel's System of Anthropo-geography* (New York: Holt, 1911), 170–77.

142. Rudolf Kjellén, *Der Staat als Lebensform* (Leipzig: S. Hirzel, 1917), 204, 215.

143. Karl Haushofer, *Weltpolitik von heute* (Berlin: Zeitgeschichte, 1934), 160.

144. Richard Hennig, *Einführung in die Geopolitik* (Leipzig: B. G. Teubner, 1935) 90.

145. Albert Speer, *Inside the Third Reich* (London: Phoenix, 1970), 97; Mark Featherstone, "Ruin Value," *Journal for Cultural Research* 9, no. 3 (2005): 301–20; Julia Hell and George Steinmetz, "Ruinopolis: Post-imperial Theory and Learning from Las Vegas," *International Journal of Urban and Regional Research* 38, no. 3 (2014): 1047–68.

146. For an overview see Mark Neocleous, "Long Live Death! Fascism, Resurrection, Immortality," *Journal of Political Ideologies* 10, no. 1 (2005): 31–49.

147. Umberto Eco, 1995. "Ur-facism," *The New York Review of Books*, June 22, 1995. http://www.nybooks.com/articles/1995/06/22/ur-fascism/, 13–14.

148. Arendt, *Origins of Totalitarianism*, 171.

CHAPTER 7

1. F. Ratzel, "Lieber Freund!" Letter to Rudolph Oldenbourg, Leipzig, 5 June 1904, Bayerische Staatsbibliothek, Ratzeliana Supplement 10.

2. For a list of obituaries, see Buttmann, *Friedrich Ratzel*, 138–47, and Ratzel and Helmolt, "Friedrich Ratzel," xxi–xxiii.

3. Paul Vidal de la Blache, "Nécrologie: Friedrich Ratzel," *Annales de Géographie* 13, no. 72 (1904): 466; Ellen Churchill Semple, "Tributes to Friedrich Ratzel," *Bulletin of the American Geographical Society* 36, no. 9 (1904): 553.

4. E. G. R., "Obituary: Friedrich Ratzel," 485.

5. F. Regel, "Ratzel, Fr., Dr., Völkerkunde. Erster Band. Die Naturvölker Afrikas. Mit 494 Abbildungen im Text, 10 Aquarelltafeln und 2 Karten. Leipzig, Bibliographisches Institut, 1885," Bayerische Staatsbibliothek, Ratzeliana II Besprechungen von

Ratzelschen Werken. Despite its lack of original empirical research, *Völkerkunde* was applauded by Anglo-American anthropologists for its comprehensiveness. F. W. Hodge, "The History of Mankind, by Professor Friedrich Ratzel, Translated from the Second German Edition by A. J. Butler, M.A," *American Anthropologist* A9, no. 11 (1896): 381–82.

6. Hermann Wagner, "Bericht über die Entwicklung der Methodik und des Studiums der Erdkunde (1889–91)," *Geographisches Jahrbuch* 14 (1891): 381.

7. Hermann Wagner, "F. Ratzels Anthropogeographie II. Oder die geographische Verbreitung des Menschen," *Zeitschrift der Gesellschaft für Erdkunde zu Berlin* 16 (1891): 465–478. See also Steinmetzler, *Anthropogeographie Friedrich Ratzels*, 14; Schultz, "Bellizistischer Raumtheoretiker," 125.

8. Friedrich Ratzel, "Erwiderung auf Hermann Wagners Besprechung der Anthropogeographie II," *Zeitschrift der Gesellschaft für Erdkunde zu Berlin* 16 (1891): 508–12.

9. Ratzel felt it necessary to mention the severity of these attacks again in the foreword to the revised *Anthropogeographie.* Ratzel, *Anthropogeographie Erster Teil, Zweite Auflage*, xi.

10. M. Ratzel, "Sehr geehrter Herr Professor!" Letter to unknown, Ammerland, 19 June 1906, Archiv für Geographie, Leibniz Institut für Länderkunde, Nachlass Friedrich Ratzel 163/657.

11. Jeremy Black, *Geopolitics and the Quest for Dominance* (Bloomington: Indiana University Press, 2016); Guzzini, *The Return of Geopolitics*; Jureit, *Das Ordnen von Räumen*; Ó Tuathail, *Critical Geopolitics*; Werber, *Geopolitik zur Einführung*; Murphy, *Heroic Earth.*

12. Stephan Günzel, *Geophilosophie: Nietzsches philosophische Geographie* (Berlin: Akademie, 2001), 191–98.

13. Ludwig Gumplowicz, *Der Rassenkampf: Sociologische Untersuchungen* (Innsbruck: Verlag der Wagner'schen Univerităts-Buchhandlung, 1883), 221–23; Ludwig Gumplowicz, *Geschichte der Staatstheorien* (Innsbruck: Verlag der Wagner'schen Univerităts-Buchhandlung, 1905), 530–40.

14. Émile Durkheim, "Monsieur et honoré collège," Letter to Friedrich Ratzel, Paris, 8 October 1899, Bayerische Staatsbibliothek, Ratzeliana Supplement 11. The only surviving letter from Durkheim to Ratzel informs the German geographer of a change of the French sociologist's address. This suggests that it may have been part of a longer exchange.

15. Émile Durkheim, "Review of Ratzel's Anthropogeographie," In *Emile Durkheim: Selected Writings*, ed. Anthony Giddens (Cambridge: Cambridge University Press, 1899 [1972]), 87.

16. Geoffrey Parker, "Ratzel, the French School and the Birth of Alternative Geopolitics," *Political Geography* 19, no. 8 (2000): 957–69.

17. Élisée Reclus, "Recent Books on the United States," *The Geographical Journal* 6, no. 5 (1895): 452. Despite their undoubted political differences, Ratzel had been generally positive about Reclus. He did criticize a lack of attention to questions of physical geography and the distribution and hierarchy of races. "E. Reclus, Nouvelle geographie universelle, Livraison 574–638, Paris 1885," *Literarisches Zentralblatt* 13

(1887): 415–16; "E. Reclus, Nouvelle geographie universelle, Tome XU, Paris 1886/87," *Literarisches Zentralblatt* 26 (1888): 882–83.

18. On Ratzel's reception by Durkheim and Vidal de la Blache and the tensions between them, see Howard F. Andrews, "The Durkheimians and Human Geography: Some Contextual Problems in the Sociology of Knowledge," *Transactions of the Institute of British Geographers* 9, no. 3 (1984): 315–36. Ratzel too had been aware of Vidal's work. Friedrich Ratzel, "P. Vidal de la Blache, Les conditions geographiques des faits sociaux, Annales de Geographie 1902," *Geographischer Literaturbericht* 310 (1902): 95.

19. David Atkinson, "Geopolitical Imaginations in Modern Italy," In *Geopolitical Traditions: A Century of Geopolitical Thought*, ed. Klaus Dodds and David Atkinson (London: Routledge, 2000), 93–117; Rosenboim, "Geopolitics and Nationalism."

20. Buttmann, *Friedrich Ratzel*, 139.

21. Sonya Gavrilova, "The Present Taxidermied: Soviet 'Common Unsaids' in Russian *Krayevedcheskiy* Museum" (DPhil diss., University of Oxford, 2019), 155; Jon D. Oldfield and Denis J. J. Shaw, "A Russian Geographical Canon? The Contested Canon of Russian and Soviet Geography, 1884–1953," *Journal of Historical Geography* 49 (2015): 75–84.

22. Mark Bassin, "Geographical Determinism in Fin-de-siècle Marxism: Georgii Plekhanov and the Environmental Basis of Russian History," *Annals of the Association of American Geographers* 82, no. 1 (1992): 3–22.

23. Hassert, "Friedrich Ratzel," 319.

24. Alfred Hettner, "Die Geographie des Menschen." *Geographische Zeitschrift* 13, no. 8 (1907): 402; Hettner, "Bücherbesprechungen: Friedrich Ratzel," 527.

25. Eckert, "Friedrich Ratzel," 256.

26. Alfred Kelly, *The Descent of Darwin: The Popularization of Darwinism in Germany, 1860–1914* (Chapel Hill: University of North Carolina Press, 1981).

27. Wilhelm II had read Mahan too. Clark, *Kaiser Wilhelm II*, 179.

28. Patrick J. Kelly, *Tirpitz and the Imperial German Navy* (Bloomington: Indiana University Press, 2011), 358.

29. Friedrich von Bernhardi, *Germany and the Next War* (London: Arnold, 1912 [1914]); Daniel Frymann, *Wenn ich der Kaiser wär': Politische Wahrheiten und Notwendigkeiten* (Leipzig: Dieterich, 1912); Friedrich Naumann, *Mitteleuropa* (Berlin: Georg Reimer, 1915).

30. Bernhardi, *Germany and the Next War*, 18.

31. Frymann, *Wenn ich der Kaiser wär'*.

32. Smith, *The Ideological Origins*, 174.

33. Jureit, *Das Ordnen von Räumen*, 161.

34. Clark, *Kaiser Wilhelm II*, 554.

35. For detailed discussions of political geographical debates before and during the First World War, see Eugen Oberhummer, "Anhang. Die politische Geographie vor Ratzel und ihre jüngste Entwicklung," in Friedrich Ratzel, *Politische Geographie* (Munich: R. Oldenbourg, 1923), 597–618. See also Schultz, "Pax Geographica," 9–11; "Fantasies of *Mitte*," 325–30). For an overview of where these debates sat within other European geopolitical discourses, see Heffernan, *The Meaning of Europe*, 71–76.

36. Alfred Hettner, *Der Friede und die deutsche Zukunft* (Stuttgart: Deutsche Verlags-Anstalt, 1917), 237.

37. Alfred Hettner, "Der Friede und die politische Geographie," *Geographische Zeitschrift* 25, no. 8/9 (1919): 233.

38. Hettner, "Der Friede und die politische Geographie," 235.

39. Müller, *Friedrich Ratzel*, 58.

40. Albrecht Penck, *Morphologie der Erdoberfläche: Erster Teil* (Stuttgart: Engelhorn, 1894), 389; Albrecht Penck, "Verehrter Freud und Kollege!" Letter to Ratzel, Vienna, 28 October 1902, Bayerische Staatsbibliothek, Ratzeliana Supplement 12.

41. Michael Heffernan, "Professor Penck's Bluff: Geography, Espionage and Hysteria in World War I." *Scottish Geographical Journal* 116, no. 4 (2008): 267–82.

42. Hans-Dietrich Schultz, "Albrecht Penck: Vorbereiter und Wegbereiter der NS-Lebensraumpolitik?" *E & G Quaternary Science Journal* 66, no. 2 (2018): 124.

43. Norman Henniges, "'Naturgesetze der Kultur': Die Wiener Geographen und die Ursprünge der 'Volks- und Kulturbodentheorie,'" *ACME: An International Journal for Critical Geographers* 14, no. 4 (2015): 1326.

44. Henniges, "Naturgesetze der Kultur," 1326.

45. Albrecht Penck, *Über politische Grenzen: Rede zum Antritt des Rektorates der Königlichen Friedrich Wilhelms-Universität in Berlin* (Berlin: Norddeutsche Buchdruckerei und Verlagsanstalt, 1917), 18.

46. Penck, *Über politische Grenzen*, 31.

47. Jureit, *Das Ordnen von Räumen*, 244.

48. Schultz, "Albrecht Penck," 127.

49. There were still Ratzelians outside of the academy too. It was thus the conservative politician and later national socialist Martin Spahn who wrote a book on great power conflict that operated with a Ratzelian concept of the state. Martin Spahn, *Die Großmächte: Richtlinien ihrer Geschichte, Maßstäbe ihres Wesens* (Berlin: Ullstein, 1918), 13.

50. Rudolf Kjellén, *Die Ideen von 1914: Eine weltgeschichtliche Perspektive* (Leipzig: S. Hirzel, 1915).

51. Rudolf Kjellén, *Grundriss zu einem System der Politik* (Leipzig: S. Hirzel, 1920), 94.

52. Kjellén, *Der Staat als Lebensform*, 46.

53. Sven Holdar, "The Ideal State and the Power of Geography: The Life-Work of Rudolf Kjellén," *Political Geography* 11, no. 3 (1921): 321. Even if Ratzel had taken notice, he would not have had sufficient time to engage with it in his 1903 second edition of *Politische Geographie.*

54. "Geopolitics" is likely to have been first used in English by the Hungarian-born historian and critic of German imperialism Emil Reich in the index of a 1905 collection of important medieval and modern legal and political documents. Emil Reich, *Selected Documents Illustrating Mediæval and Modern History* (London: P. S. King and Son, 1905), 747, 750. The term appears again in a 1908 handbook of descriptive and mathematical geography. Without recourse to either Ratzel or Kjellén but drawing quite liberally, the author admitted, on Ritter and Reclus, he used it to denote "the immense importance of geographical factors both on past history and present life." Emil Reich,

Handbook of Geography: Descriptive & Mathematical (London: Duckworth, 1908), 1. Whilst the book seems to have had little impact upon its contemporaries, its author was accused of plagiarism by the Oxford geographer A. J. Herbertson. A. J. Herbertson, "Review: Dr. Reich's Handbook of Geography," *Geographical Journal* 34 (1909): 203–6.

55. Holdar, "The Ideal State," 319.

56. Werber, *Geopolitik zur Einführung*, 59; Holdar, "The Ideal State," 316.

57. Kjellén, *Der Staat als Lebensform*, 57.

58. Kjellén, *Der Staat als Lebensform*, 213.

59. Kjellén, *Grundriss zu einem System*, 61.

60. Abrahamsson, "Genealogy of Lebensraum," 40.

61. Holdar, "The Ideal State."

62. Mildenberger notes that it would have been increasingly difficult to sustain this tension in the years that followed. Mildenberger "Lebensraum oder Umwelt?" 128.

63. Anne Harrington, *Reenchanted Science: Holism in German Culture from Wilhelm II to Hitler* (Princeton, NJ: Princeton University Press, 1996).

64. Jakob von Uexküll, *Staatsbiologie: Anatomie—Physiologie—Pathologie des Staates* (Berlin: Gebrüder Paetel, 1920), 37.

65. Mildenberger, "Lebensraum oder Umwelt?"

66. Jakob von Uexküll, *Nie geschaute Welten* (München: Paul List, 1936 [1957]), 184.

67. Francesca Michelini and Kristian Köchy, *Jakob von Uexküll and Philosophy: Life, Environments, Anthropology* (Abingdon: Routledge, 2019); Boria Sax and Peter H. Klopfer "Jakob von Uexküll and the Anticipation of Sociobiology," *Semiotica* 134, no. 1 (2001): 777.

68. When the British biologist Richard Dawkins (1941–) employs the term "living space" as a synonym for habitat in his 1976 bestseller *The Selfish Gene*, he is not doing so because he is likely to have read Ratzel. Richard Dawkins, *The Selfish Gene* (Oxford: Oxford University Press, 1976 [2006]), 67. The Austrian ethologist and erstwhile national socialist Konrad Lorenz (1903–1989), in later years a critic of von Uexküll, used the German term *Lebensraum* frequently, as do many German biologists to this day. Konrad Lorenz, "Biologische Fragestellung in der Tierpsychologie," *Zeitschrift für Tierpsychologie* 1, no. 1 (1937): 24–32; Konrad Lorenz, "Ein neuer Wissenschaftszweig—die Kulturethologie," in *Otto Koenig 70 Jahre: Kulturwissenschaftliche Beiträge zur Verhaltensforschung*, ed. Gesellschaft der Freunde der Forschungsgemeinschaft Wilhelminenberg (Vienna: Ueberreuter, 1984), 7–9.

69. Ellen Churchill Semple, "Slavery: A Study in Sociology" (MA Thesis, Vassar College, 1891).

70. Ellen Churchill Semple, "Dear Dr Keltie," Letter to John Scott Keltie, Piccadilly, 30 October 1912, Correspondence Block 1911–1920, Royal Geographical Society. See also Ellen E. Adams, "Ellen Churchill Semple and American Geography in an Era of Imperialism" (PhD diss., College of William and Mary, 2011), 56; Keighren, *Bringing Geography to Book*, 14.

71. Ellen Churchill Semple, *American History and its Geographic Conditions* (Boston and New York: Houghton, Mifflin and Company, 1903); Semple, *Influences of Geographic Environment*.

72. Richard Peet, "The Social Origins of Environmental Determinism," *Annals of the Association of American Geographers* 75, no. 3 (1985): 319. On Turner and Ratzel, see Mark Bassin, "Turner, Solov'ev, and the 'Frontier Hypothesis': The Nationalist Signification of Open Spaces," *Journal of Modern History* 65, no. 3 (1993): 473–511.

73. Ellen Churchill Semple, "An Abridgement of the Partition of Asiatic Turkey," 30 April 1918, Records of the American Commission to Negotiate Peace, Record Group 256, Inquiry document 500, Records of the Inquiry 1914–19, National Archives.

74. Allen D. Bushong, "Ellen Churchill Semple, 1863–1932," In *Geographers: Biobibliographical Studies 8*, T. W. Freeman (London: Mansell, 1984), 89.

75. Ratzel had also encouraged another student, though unsuccessfully, to publish an English-language translation of his pamphlet *Das Meer als Quelle der Volksgröße*. W. C. Henkel, "Sir", Letter to Clements R. Markham, Cassel, 9 August 1901, Royal Geographical Society, Correspondence Blocks, Correspondence 1881–1910.

76. Ratzel, *Vereinigten Staaten, Zweiter Band, Zweite Auflage*, vii.

77. E. C. Semple, "Dear Professor Ratzel," Letter, Louisville, 9 December 1902, Bayerische Staatsbibliothek, Ratzeliana I Supplement 13.

78. E. Semple, "Geehrter Herr Professor," Letter to Friedrich Ratzel, n.p., 30 January 1893, Archiv für Geographie, Leibniz Institut für Länderkunde, Nachlass Friedrich Ratzel 163/686.

79. In 1902, he had noted Semple's clarity and talent for good description but also observed, somewhat patronizingly, a lack of scientific fundament that statistical material or cartography would have provided. F. Ratzel, "Diary entry 10 March 1902," *Tagebuch 4.7.1900–19.7.1903, Abschrift angefertigt von Günter Buttmann* Bayerische Staatsbibliothek, Ratzeliana Supplement 17a VI 20. He seems to have enjoyed the attention he received from Semple, remarking in 1896 how he had been delighted to receive a letter from "Miss Semple with an attractive keyring" and that he had responded by sending her some of his writings. F. Ratzel, "Diary entry 21 December 1896," *Tagebuch 4.10.1896–6.1.1889, Abschrift angefertigt von Günter Buttmann* Bayerische Staatsbibliothek, Ratzeliana Supplement 17a VI 15a. As we have seen in chapter 1, Ratzel had discussed Semple's translations of his work with her. He undoubtedly thought of her as a worthy scholar to carry his ideas across the Atlantic. During hyperinflation of the 1920s, Semple helped Ratzel's widow and daughters out financially. Wanklyn, *Friedrich Ratzel*, 51.

80. F. Ratzel, "Diary entry 10 March 1902," *Tagebuch 4.7.1900–19.7.1903, Abschrift angefertigt von Günter Buttmann* Bayerische Staatsbibliothek, Ratzeliana Supplement 17a VI 20; F. Ratzel, "Diary entry 13 May 1902," *Tagebuch 4.7.1900–19.7.1903, Abschrift angefertigt von Günter Buttmann* Bayerische Staatsbibliothek, Ratzeliana Supplement 17a VI 20.

81. Ratzel, *Politische Geographie: Geographie der Staaten*, 827; Friedrich Ratzel, "E. C. Semple, The Anglo-Saxons of the Kentucky Mountains, Geographical Journal 1901," *Geographischer Literaturbericht* 226 (1902): 67–68.

82. Ellen Churchill Semple, "The North-Shore Villages of the Lower St Lawrence," in *Zu Friedrich Ratzels Gedächtnis. Geplant als Festschrift zum 60. Geburtstag, nun als Grabspende dargebracht von Fachgenossen und Schülern, Freunden und Verehrern*, ed. Hans Helmolt (Leipzig: Seele, 1904), 349–60.

83. The book drew heavily on Ratzel, particularly *Anthropogeographie* and his

regional geography of the United States, concurring with the basic Ratzelian assumption that "the struggle for existence is the struggle for space." Semple, *American History*, 244.

84. Semple, *American History*, 280.

85. Semple, *Influences of Geographic Environment.*

86. F. Ratzel, "Diary entry 25 October 1894," *Tagebuch 1894, Abschrift angefertigt von Günter Buttmann* Bayerische Staatsbibliothek, Ratzeliana Supplement 17a VI 12.

87. F. Ratzel, "Liebe Miss Semple!", Letter to Ellen Churchill Semple, Riva, 4 April 1903, University of Kentucky, Special Collections Research Center, Ellen Churchill Semple papers, Correspondence, 1900–1952, box 6, folder 4.

88. Keighren, *Bringing Geography to Book*, 50.

89. Semple, *Influences of Geographic Environment*, 1.

90. Lucian Ashworth, "Mapping a New World: Geography and the Interwar Study of International Relations," *International Studies Quarterly* 57, no. 1 (2013): 140.

91. Semple, *Influences of Geographic Environment*, viii.

92. Semple, *Influences of Geographic Environment*, v.

93. Semple, *Influences of Geographic Environment*, vi.

94. Keighren, *Bringing Geography to Book.*

95. D. C. Munro, "Criticism of 'Partition of Asiatic Turkey,'" 6 August 1918, Records of the American Commission to Negotiate Peace, Record Group 256, Inquiry document 501, Records of the Inquiry 1914–19, National Archives.

96. H. Hawks, "Manchuria Needs Japan, Says Eminent Woman Scientists Here," *Palm Beach Post*, 17 January 1932, University of Kentucky, Special Collections Research Center, Ellen Churchill Semple papers, scrapbook, 1901–1932, box 6, folder 8.

97. Ellen Churchill Semple, "Two Works on Political Geography," *Geographical Review* 14, no. 4 (1924): 666.

98. Semple, *American History*, 62, 96; Semple, *Influences of Geographic Environment*, 22–23, 199.

99. William A. Koelsch, "Franz Boas, Geographer, and the Problem of Disciplinary Identity," *Journal of the History of the Behavioural Sciences* 40, no. 1 (2004): 1–22.

100. Franz Boas, "Werter Herr Professor," Letter to Friedrich Ratzel, Worcester (Mass.), 5 December 1889, Bayerische Staatsbibliothek, Ratzeliana Supplement 13.

101. Leonard B. Glick, "Types Distinct from Our Own: Franz Boas on Jewish Identity and Assimilation," *American Anthropologist* 84, no. 3 (1982): 553.

102. Mark Anderson, "Ruth Benedict, Boasian Anthropology, and the Problem of the Colour Line," *History and Anthropology* 25, no. 3 (2014): 395–414.

103. Richard Powell, "The Study of Geography? Franz Boas and His Canonical Returns," *Journal of Historical Geography* 49 (2015): 21–30.

104. Franz Boas, "The History of Anthropology," *Science* 20, no. 512 (1904): 519; Matti Bunzl, "Franz Boas and the Humboldtian Tradition: From *Volksgeist* und *Nationalcharacter* to an Anthropological Concept of Culture," in *Volksgeist as Method and Ethic: Essays on Boasian Ethnography and the German Anthropological Tradition*, ed. George W. Stocking Jr. (Madison: University of Wisconsin Press, 1996), 60.

105. Koelsch, "Franz Boas, Geographer," 18.

106. Franz Boas, "Methods of Research," in *General Anthropology*, ed. Franz Boas (Boston: DC Health and Company, 1938), 669.

107. Ruth Benedict, "Obituary: Franz Boas," *Science* 97, no. 25 (1943): 61.

108. Friedrich Ratzel, "F. Boas, Dissemination of Tales among the Natives of North America, NewYork 1891," *Geographischer Literaturbericht* 1398 (1891): 102–3.

109. Daniel Levering Lewis, *W. E. B. Du Bois: A Biography of Race 1868–1919* (New York: Henry Holt and Company, 1993), 136.

110. Kendi, *Stamped from the Beginning*, 296.

111. Kenneth Barkin, "W. E. B. Du Bois' Love Affair with Imperial Germany," *German Studies Review* 28, no. 2 (2005): 285–302.

112. Russell, A. Berman, "Du Bois and Wagner: Race, Nation, and Culture between the United States and Germany," *The German Quarterly* 70, no. 2 (1997): 123–135.

113. Robert Gooding-Williams, "Outlaw, Appiah, and Du Bois's 'The Conservation of Races,'" in *W. E. B. Du Bois on Race and Culture*, ed. Bernard W. Bell, Emily Grosholz, and James B. Stewart (London: Routledge, 1996), 46; Kimberly Ann Harris, "W. E. B. Du Bois's 'Conservation of Races': A Metaphilosophical Text," *Metaphilosophy* 50, no. 5 (2019): 675.

114. W. E. B. Du Bois, *The Negro American Family* (Atlanta: Atlanta University Press, 1908), 10–17, 48.

115. W. E. B. Du Bois, *The Negro* (New York: Henry Holt and Company, 1915), 105, 139; see also W. E. B. Du Bois, "The Conservation of Races," *The American Negro Academy Occasional Papers* 2 (1897): 1–15.

116. Du Bois, *The Negro*, 244.

117. Indeed, Du Bois had included even Carl Peters and Henry Morton Stanley in his recommendations, although their texts were not starred.

118. Du Bois, *Black Reconstruction*.

119. Sauer, "The Formative Years of Ratzel;" Stehlin, "Introduction."

120. Smith, *American Empire*.

121. Smith, *American Empire*, 246.

122. Smith, *American Empire*, 38, 71.

123. Isaiah Bowman, *The New World: Problems in Political Geography*. (New York: World Book Company, 1921 [1928]).

124. Ashworth, "Mapping a New World," 209. In later years, Bowman sought to develop a global perspective on the frontier that universalized the American experience. Isaiah Bowman, "The Pioneer Fringe," *Foreign Affairs* 6, no. 1 (1927): 49–66.

125. Keighren, *Bringing Geography to Book*, 139.

126. Clarence G. Contee, "Du Bois, the NAACP, and the Pan-African Congress of 1919," *Journal of Negro History* 57, no. 1 (1972): 13–28.

127. After a stint as *Privatdozent*, Haushofer was in 1921 made *Honorarprofessor*. In 1933 he was appointed *ordentlicher Professor*. Christian Spang, *Karl Haushofer und Japan: Die Rezeption seiner geopolitischen Theorien in der deutschen und japanischen Politik* (Munich: Iudicium, 2013), 158.

128. While Haushofer only occasionally made reference to Mackinder, he frequently invoked Kjellén. He observed in private correspondence that Kjellén had taken "everything" from Ratzel. Karl Haushofer, "Karl Haushofer an Max-Hildebert Böhm, 8. Feb-

ruar 1931," in *Karl Haushofer: Leben und Werk, Band II*, ed. Hans-Adolf Jacobsen (Boppard am Rhein: Boldt, 1931 [1979]), 104.

129. Haushofer claimed to have first met Ratzel through his father Max H. Haushofer. Ratzel mentions meetings with Max Haushofer in his diary from the mid-1870s onwards. Karl Haushofer, "Friedrich Ratzel im raumpolitischen Umbruch unserer Zeit," 81–86; F. Ratzel, "Diary entry 24 November 1875," *Tagebuch begonnen 5.10.1875, Abschrift angefertigt von Günter Buttmann* Bayerische Staatsbibliothek, Ratzeliana Supplement 17a VI 3. This of course does not imply a direct intellectual transfer from Ratzel to Karl Haushofer at an early stage in the latter's life, however much Karl would in later years play up such a personal connection. Spang, *Karl Haushofer und Japan*, 220–21. But it would perhaps explain why Haushofer was so committed to upholding Ratzel's legacy.

130. Karl Haushofer, "Politische Erdkunde und Geopolitik," in *Freie Wege vergleichender Erdkunde*, ed. Erich von Drygalski (Munich: R. Oldenbourg, 1925), 89.

131. Haushofer, *Wehrgeopolitik*.

132. Karl Haushofer, *Ratzel: Erdenmacht und Völkerschicksal, herausgegeben und eingeleitet von Generalmajor a. D. Prof. Dr. Karl Haushofer* (Stuttgart: Alfred Kröner, 1941).

133. Karl Haushofer, "Geopolitische Einflüsse bei den Verkörperungsversuchen von nationalem Sozialismus und sozialer Aristokratie," *Zeitschrift für Geopolitik* 1 (1924): 127–34.

134. Karl Haushofer, *Der Nationalsozialistische Gedanke in der Welt* (Munich: Georg Callwey, 1933).

135. Spang, *Karl Haushofer und Japan*, 375. A link is sometimes made between Ratzel's idealization of the German peasantry and a similar romanticization which could later be found in national socialism. But it is important to note that *Reichsbauernführer* (Reichs peasant leader) Richard Walter Darré, a crucial blood-and-soil ideologue, did not engage with Ratzel's writings in his books, although he did cite other geographers (see for instance Walther Darré, *Das Bauerntum als Lebensquell des nordischen Rasse* (Munich: J. F. Lehmann, 1928 [1942])).

136. Bassin, "Imperialism and the Nation State," 126–29.

137. In the spring of 1941, Hitler is said to have accused Haushofer of inspiring Rudolf Hess's unsuccessful flight to England in the hope of brokering a peace with London. Jacobsen, *Karl Haushofer I*, 451. Haushofer's marriage to a person of Jewish descent was adding to the pressure on him. But it was not until 1944, when Haushofer's son Albrecht (a geopolitician in his own right) was associated with the failed assassination attempt on Hitler that Haushofer fell from grace. Spang, *Karl Haushofer und Japan*, 381.

138. See for instance Haushofer, *Geopolitik der Pan-Ideen*; Haushofer, *Weltpolitik von heute*. See also Murphy, *Heroic Earth*, 107.

139. Leslie W. Hepple, "Dudley Stamp and the *Zeitschrift für Geopolitik*," *Geopolitics* 13, no. 2 (2009): 388; see also Murphy, *Heroic Earth*, ix–xi.

140. Werber, *Geopolitik zur Einführung*, 111.

141. Fritz Hesse, "Das Gesetz der wachsenden Räume," *Zeitschrift für Geopolitik* 1 (1924): 1–4; Werber, *Geopolitik zur Einführung*, 111.

142. Karl Haushofer, "Die Einheit der Monsoonländer," *Zeitschrift für Geopolitik* 1 (1924): 25, 27.

143. Erich Obst, "Die geopolitischen Leitlinien des europäischen Russlands," *Zeitschrift für Geopolitik* 1 (1924): 5; Franz Termer, "Die natürlichen Grundlagen amerikanischer Staatsentwicklung," *Zeitschrift für Geopolitik* 1 (1924): 28–35.

144. Hermann Lautensach, "Die Mittelmeere als geopolitische Kraftfelder," *Zeitschrift für Geopolitik* 1 (1924): 36–47.

145. See for instance Otto Maull, "Friedrich Ratzel zum Gedächtnis," *Zeitschrift für Geopolitik* 5 (1929): 628; Kurt Vowinckel, "Zum Begriff Lebensraum," *Zeitschrift für Geopolitik* 16 (1939): 638–39.

146. David T. Murphy, "Space, Race and Geopolitical Necessity: Geopolitical Rhetoric in German Colonial Revanchism, 1919–1933," in *Geography and Empire*, ed. Anne Godlewska and Neil Smith (Oxford: Blackwell, 1994), 176–77.

147. Karl Haushofer, "Einleitung: Friedrich Ratzel als raum- und volkspolitischer Gestalter," in *Erdenmacht und Völkerschicksal, herausgegeben und eingeleitet von Generalmajor a. D. Prof. Dr. Karl Haushofer*, Friedrich Ratzel (Stuttgart: Alfred Kröner, 1941), xxv.

148. Hitler, *Mein Kampf*, 8, 732.

149. Hitler, *Hitler's Second Book*.

150. Bruno Hippler, *Hitlers Lehrmeister: Karl Haushofer als Vater der NS Ideologie* (St. Ottilien: EOS, 1996); Herwig, *The Demon of Geopolitics*.

151. Othmar Plöckinger, *Geschichte eines Buches: Adolf Hitlers Mein Kampf: 1922–1945* (Munich: Oldenbourg, 2011), 145.

152. David T. Murphy, "Hitler's Geostrategist? The Myth of Karl Haushofer and the 'Institut für Geopolitik,'" *The Historian* 76, no. 1 (2014): 1–25; Ó Tuathail, *Critical Geopolitics*.

153. Spang, *Karl Haushofer und Japan*, 369–72; see also Jacobsen, *Karl Haushofer I*, 241–46.

154. Plöckinger, *Geschichte eines Buches*, 52.

155. Spang, *Karl Haushofer und Japan*, 386.

156. Haushofer, "Einleitung," xxvi. Lange also highlights the degree to which Hitler's political language drew on the discourses of the 1890s pan-German movement, of which Ratzel had of course been a crucial proponent. Lange, "Der Terminus Lebensraum," 428.

157. Edmund A. Walsh, "The Mystery of Haushofer," *Life*, September 16, 1946, 108. Rudolf Hess claimed in a 1924 letter to Haushofer that Hitler had been impressed with Haushofer, one of whose books (on Japan) he had read. And yet, there were a number of crucial dissonances between Hitler and Haushofer's geopolitics, particular in relation to the question as to how Germany should approach the Soviet Union on the international stage. Spang, *Karl Haushofer und Japan*, 387.

158. Hitler did not acknowledge Ratzel in *Mein Kampf*, but then he did not cite Haushofer or the vast majority of his other influences either. *Mein Kampf* had many influences, including Oswald Spengler, Houston Stewart Chamberlain, Hans F. K. Günther, and Heinrich Claß (for an overview, see Roman Töppel, "'Volk und Rasse': Hitlers Quellen auf der Spur," *Vierteljahrshefte für Zeitgeschichte* 64, no. 1 (2016): 1–35.

159. Töppel, "Volk und Rasse," 14. I have some sympathy with Töppel's view that Hitler was prone to a selective reading. But I would add that the addition of political geographical lines of argument into Hitler's ideological project reveals that his worldview was not entirely settled at this stage.

160. Lange notes that many passages in *Politische Geographie*, including on Napoleon's Russian campaign, could in fact have been interpreted as a warning to Hitler. Lange, "Der Terminus Lebensraum," 434. But Schultz counters that Ratzel's warnings were always too vague to serve as such. Schultz, "Hätte doch die Erde," 38.

161. Jacobsen, *Karl Haushofer I*, 241; Plöckinger, *Geschichte eines Buches*, 145.

162. Lange, "Der Terminus Lebensraum."

163. Jacobsen, *Karl Haushofer I*, 243–44.

164. Spang, *Karl Haushofer und Japan*, 223.

165. Hitler, *Mein Kampf*, 360.

166. Hans Grimm, *Volk ohne Raum* (Munich: Albert Langen, 1926); Schultz, "Hätte doch die Erde," 38.

167. Hitler, *Mein Kampf*, 436. Note that the relationship between *Volk* and *Staat* is much more direct in Hitler than it is in Ratzel. Hitler insisted on racial purity as the basis of a strong state in ways that Ratzel was far less keen to do. We have already discussed in chapter 3 the relative absence of antisemitism in Ratzel's writings.

168. James Q. Whitman, *Hitler's American Model: The United States and the Making of Nazi Race Law* (Princeton, NJ: Princeton University Press, 2017); Brendan Simms, *Hitler: Only the World Was Enough* (London: Penguin, 2019).

169. Hitler, *Mein Kampf*, 25.

170. Simms, *Hitler*, 99, 120.

171. Termer, "Die natürlichen Grundlagen;" Töppel, "Volk und Rasse," 13.

172. Hitler, *Mein Kampf*, 313.

173. Franz Termer, "Einführung," in *Aus Mexico: Reiseskizzen aus den Jahren 1874 und 1875*, Friedrich Ratzel (Stuttgart: Brockhaus, 1969), v–xii. *Zeitschrift für Geopolitik* also included in 1924 an article on the KKK that explained that the second Klan was the expression of a naïve, fanatical, anti-German, and ultimately ridiculous "American form of fascism," now no longer limited to the South. Herman George Scheffauer, "Der Ku Klux Klan," *Zeitschrift für Geopolitik* 1 (1924): 700–704. There is no evidence that Hitler had read Ratzel's earlier books on the United States. But he did display similar positions to Ratzel when it came to the American South, which combined a rejection of slavery in principle both with antiblackness and a fascination with segregation and miscegenation laws. Johnpeter Horst Grill and Robert L. Jenkins, "The Nazis and the American South in the 1930s: A Mirror Image?" *Journal of Southern History* 58, no. 4 (1992): 671; Simms, *Hitler*, 121.

174. Schlögel, *Im Raume lesen wir*.

175. Richard Dawkins, *The Greatest Show on Earth: The Evidence of Evolution* (New York: Free Press, 2009), 62.

176. Weikart, *From Darwin to Hitler*.

177. *Bekenntnis der Professoren an den deutschen Universitäten und Hochschulen zu Adolf Hitler und dem nationalsozialistischen Staat*. Dresden: Nationalsozialistischer Lehrerbund Deutschland/Sachsen, 1933, 132.

178. Heinrich Schmitthenner, *Lebensräume im Kampf der Kulturen* (Leipzig: Quelle and Meyer, 1938).

179. Ewald Banse, *Große Forschungsreisende* (Munich: Lehmann, 1933); Friedrich Metz, "Friedrich Ratzel, Begründer der politischen Geographie," *Mein Heimatland* 25 (1938): 314–18. On the eccentric "aesthetic geographer" and orientalist travel writer Banse, see Murphy, *Heroic Earth*, 75–85. Friedrich Metz has thus far only received limited attention in the history of geography. For a more general list of geographers with Nazi biographies, see Arbeitskreis Geographie und Faschismus der Bundesfachschaftentagung Geographie. "Beiträge deutscher Geographieprofessoren zum Faschismus," *Geographische Sumpfschau* 4 (1996): 1–124. Geopolitics itself had featured from the 1920s onwards a number of non-Ratzelian strands, most prominently that of the pan-Europeanist Richard Coudenhove-Kalergi. Heffernan, *The Meaning of Europe*, 125–27; Benjamin J. Thorpe, "The Time and Space of Richard Coudenhove-Kalergi's Pan-Europe, 1923–1939" (PhD diss., University of Nottingham, 2018).

180. Trevor Barnes and Claudio Minca, "Nazi Spatial Theory: The Dark Geographies of Carl Schmitt and Walter Christaller," *Annals of the Association of American Geographers* 103, no. 3 (2013): 669–87. Christaller's 1933 *Die zentralen Orte in Süddeutschland* (Central places in southern Germany) was undoubtedly in a dialogue with Ratzel and his followers. But Christaller stated unambiguously in a footnote that Ratzel's conception of a town was "useless" for his purposes. Walter Christaller, *Central Places in Southern Germany* (New Jersey: Prentice-Hall, 1933 [1966]), 24.

181. Alfred Rosenberg, *Der Mythus des 20. Jahrhunderts* (Munich: Hoheneichen, 1930).

182. Franz Knieper, *Geopolitik für die Unterrichtspraxis* (Bochum: Verlags- und Lehrmittel-Anstalt Ferdinand Kamp, 1937), 9. The 1937 fifth edition of this book included a foreword by Karl Haushofer that highlighted the pivotal contributions of Ratzel and Kjellén. Knieper nonetheless noted that Ratzel's *Politische Geographie* had overstated the role of environmental factors. He also highlighted what he deemed to be "the sharpest of oppositions" between the "Jewish race" and the "Aryans" in the field of cultural and economic life. Knieper, *Geopolitik für die Unterrichtspraxis*, 9, 76.

183. Lange, Der Terminus Lebensraum, 436.

184. Karsten Uhl, "The Idea of Lebensraum and the Spatial Order of Power at German Factories, 1900–1945," *European Review of History* 20, no. 2 (2013): 287–307.

185. Zoltán Hajdú, "The Influence of Friedrich Ratzel on Hungarian Geography," in *Europe between Political Geography and Geopolitics: On the Centenary of Ratzel's Politische Geographie*, ed. Marco Antonsich, Vladimir Kolossov, and Maria Paola Pagnini (Rome: Società Geografica Italiana, 2001), 129–43; Steven Seegel, *Map Men: Transnational Lives and Deaths of Geographers in the Making of East Central Europe* (Chicago: University of Chicago Press, 2018), 128; Keiichi Takeuchi, "Japanese Geopolitics in the 1930s and 40s," In *Geopolitical Traditions: A Century of Geopolitical Thought*, ed. Klaus Dodds and David Atkinson (London: Routledge, 2000), 81.

186. Ernst Friedrich Flohr, "Versuch einer Klärung des Begriffs Lebensraum," *Geographische Zeitschrift* 48, 11/12 (1942): 393–404; Heinrich Schmitthenner, "Zum Begriff Lebensraum," *Geographische Zeitschrift* 48, no. 11/12 (1942): 405–17. Schrepfer, "Was heißt Lebensraum?"

187. Werner Daitz, *Lebensraum und gerechte Weltordnung: Grundlagen einer Anti-Atlantikcharta* (Amsterdam: De Amsterdamsche Keurkamer, 1943).

188. Buttmann, *Friedrich Ratzel*, 138.

189. Kurt Vowinckel, "Ein Seher vom Oberrhein. Fr. Ratzel, der Schöpfer der deutschen Geopolitik." *Straßburger Neueste Nachrichten*, April 13, 1944.

190. Haushofer, "Friedrich Ratzel im raumpolitischen Umbruch," 82, 87.

191. Carl Schmitt, *Land and Sea* (Washington DC: Plutarch, 1942 [1997]), 57.

192. Carl Schmitt, *The Nomos of the Earth in the International Law of the jus publicum Europaeum* (New York: Telos, 1950 [2006]).

193. Stuart Elden, "Reading Schmitt Geopolitically," *Radical Philosophy* 161 (2010): 18–26. This is crucial not least because it challenges the notion that Schmitt's theorizing sits on a higher plane than the jingoistic oversimplicities of Ratzel and his followers.

194. Schmitt, *Land and Sea*, 52.

195. G. L. Ulmen, "Translator's Introduction," In *The Nomos of the Earth in the International Law of the Jus Publicum Europaeum*, Carl Schmitt, 9–34 (New York: Telos, 2006), 23.

196. Patricia Chiantera-Stutte, " Space, *Großraum* and *Mitteleuropa* in Some Debates of the Early Twentieth Century," *European Journal of Social Theory* 11, no. 2 (2008): 194–96; Elden, "Reading Schmitt Geopolitically," 24; Matthew Specter, "Grossraum and Geopolitics: Resituating Schmitt in an Atlantic Context," *History and Theory* 56, no. 3 (2017): 403–5. Ratzel, it should be mentioned, was not the only who appeared in Schmitt's writings from the 1940s and 1950s. Indeed, he singled out Mackinder as his inspiration at the outset of Nomos of the Earth. Schmitt, *Nomos of the Earth*, 37.

197. Carl Schmitt, "Die Bibliothek Carl Schmitt," accessed 10 February 2023. Availableathttps://www.carl-schmitt.de/wp-content/uploads/CSG_Bibliothek_H2_2021.pdf.

198. Carl Schmitt, "Dialogue on New Space;" in *Dialogues on Space and Power*, eds. Andreas Kalyvas and Federico Finchelstein (Cambridge: Polity, 2015 [1958]), p.54.

199. Schmitt's original essay uses the Ratzelian term *Raumbewältigung*, which Nunan translates as "coming to terms with space," but which is in fact better translated as "the mastering of space." Carl Schmitt, "The *Großraum* Order of International Law with a Ban on Intervention for Spatially Foreign Powers: A Contribution to the Concept of *Reich* in International Law (1939–1941)," in *Carl Schmitt: Writings on War*, ed. Timothy Nunan (Cambridge: Polity, 1941 [2011]), 122; Schmitt, *Völkerrechtliche Großraumordnung*, 79. It should be noted too that Schmitt borrowed from Kjellen. *Geopolitik zur Einführung*, 83.

200. "The *Großraum* Order," 122.

201. "The *Großraum* Order," 119.

202. Schmitt, *Nomos of the Earth*, 283.

203. Schmitt, *Land and Sea*, 39.

204. Ratzel, *Politische Geographie: Geographie der Staaten*, 541.

205. Ratzel used the term "land appropriation" (*Landnahme*) in passing both in *Anthropogeographie* and *Politische Geographie* but did not make it a central concept in the way that Schmitt would in *Nomos of the Earth*, where the latter argued that it was an original act of spatial ordering that made the law appear. Ratzel, *Anthropogeographie Erster Teil*, 173; *Politische Geographie: Geographie der Staaten*, 348; Schmitt,

Nomos of the Earth, 80–81. Schmitt recognized that his own concept of *Landnahme* was different from older usages of the term as a synonym for land divisions that had appeared in colonial discourses at the turn of the century. That said, Schmitt's idea of a taking of land that is as yet unmediated by the law was not in fundamental contradiction with Ratzel's political geography.

206. Oliver Simons, "Carl Schmitt's Spatial Rhetoric," in *The Oxford Handbook of Carl Schmitt*, ed. Jens Meierhenrich and Oliver Simons (Oxford: Oxford University Press, 2016), 788.

207. Michael Heffernan, "Mapping Schmitt," in *Spatiality, Sovereignty and Carl Schmitt: Geographies of the Nomos*, ed. Stephen Legg (London: Routledge, 2011), 237; Claudio Minca and Nick Vaughan-Williams, "Carl Schmitt and the Concept of Border," *Geopolitics* 17, no. 4 (2012): 759. In the light of the above, I am not convinced that Schmitt's treatment of Ratzel's political geography amounts to a "dismissal." But what is certainly true is that Schmitt's geography was fundamentally "abstract" and "rhetorical" and devoid of the empirical richness that can be found in Ratzel's writings. Heffernan, "Mapping Schmitt," 237. Schmitt also rejected Ratzel's *halogäische Erdauffassung*. Schmitt, *Nomos of the Earth*, 89.

208. Minca and Rowan, *On Schmitt and Space*, 176.

209. Carl Schmitt, "Zum Gedächtnis von Sege Maiwald," *Zeitschrift für Geopolitik* 23 (1952): 447–48.

210. Isaiah Bowman, "Geography vs. Geopolitics," *Geographical Review* 32, no. 4 (1942): 655.

211. Bowman, "Geography vs. Geopolitics," 653; see also Bowman, *The New World*, 266.

212. Bowman, "Geography vs. Geopolitics," 655.

213. Hans Weikart, "Review: Haushofer and the Pacific," *Foreign Affairs* 20 (1942): 736.

214. Sigmund Neumann, "Fashions in Space," *Foreign Affairs* 21, no. 2 (1943): 279.

215. Klinke, *Cryptic Concrete*, 45–46.

216. Heinrich Schmitthenner, *Lebensräume im Kampf der Kulturen* (Heidelberg: Quelle and Meyer, 1938 [1951]), 6.

217. Schmitthenner, *Lebensräume im Kampf der Kulturen*, 210. Other interwar geopoliticians sought to salvage what was left of *Lebensraum*. Otto Maull, *Politische Geographie* (Berlin: Safari, 1956), 17.

218. Franz Knieper, *Politische Geographie (Geopolitik) für die Unterrichtspraxis* (Ahlen: Sommer, 1957), 20–21.

219. Werner Seegraef, "Politische Geographie in Schulbüchern für den Erdkundeunterricht seit 1945—ein Überblick," *Internationale Schulbuchforschung* 5, no. 3 (1983): 288–89.

220. Carl Troll, "Geographic Science in Germany during the Period 1933–1945: A Critique and Justification," *Annals of the Association of American Geographers* 39, no. 2 (1949): 134. For a further nuanced defense of Ratzel's anthropogeographical approach, see Overbeck, "Das politisch geographische Lehrgebäude."

221. Günther Heyden, *Kritik der deutschen Geopolitik: Wesen und soziale Funktion einer reaktionären soziologischen Studie* (Berlin: Dietz, 1958).

222. Ingrid Hönsch, "Friedrich Ratzel's Collected Papers in the Archive for Geogra-

phy in Leipzig," in *Europe between Political Geography and Geopolitics: On the Centenary of Ratzel's Politische Geographie*, ed. Marco Antonsich, Vladimir Kolossov, and Maria Paola Pagnini (Rome: Società Geografica Italiana, 2001), 98–100. A Leipzig street and a public building that had been named after Ratzel did not change their name. In 1994, a Leipzig school was named after Ratzel. It closed a decade later.

223. Schöller, "Wege und Irrwege."

224. Karl Heinz Pfeffer and Kurt Vowinckel, "Was will Geopolitik heute?" *Zeitschrift für Geopolitik* 24 (1954): 193–200; Schöller "Wege und Irrwege," 4.

225. I discuss the revival of geopolitics after 1945 in more detail in chapter 3 of Klinke, *Cryptic Concrete*.

226. Friedrich Ruge, *In vier Marinen. Lebenserinnerungen als Beitrag zur Zeitgeschichte* (Munich: Bernard & Graefe, 1979), 16.

227. Ratzel mentions Friedrich Ruge's father Walter in his diary. F. Ratzel, "Diary entry 29 December 1902," *Tagebuch 4.7.1900–19.7.1903, Abschrift angefertigt von Günter Buttmann* Bayerische Staatsbibliothek, Ratzeliana Supplement 17a VI 20. The grandfather Sophus Ruge was an economic geographer at the precursor of the Technische Universität Dresden.

228. Ruge, *In vier Marinen*, 345.

229. Friedrich Ruge, "Die vergessene See," *Zeitschrift für Geopolitik* 26 (1955): 355–60.

230. Friedrich Ruge, *Seemacht und Sicherheit: Eine Schicksalsfrage für alle Deutschen* (Tübingen: Fritz Schlichtenmeyer, 1955), 64.

231. Ruge, *In vier Marinen*, 429.

232. Erich Hampe, *Im Spannungsfeld der Luftmächte: Eine Einführung in Luftgefahr und Bevölkerungsschutz* (Cologne: Maximilian, 1956), 10.

233. Alexander Löfken, "Warum 'Baulicher Luftschutz'?" *Ziviler Luftschutz* 17, no. 2 (1954): 34; see also Alexander Löfken, "Überleben in Schutzraumbauten." *Ziviler Luftschutz* 24, no. 5 (1960): 162–68.

234. Morgenthau, *Politics among Nations*, 116, 479; see also Curran Flynn, "Political Geography and Morgenthau's Early American Works," *Cambridge Review of International Affairs* 29, no. 4 (2016): 1592.

235. Lucian Ashworth, "A New World? Isaiah Bowman and the Americanization of Ratzel's Political Geography" (Unpublished manuscript, 2018), 9. Given that Semple's geography too had an afterlife in American environmentalist geography throughout the twentieth century, Ratzel continued to exert an influence in the United States until at least the 1990s. Jared Diamond, *Guns, Germs and Steel: The Fates of Human Societies* (London: W. W. Norton, 1997), 431. Livingstone, "Environmental Determinism," 376.

236. Herman Kahn, *On Thermonuclear War* (New Brunswick, NJ: Transaction, 1960 [2007]), 517.

237. Saul B. Cohen, *Geography and Politics in a World Divided, 2nd edition* (New York: Oxford University Press, 1963 [1973]). Richard Pipes, "Militarism and the Soviet State," *Daedalus* 109, no, 4 (1980): 11; see also Dalby, *Creating the Second Cold* War, 76–77.

238. Leslie W. Hepple, "Metaphor, Geopolitical Discourse and the Military in South America," in *Writing Worlds: Discourse, Text and the Metaphor in the Representations*

of Landscape, ed. Trevor J. Barnes and James S. Duncan (London: Routledge, 1992), 136–54. "Möge Gott helfen." *Der Spiegel*, 17 March 1980. https://www.spiegel.de/spiegel/print/d-14319788.html.

239. Bassin, "Imperialism and the Nation State;" Bassin, "Race contra Space."

240. Ó Tuathail, *Critical Geopolitics*.

241. Karlheinz Weißmann, *Rückruf in die Geschichte. Die deutsche Herausforderung: Alte Gefahren—neue Chancen* (Berlin: Ullstein, 1992), 129; Karlheinz Weißmann, *Rubikon: Deutschland vor der Entscheidung* (Berlin: JF Edition, 2016).

242. Nationaldemokratische Partei Deutschlands, "Portfolio: Identität." Accessed September 20, 2019. Available at https://npd.de/portfolio/identitaet/.

243. Behlül Ozkan, "Turkey, Davutoglu and the Idea of Pan-Islamism," *Survival* 56, no. 4 (2014): 119–40.

244. Klinke and Bassin, "Lebensraum and Its Discontents;" Aleksandr Dugin, *Osnovy geopolitiki: Geopoliticheskoe budushchee Rossii* (Moscow: Arktogeia, 1997); Aleksandr Dugin, *Last War of the World-Island: The Geopolitics of Contemporary Russia* (London: Arktos, 2015), 4.

245. Ronald Beiner, "Stephen Bannon's World: Dangerous Minds in Dangerous Times." *The Conversation*, 25 July 2018. https://theconversation.com/stephen-bannons-world-dangerous-minds-in-dangerous-times-100373.

246. Weißmann, *Rubikon*, 70; see also Klinke, "Geopolitics and the Political Right," 511–13. Weißmann was not the only admirer of Ratzel's who was allied with the *Alternative für Deutschland*. Bernhard Streck, "Diffusionism and Geopolitics in the Work of Friedrich Ratzel," in *Europe Between Political Geography and Geopolitics: On the Centenary of Ratzel's Politische Geographie*, ed. Marco Antonsich, Vladimir Kolossov, and Maria Paola Pagnini (Rome: Società Geografica Italiana, 2001), 51–66; Bernhard Streck, "Gutachten zu Wolfgang Gedeon: 'Christlich-europäische Leitkultur' (3 Bd.) unter besonderer Berücksichtigung des Antisemitismusfrage." Accessed 7 October 7 2020. http://www.wolfgang-gedeon.de/wp-content/uploads/2016/11/Gutachten_Prof_Streck_zu_Gedeon.pdf.

247. Osterhammel, *Die Verwandlung der Welt*, 191; Weißmann, *Rubikon*, 73. *Kulturkreis* may not have been a central concept of Ratzel's and was already in circulation before his day, but the Leipzig geographer had certainly used it. Paul Leser, "Zur Geschichte des Wortes Kulturkreis," *Anthropos* 58, no. 1/2 (1963): 1–36.

248. Haruyo Yoshida, "Streitfall 'Kulturkreise': Zur Aktualität der Positionen von Spengler, Frobenius und Bloch," *Neue Beiträge zur Germanistik* 3, no. 3 (2004): 61.

249. Richard Hartshorne, "Recent Developments in Political Geography I," *American Political Science Review* 29, no. 5 (1935): 789; Sprengel, *Kritik der Geopolitik*, 26; Werber, *Geopolitik zur Einführung*, 26–27.

250. Smith, "Friedrich Ratzel," 68.

CHAPTER 8

1. Abrahamsson, "Genealogy of Lebensraum;" Barua, "Ratzel, Bio-geography;" Keighren, "History and Philosophy of Geography III"; Mitch Rose, "The Question of

Culture in Cultural Geography: Latent Legacies and Potential Futures," *Progress in Human Geography* 45, no. 5 (2021): 951–71; Usher, "Territory *incognita*."

2. Verne "The Neglected 'Gift,'" 87.

3. Usher, "Territory *Incognita*," 1021, 1023; see also Rose, "The Question of Culture."

4. Usher, "Territory *Incognita*," 1034.

5. Rose argues the exact opposite here, namely that in Ratzel's later work "imagined hierarchies of *volk* culture where different cultural 'strengths' compete for dominance" were "disbanded in favour of a more interactionist tone." This too, I hope to have shown, does not hold up to scrutiny. Ratzel remained committed to a hierarchy of *Völker* and races in his very last writings. Rose, "The Question of Culture," 955.

6. Elizabeth A. Povinelli, *Geontologies: A Requiem to Late Liberalism*. (Durham, NC: Duke University Press, 2016); Juanita Sundberg, "Decolonizing Posthumanist Geographies," *Cultural Geographies* 21, no. 1 (2014): 33–47.

7. Povinelli, *Geontologies*, 176.

8. Bennett, *Vibrant Matter*, 120.

9. Thomas Lemke, "An Alternative Model of Politics? Prospects and Problems of Jane Bennett's Vital Materialism," *Theory, Culture and Society* 35, no. 6 (2018): 31–54.

10. Jane Bennett, "Author Response," *Dialogues in Human Geography* 1, no. 3 (2011): 404–6.

11. Nigel Clark, "Geopolitics of the Threshold," *Political Geography* 37 (2013): 48–50; Deborah Dixon, "The Way of the Flesh: Life, Geopolitics and the Weight of the Future," *Gender, Place & Culture* 21, no. 2 (2014): 136–51.

12. Mbembe, *Necropolitics*, 77.

13. Mbembe, *Necropolitics*, 20.

14. Mbembe, *Necropolitics*, 1.

15. Kundrus, "German Colonialism," 25.

16. Bauman, *Modernity and the Holocaust*, 13.

17. Ratzel, "Nationalitäten und Rassen," 487.

18. Mbembe, *Necropolitics*, 138.

Bibliography

WORKS BY FRIEDRICH RATZEL

"America's Race Problems. Addresses at the Annual Meeting of the American Academy of Political and Social Science, Philadelphia, April Twelfth and Thirteenth, Philadelphia 1901." *Geographischer Literaturbericht* 521 (1902): 149–50.

Anthropogeographie Erster Teil, Dritte Auflage: Grundzüge der Anwendung der Erkunde auf die Geschichte. Stuttgart: J. Engelhorn, 1882 (1909).

Anthropogeographie Erster Teil: Grundzüge der Anwendung der Erkunde auf die Geschichte. Stuttgart: J. Engelhorn, 1882.

Anthropogeographie Erster Teil, Zweite Auflage: Grundzüge der Anwendung der Erkunde auf die Geschichte. Stuttgart: J. Engelhorn, 1882 (1899).

Anthropogeographie Zweiter Teil, Dritte Auflage: Die geographische Verbreitung des Menschen. Stuttgart: J. Engelhorn, 1891 (1912).

Anthropogeographie Zweiter Teil, Zweite Auflage: Die geographische Verbreitung des Menschen. Stuttgart: J. Engelhorn, 1891 (1899).

"Aus den Baumwollenstaaten I." *Globus* 30 (1876): 314–18.

"Aus den Baumwollenstaaten II." *Globus* 30 (1876): 344–46.

"Aus Deutschamerika." *Die Grenzboten* 54 (1895): 403–4.

Aus Mexico: Reiseskizzen aus den Jahren 1874 und 1875. Stuttgart: Brockhaus, 1878 (1969).

"Beiträge zur Anatomie von Enchytraeus vermicularis Henle." *Zeitschrift für Wissenschaftliche Zoologie* 18 (1867): 99–108.

"Beiträge zur anatomischen und systematischen Kenntnis der Oligochaeten." *Zeitschrift für Wissenschaftliche Zoologie* 18 (1868): 563–91.

"Briefe und Berichte Emin Paschas." *Beilage zur Allgemeinen Zeitung* 81, 1888: 1185–87.

"Brutal." *Die Grenzboten* 56 (1897): 444.

"B. T. Washington: Vom Sklaven empor! Berlin 1902." *Geographischer Literaturbericht* 315 (1903): 97.

"C. Pearson, The Chances of Death and Other Studies in Evolution." *Literarisches Zentralblatt* 8 (1898): 254–55.

"Das Buch des Dr. Karl Peters." *Die Grenzboten* 50 (1891): 600–610.

"Das Deutschtum in Nordamerika." In *Kleine Schriften von Friedrich Ratzel, herausgegeben von Hans Helmolt, 1. Band*, edited by Hans Helmolt, 358–60, Munich: R. Oldenbourg, 1904 (1906).

Das Meer als Quelle der Volksgröße: Eine politisch–geographische Studie. Munich: R. Oldenbourg, 1900 (1911).

"Der Geist, der über den Wassern schwebt." In *Kleine Schriften von Friedrich Ratzel, herausgegeben von Hans Helmolt, 1. Band*, edited by Hans Helmolt, 237–49. Munich: R. Oldenbourg, 1901 (1906).

Der Lebensraum: Eine biogeographische Studie. Tübingen: Verlag der Laupp'schen Buchhandlung, 1901.

"Der Lebensraum: Eine biogeographische Studie." In *Festgaben für Albert Schäffle zur siebenzigsten Wiederkehr seines Geburtstages a, 24. Februar 1901*, edited by Karl Bücher et al., 102–89. Tübingen: Verlag der Laupp'schen Buchhandlung, 1901.

"Der mitteleuropäische Wirtschaftsverein." *Die Grenzboten* 63 (1904): 253–59.

"Der Nicaraguakanal und die Monroe-Doktrin." *Beilage zur Allgemeinen Zeitung* 227 (1897): 1–7.

"Der Staat als Organismus." *Die Grenzboten* 55 (1896): 614–23.

"Der Ursprung der Arier in geographischem Licht." In *Kleine Schriften von Friedrich Ratzel, herausgegeben von Hans Helmolt, 2. Band*, edited by Hans Helmolt, 391–401. Munich: R. Oldenbourg, 1899 (1906).

"Der Verfall der Nekrologie." *Die Grenzboten* 53 (1894): 238–39.

"Der Verschönerungsverein." *Die Grenzboten* 52 (1893): 574–75.

"Deutschland und Frankreich." *Die Grenzboten* 52 (1893): 289–94.

"Deutsch-Chinesisch." *Die Grenzboten* 55 (1896): 622–26.

"Deutscher Kolonien- und Zeitungsklatsch." *Die Grenzboten* 55 (1896): 527.

Deutschland: Eine Einführung in die Heimatkunde. Leipzig: Wilhelm Grunow, 1898.

"Deutschlands Lage." *Die Grenzboten* 55 (1896): 105–8.

"Deutschands Stellung und Rechte am Niger." *Die Grenzboten* 56 (1897): 545–51.

"Die Aussichten unsers südwestafrikanischen Schutzgebietes." *Die Grenzboten* 51 (1892): 171–75.

"Die Beurteilung der Neger." *Die Grenzboten* 51 (1892): 20–24.

Die chinesische Auswanderung: Ein Beitrag zur Cultur- und Handelsgeographie. Breslau: J. U. Kern, 1876.

Die deutsche Tiefsee-Expedition." In *Kleine Schriften von Friedrich Ratzel, herausgegeben von Hans Helmolt, 2. Band*, edited by Hans Helmolt, 382–390. Munich: R. Oldenbourg, 1898 (1906).

Die Erde in vierundzwanzig gemeinverständlichen Vorträgen über Allgemeine Erdkunde. Stuttgart: J. Engelhorn, 1881.

Die Erde und das Leben, 1. Band. Leipzig: Bibliographisches Institut, 1902.

Die Erde und das Leben, 2. Band. Leipzig: Bibliographisches Institut, 1902.

"Die Gegenwart und Zukunft der Siebenbürger Sachsen." *Die Grenzboten* 51 (1892): 449–57.

"Die geographische Lage Deutschlands." *Die Grenzboten* 55 (1896): 390–97; 450–56.

"Die geographische Methode in der Frage nach der Urheimat der Indogermanen." In *Kleine Schriften von Friedrich Ratzel, herausgegeben von Hans Helmolt, 2. Band*, edited by Hans Helmolt, 528–36. Munich: R. Oldenbourg, 1904 (1906).

"Die Gesetze des räumlichen Wachstums der Staaten. Ein Beitrag zur wissenschaftlichen politischen Geographie." *Petermanns Mitteilungen* 42, no. 5 (1896): 97–107.

"Die Kant-Lapclacesche Hypothese und die Geographie." In *Kleine Schriften von Friedrich Ratzel, herausgegeben von Hans Helmolt, 2. Band*, edited by Hans Helmolt, 420–36. Munich: R. Oldenbourg, 1901 (1906).
"Die Maske ab!" *Die Grenzboten* 53 (1894): 337–40.
"Die nordatlantischen Mächte. Eine politisch-geographische Studie." *Marine-Rundschau* 7–12 (1903): 911–39.
"Die nordatlantischen Mächte. Eine politisch-geographische Studie." *Marine-Rundschau* 7–12 (1903): 1047–62.
Die Schneedecke, besonders in Gebirgen. Stuttgart: Engelhorn, 1889.
"Die Tagesansicht Gustav Theodor Fechners." *Die Grenzboten* 60 (1901): 169–78.
Die Vereinigten Staaten von Nord-Amerika, Zweiter Band: Culturgeographie unter besonderer Berücksichtigung der wirthschaftlichen Verhältnisse. Munich: R. Oldenbourg, 1880.
Die Vereinigten Staaten von Nord-Amerika, Zweiter Band, Zweite Auflage: Politische und Wirthschaftsgeographie. Munich: R. Oldenbourg, 1880 (1893).
Die Vereinigten Staaten von Nord-Amerika, Erster Band, Physikalische Geographie und Naturcharakter. Munich: R. Oldenbourg, 1878.
Die Vorgeschichte des europäischen Menschen. Munich: R. Oldenbourg, 1874.
"Die Zeichen des cubanischen Aufstandes." *Die Grenzboten* 54 (1895): 479–81.
"Die Zeitforderung in den Entwicklungswissenschaften." *Annalen der Naturphilosophie* 1/2 (1902): 40–98.
"Dr Karl Peters." *Die Grenzboten* 56 (1897): 252–56.
"Durch den dunkeln Weltteil, Bd. 1, Leipzig, 1878." *Literarisches Zentralblatt* (1878): 1570–1571.
"E. C. Semple, The Anglo-Saxons of the Kentucky Mountains, Geographical Journal 1901." *Geographischer Literaturbericht* 226 (1902): 67–68.
"1860er Antisemitismus." *Die Grenzboten* 53 (1894): 282–83.
"Eine 'Ehrung' für Virchow." *Die Grenzboten* 53 (1894): 335–36.
"Eine vergessene Stadt in Südfrankreich." In *Kleine Schriften von Friedrich Ratzel, herausgegeben von Hans Helmolt, 1. Band*, edited by Hans Helmolt, 10–15. Munich: R. Oldenbourg, 1869 (1906).
"Eingabe an das Reichskanzleramt in Sachen Emin Pascha." In *Kleine Schriften von Friedrich Ratzel, herausgegeben von Hans Helmolt, 1. Band*, edited by Hans Helmolt, 519–21. Munich: R. Oldenbourg, 1891 (1906).
"Ein gutes Ziel für deutsche Auswanderung I." *Allgemeine Zeitung* 118 (1880): 1713–14.
"Ein gutes Ziel für deutsche Auswanderung II." *Allgemeine Zeitung* 119 (1880): 1729–30.
"Einige Aufgaben der politischen Ethnographie." In *Kleine Schriften von Friedrich Ratzel, herausgegeben von Hans Helmolt, 2. Band*, edited by Hans Helmolt, 402–19. Munich: R. Oldenbourg, 1900 (1906).
"Emin Pascha." *Deutsche Revue* 17 (1892): 211–23.
"Emin Pascha: Eine Darstellung seines Lebens und Wirkens mit Benutzung seiner Tagebücher, Briefe und wissenschaftlichen Aufzeichungen von Georg Schweitzer." *Geographische Zeitschrift* 4, no. 5 (1898): 301–3.

"Emin Pascha." In *Kleine Schriften von Friedrich Ratzel, herausgegeben von Hans Helmolt, 1. Band*, edited by Hans Helmolt, 511–18. Munich: R. Oldenbourg, 1904 (1906).

"Entwurf einer neuen politischen Karte von Afrika. Nebst einigen allgemeinen Bemerkungen über die Grundsätze der politischen Geographie." *Petermanns Mitteilungen* 31 (1885): 245–250.

"E. Reclus, Nouvelle geographie universelle, Livraison 574–638, Paris 1885." *Literarisches Zentralblatt* 13 (1887): 415–16.

"E. Reclus, Nouvelle geographie universelle, Tome XU, Paris 1886/87." *Literarisches Zentralblatt* 26 (1888): 882–83.

"Ernst Haeckel." In *Kleine Schriften von Friedrich Ratzel, herausgegeben von Hans Helmolt, 1. Band*, edited by Hans Helmolt, 373–76. Munich: R. Oldenbourg, 1872 (1906).

"Erwiderung auf Hermann Wagners Besprechung der Anthropogeographie II." *Zeitschrift der Gesellschaft für Erdkunde zu Berlin* 16 (1891): 508–12.

"Ethnographie und Geschichtswissenschaft in Amerika." In *Kleine Schriften von Friedrich Ratzel, herausgegeben von Hans Helmolt, 1. Band*, edited by Hans Helmolt, 347–54. Munich: R. Oldenbourg, 1897 (1906).

"Europa." In *Brockhaus Konversationslexikon, 14. Auflage, Band VI*, no editor, 422–35. Leipzig: Brockhaus, 1893.

"F. Boas, Dissemination of Tales among the Natives of North America, New York 1891." *Geographischer Literaturbericht* 1398 (1891): 102–3.

"F. Boas, Human Faculty as Determined by Race, New York 1894." *Geographischer Literaturbericht* 40 (1895): 12–13.

"5 Jahre deutscher Kolonialpolitik." *Allgemeine Zeitung* 157 (1889): 2377–78.

"Flottenfrage und Weltlage." In *Kleine Schriften von Friedrich Ratzel, herausgegeben von Hans Helmolt, 2. Band*, edited by Hans Helmolt, 375–81. Munich: R. Oldenbourg, 1898 (1906).

Frommanns Taschenbuch für Fussreisende, herausgegeben und ergänzt von Dr. Friedrich Ratzel. Stuttgart: Frommanns, 1880.

"F. Turner, The Significance of the Frontier in American History, Washington 1894." *Geographischer Literaturbericht* 574 (1895): 122.

Géographie politique. Paris: Éditions régionales européennes, 1897 (1988).

"Geographische Handbücher." *Die Grenzboten* 48 (1889): 331–36.

"Geschichte, Völkerkunde und historische Perspektive." *Historische Zeitschrift* 93, no. 1 (1904): 1–46.

"Geschichte, Völkerkunde und historische Perspektive." In *Kleine Schriften von Friedrich Ratzel, herausgegeben von Hans Helmolt, 2. Band*, edited by Hans Helmolt, 488–525. Munich: R. Oldenbourg, 1904 (1906).

"Gummiräder." *Die Grenzboten* 53 (1894): 110–11

"Gustav Nachtigal." In *Allgemeine Deutsche Biographie XXIII*, edited by Historische Commission bei der Königl. Akademie der Wissenschaften, 193–99. Leipzig: Duncker und Humblot, 1886.

"Heinrich Driesmans: Rasse und Milieu, Berlin 1902." *Historische Vierteljahrschrift* 7 (1904): 391–97.

"H. F. v. Behr, Kriegsbilder aus dem Araberaufstand in Deutsch-Ostafrika, Leipzig 1891." *Literarisches Zentralblatt* 14 (1891): 453.

The History of Mankind, Volume I. London: MacMillan, 1885 (1896).

The History of Mankind, Volume II. London: MacMillan, 1886 (1897).

The History of Mankind, Volume III. London: MacMillan, 1888 (1898).

"H. M. Stanley, Durch den dunkeln Weltteil, Bd. I, Leipzig 1878." *Literarisches Zentralblatt* (1878): 1570–71.

"H. Spencer, Prinzipien der Biologie, Bd. I, Stuttgart, 1876." *Archiv für Anthropologie* 10 (1878): 339–41.

"H. Spencer, Prinzipien der Biologie, Bd. II, Stuttgart, 1876." *Archiv für Anthropologie* 11 (1879): 181–82.

"H. Wißmann, L. Wolf, C. v. Francois und H. Müller, Im Innern Afrikas, Leipzig 1888." *Geographischer Literaturbericht* 359 (1888): 80.

"H. Wißmann, Unter deutscher Flagge quer durch Afrika von West nach Ost, Berlin 1889." *Geographischer Literaturbericht* 135 (1890): 19.

"In welcher Richtung beeinflussen die afrikanischen Ereignisse die Tätigkeit des Kolonialvereins?" *Deutsche Kolonialzeitung* 2 (1885): 38–44.

"Koloniale Biergespräche." *Die Grenzboten* 54 (1895): 246–47.

"Land und Landschaft in der nordamerikanischen Volksseele." In *Kleine Schriften von Friedrich Ratzel, herausgegeben von Hans Helmolt, 1. Band*, edited by Hans Helmolt, 277–92. Munich: R. Oldenbourg, 1902 (1906).

"Laws of the Spatial Growth of States." In *The Structure of Political Geography*, edited by Roger E. Kasperson and Julian V. Minghi, 17–28. London: Transactions, 1896 (1969).

"Lebensraum—a Biogeographical Study. Translated by Tul'si (Tuesday) Bhambry." *Journal of Historical Geography* 61 (1901 [2018]): 59–80.

"Man as a Life Phenomenon on the Earth's Surface." In *The History of the World*, vol. I, edited by Hans F. Helmolt, 61–106. New York: Dodd, Mead and Co., 1902.

"Moritz Wagner." In *Kleine Schriften von Friedrich Ratzel, herausgegeben von Hans Helmolt, 1. Band*, edited by Hans Helmolt, 461–75. Munich: R. Oldenbourg, 1895 (1906).

"Nationalitäten und Rassen." In *Kleine Schriften von Friedrich Ratzel, herausgegeben von Hans Helmolt, 2. Band*, edited by Hans Helmolt, 462–87. Munich: R. Oldenbourg, 1904 (1906).

"Neger." In *Brockhaus Konversationslexikon, 14. Auflage, Bd XII*, no editor, 228. Leipzig: Brockhaus, 1884.

"Neue Untersuchungen über die Vogelnester." *Meyers Ergänzungsblätter zur Kenntnis der Gegenwart* 2 (1870): 496–500.

"Nicht schielen!" *Die Grenzboten* 51 (1892): 411–12.

"Nochmals die englische Heuchelei." *Die Grenzboten* 53 (1894): 472–73.

"Oskar Peschel." In *Kleine Schriften von Friedrich Ratzel, herausgegeben von Hans Helmolt, 1. Band*, edited by Hans Helmolt, 430–47. Munich: R. Oldenbourg, 1887 (1906).

Politische Geographie: Geographie der Staaten, des Verkehrs und des Krieges. Munich: R. Oldenbourg, 1897 (1903).

"P. Vidal de la Blache, Les conditions geographiques des faits sociaux, Annales de Geographie 1902." *Geographischer Literaturbericht* 310 (1902): 95.
Raum und Zeit in Geographie und Geologie: Naturphilosophische Betrachtungen, herausgegeben von Dr Paul Barth. Leipzig: Johann Ambrosius Barth, 1907.
Sein und Werden: Eine populäre Schöpfungsgeschichte. Leipzig: Gebhardt und Reisland, 1869.
Sketches of Urban and Cultural Life in North America. New Brunswick, NJ: Rutgers University Press, 1876 (1988).
"Stabsarzt Dr. Ludwig Wolf." *Beilage zur Allgemeinen Zeitung* 71 (1890): 1–2.
"Stabsarzt Dr. Ludwig Wolf." *Mitteilungen des Vereins für Erdkunde zu Leipzig* 1889 (1890): 104–14.
"Stanleys neues Buch." *Beilage zur Allgemeinen Zeitung* 168 (1890): 1–3.
"Studies in Political Areas: The Political Territory in Relation to Earth and Continent." *American Journal of Sociology* 3, no. 3 (1897): 449–63.
"Studies in Political Areas II: Intellectual, Political, and Economic Effects of Large Areas." *American Journal of Sociology* 3, no. 4 (1898): 449–63.
"Studies in Political Areas III: The Small Political Area." *American Journal of Sociology* 4, no. 3 (1898): 366–79.
"The Territorial Growth of States." *Scottish Geographical Magazine* 12, no. 7 (1896): 351–61.
"Th. Roosevelt, The Winning of the West, New York 1895." *Geographischer Literaturbericht* 816 (1895): 176.
"Über den Lebensraum." *Die Umschau* 1, no. 21 (1897): 363–67.
"Über die anthropogeographischen Begriffe geschichtliche Tiefe und Tiefe der Menschheit." In *Kleine Schriften von Friedrich Ratzel, herausgegeben von Hans Helmolt, 2. Band*, edited by Hans Helmolt, 215–236. Munich: R. Oldenbourg, 1890 (1906).
"Über die Anwendung des Begriffs Oekumene auf geographische Probleme der Gegenwart (Mit einer Karte)." *Berichte über die Verhandlungen der Königlich Sächsischen Gesellschaft der Wissenschaften zu Leipzig, philologisch-historische Klasse* 40 (1888): 137–80.
"Über Kalifornien." In *Kleine Schriften von Friedrich Ratzel, herausgegeben von Hans Helmolt, 2. Band*, edited by Hans Helmolt, 1–18. Munich: R. Oldenbourg, 1877 (1906).
Über Naturschilderung. Munich: R. Oldenbourg, 1904 (1923).
"Über politische Verhältnisse in Innerafrika." In *Kleine Schriften von Friedrich Ratzel, herausgegeben von Hans Helmolt, 2. Band*, edited by Hans Helmolt, 158–72. Munich: R. Oldenbourg, 1888 (1906).
"Unsere Bahnhofskneipen." *Die Grenzboten* 52 (1893): 190.
"Unsere Volkstrachten." *Die Grenzboten* 56 (1896): 357–64.
"Verkehrter Bismarckkultus." *Die Grenzboten* 52 (1893): 141–42.
"Vorwort" In *Durch Krieg zum Frieden: Stimmungsbilder aus den Jahren 1870–1871*, by Karl Stieler, i–ii. Stuttgart: Bonz und Comp., 1886 (1895).
Wandertage eines Naturforschers. 1. Theil: Zoologische Briefe vom Mittelmeer. Briefe aus Süditalien. Leipzig: Brockhaus, 1873.
Wandertage eines Naturforschers. 2. Theil: Schilderungen aus Siebenbürgen und den Alpen. Leipzig: Brockhaus, 1874.

"Weiße und Farbige im indischen Archipelagus." *Globus* 16 (1869): 375–78.
"Weltentwicklung und Weltschöpfung. Mit einem Anhang über Lyells und Darwins Gottesideen." *Die Grenzboten* 61 (1902): 669–84.
Wider die Reichsnörgler: Ein Wort zur Kolonialfrage aus Wählerkreisen. Munich: Oldenbourg, 1884.
"Zu Karl Ritters hundertjährigem Geburtstage." In *Kleine Schriften von Friedrich Ratzel, herausgegeben von Hans Helmolt, 1. Band*, edited by Hans Helmolt, 377–428. Munich: R. Oldenbourg, 1879 (1906).
"Zukunft und Beurteilung der Neger." *Deutsche Revue* 4, no. 4 (1880): 97–111.
"Zur Entwicklungsgeschichte des Regenwurms (Lumbricus Agricola Hoffm.)." *Zeitschrift für Wissenschaftliche Zoologie* 18 (1868): 547–62.
"Zur Frage der Indogermanen-Heimat." *Archiv für Rassen- und Gesellschafts-Biologie* 1, no. 4 (1904): 579–80.
"Zur Frage der Indogermanen-Heimat." In *Kleine Schriften von Friedrich Ratzel, herausgegeben von Hans Helmolt, 2. Band*, edited by Hans Helmolt, 541–42. Munich: R. Oldenbourg, 1904 (1906).
"Zur Geschichte des Deutschtums in Nordamerika." *Die Grenzboten* 56 (1897): 519–22.
"Zur Kenntnis der englischen Weltpolitik: 1." *Die Grenzboten* 54 (1895): 49–59.
"Zur Kenntnis der englischen Weltpolitik: 6. England in Südafrika." *Die Grenzboten* 54 (1895): 7–21.
, and Hans Helmolt. "Friedrich Ratzel: Ein Lebensabriß von ihm selbst und vom Herausgeber." In *Kleine Schriften von Friedrich Ratzel, herausgegeben von Hans Helmolt, 1. Band*, edited by Hans Helmolt, xxi–xxxiii. Munich: R. Oldenbourg, 1906.

LETTERS, DIARIES, AND OTHER ARCHIVAL SOURCES

Baumann, Oscar. "Hochgeehrter Herr Professor." Letter to Friedrich Ratzel, Vienna, 11 May 1887, Bayerische Staatsbibliothek, Ratzeliana Supplement 11.
Baumann, Oscar. "Sehr geehrter Herr Professor." Letter to Friedrich Ratzel, Berlin, 8 December 1889, Bayerische Staatsbibliothek, Ratzeliana Supplement 11.
Boas, Franz. "Werter Herr Professor." Letter to Friedrich Ratzel, Worcester (Mass.), 5 December 1889, Bayerische Staatsbibliothek, Ratzeliana Supplement 13.
Drygalski, Erich von. "Sehr geehrter Herr Doktor!" Letter to unknown, Königsberg, 5 January 1899, Bayerische Staatsbibliothek, Ratzeliana Supplement 11.
Drygalski, Erich von. "Hochverehrter Herr Geheimrat!" Letter to Friedrich Ratzel, Potsdam, 5 January 1899, Bayerische Staatsbibliothek, Ratzeliana Supplement 11.
Durkheim, Émile. "Monsieur et honoré college." Letter to Friedrich Ratzel, Paris, 8 October 1899, Bayerische Staatsbibliothek, Ratzeliana Supplement 11.
Fabri, Friedrich. "Wertester Herr Professor!" Letter to Friedrich Ratzel, Berlin, 15 November 1885, Bayerische Staatsbibliothek, Ratzeliana Supplement 11.
Fabri, Friedrich. "Wertester Herr und Freund!" Letter to Friedrich Ratzel, Godesberg, 9 May 1889, Bayerische Staatsbibliothek, Ratzeliana Supplement 11.

Grothe, Hugo. "Hochverehrter Herr Professor!" Letter to Friedrich Ratzel, Munich, 30 August 1903, Bayerische Staatsbibliothek, Ratzeliana Supplement 12.

Hartmann, Georg. "Hochgeehrter Herr Professor!" Letter to Friedrich Ratzel, Chorab, 6 October 1900, Bayerische Staatsbibliothek, Ratzeliana Supplement 12.

Hawks, Helen. "Manchuria Needs Japan, Says Eminent Woman Scientists Here." *Palm Beach Post*, 17 January 1932, University of Kentucky, Special Collections Research Center, Ellen Churchill Semple papers, scrapbook, 1901–1932, box 6, folder 8.

Heger, Franz. "Hochgeehrter Herr Professor." Letter to Friedrich Ratzel, Vienna, 2 May 1894, Archiv für Geographie, Leibniz Institut für Länderkunde, Nachlass Friedrich Ratzel 163/658.

Henkel, W. C. "Sir." Letter to Clements R. Markham, Cassel, 9 August 1901, Royal Geographical Society, Correspondence Blocks, Correspondence 1881–1910.

Hübbe-Schleiden, Wilhelm. "Verehrter Herr Professor!" Letter to Friedrich Ratzel, Barmen, 18 July 1881, Bayerische Staatsbibliothek, Ratzeliana Supplement 12.

Kaiser-Wilhelm-Institut für Genealogie und Demographie. "Akte Friedrich Ratzel." Historisches Archiv des Max-Planck-Instituts für Psychiatrie, MPIP-HA, GDA 560.

Klutschak, Heinrich. (1881) "Sehr geehrter Herr!" Letter to Friedrich Ratzel, Bremen, 29 November 1881, Bayerische Staatsbibliothek, Ratzeliana Supplement 12.

Knauer, Friedrich. (1902) "Naturbeobachtung" Bayerische Staatsbibliothek, Ratzeliana II Besprechungen von Ratzelschen Werken.

London Assurance Corporation, United States branch. Legal advice, New York, 5 January 1893, Archiv für Geographie, Leibniz Institut für Länderkunde, Nachlass Friedrich Ratzel 163/646.

Lüdecke, Richard. "Sehr geehrter Herr Professor!" Letter to Friedrich Ratzel, Gotha, 29 August 1892, Archiv für Geographie, Leibniz Institut für Länderkunde, Nachlass Friedrich Ratzel 155/319.

Munro, Dana C. "Criticism of 'Partition of Asiatic Turkey.'" 6 August 1918, Records of the American Commission to Negotiate Peace, Record Group 256, Inquiry document 501, Records of the Inquiry 1914–19, National Archives.

n.a. "Deutsche Amerikaner." Bayerische Staatsbibliothek, Ratzeliana II Besprechungen von Ratzelschen Werken.

Nachtigal, Gustav. "Hochgeehrter Herr Professor." Letter to Friedrich Ratzel, Berlin, 28 June 1880, Bayerische Staatsbibliothek, Ratzeliana Supplement 12.

Penck, Albrecht. "Verehrter Freud und Kollege!" Letter to Ratzel, Vienna, 28 October 1902, Bayerische Staatsbibliothek, Ratzeliana Supplement 12.

Ratzel, Friedrich. "Bismarck und die Sueddeutschen" Unpublished manuscript, Archiv für Geographie, Leibniz Institut für Länderkunde, Nachlass Friedrich Ratzel 160/476.

Ratzel, Friedrich. *Tagebuch, 1879–1881, Abschrift angefertigt von Günter Buttmann* Bayerische Staatsbibliothek, Ratzeliana Supplement 17a VI 7.

Ratzel, Friedrich. *Tagebuch 4.10.1896–6.1.1898, Abschrift angefertigt von Günter Buttmann*, Bayerische Staatsbibliothek, Ratzeliana Supplement 17a VI 15a.

Ratzel, Friedrich. *Tagebuch 1894, Abschrift angefertigt von Günter Buttmann*, Bayerische Staatsbibliothek, Ratzeliana Supplement 17a VI 12.

Ratzel, Friedrich. *Tagebuch 1895, Abschrift angefertigt von Günter Buttmann*, Bayerische Staatsbibliothek, Ratzeliana Supplement 17a VI 13.

Ratzel, Friedrich. *Tagebuch begonnen 5.10.1875, Abschrift angefertigt von Günter Buttmann*, Bayerische Staatsbibliothek, Ratzeliana Supplement 17a VI 3.

Ratzel, Friedrich. *Tagebuch 20.7.1903–9.8.1904, Abschrift angefertigt von Günter Buttmann*, Bayerische Staatsbibliothek, Ratzeliana Supplement 17a VI 21.

Ratzel, Friedrich. *Tagebuch 4.7.1900–19.7.1903, Abschrift angefertigt von Günter Buttmann*, Bayerische Staatsbibliothek, Ratzeliana Supplement 17a VI 20.

Ratzel, Friedrich. *Tagebuch, 5.10.1876–9.3.1878, Abschrift angefertigt von Günter Buttmann*, Bayerische Staatsbibliothek Ratzeliana, Supplement 17a VI 4.

Ratzel, Friedrich. *Tagebuch über Englandreise im Aug/Sep 1881, Abschrift angefertigt von Günter Buttmann*, Bayerische Staatsbibliothek, Ratzeliana Supplement 17a VI 9a.

Ratzel, Friedrich. *Tagebuch 24.7.1873–4.5.1875, Abschrift angefertigt von Günter Buttmann*, Bayerische Staatsbibliothek, Ratzeliana Supplement 17a VI 1.

Ratzel, Friedrich. "Geehrter Herr Professor." Letter to Ernst Haeckel, Carlsruhe, 25 September 1869. Transliterated by Peter Zigman. *Würzburger medizinhistorische Mitteilungen* 28: 387–388.

Ratzel, Friedrich. "Hochgeehrter Herr Professor." Letter to Ernst Haeckel, Munich, 24 March 1872. Transliterated by Peter Zigman. *Würzburger medizinhistorische Mitteilungen* 28: 398–401.

Ratzel, Friedrich. "Hochverehrter Herr Kollege!" Letter to Wilhelm Wundt, 6 February 1887, Leipzig, Universitätsarchiv Leipzig, Nachlass Wilhelm Wundt; NA Wundt/III/1601–1700/1601/187/489–490.

Ratzel, Friedrich. "Liebe Hedwig!" Letter to Hedwig Ratzel, Frankfurt, 14 October 1895, Bayerische Staatsbibliothek, Ratzeliana Supplement 10.

Ratzel, Friedrich. "Liebe Loll!" Letter to Lila Ratzel, Leipzig, 24 May 1902, Bayerische Staatsbibliothek, Ratzeliana Supplement 10.

Ratzel, Friedrich. "Liebe Miss Semple!" Letter to Ellen Churchill Semple, Riva, 4 April 1903, University of Kentucky, Special Collections Research Center, Ellen Churchill Semple papers, Correspondence, 1900–1952, box 6, folder 4.

Ratzel, Friedrich. "Lieber Freund!" Letter to Karl Lamprecht, Leipzig, 18 July 1899, Universitäts- und Landesbibliothek Bonn, Nachlass Lamprecht, S 2713 UL 4: 3: 28.

Ratzel, Friedrich. "Lieber Freund!" Letter to Karl Lamprecht, Leipzig, 29 January 1897, Universitäts- und Landesbibliothek Bonn, Nachlass Lamprecht, S2714 Korr. 44.

Ratzel, Friedrich. "Lieber Kollege!" Letter to Kurt Hassert, Leipzig, 6 January 1902, Archiv für Geographie, Leibniz Institut für Länderkunde, Nachlass Friedrich Ratzel 163/657.

Ratzel, Friedrich. "Lieber Freund!" Letter to Rudolph Oldenbourg, Leipzig, 5 June 1904, Bayerische Staatsbibliothek, Ratzeliana Supplement 10.

Ratzel, Friedrich. "Lieber Freund!" Letter to Rudolph Oldenbourg, Leipzig, 10 February 1900, Bayerische Staatsbibliothek Ratzeliana Supplement 10.

Ratzel, Friedrich. "Lieber Freund!" Letter to Rudolph Oldenbourg, Leipzig, 17 February 1901, Bayerische Staatsbibliothek Ratzeliana Supplement 10.

Ratzel, Friedrich. "Lieber Freund." Letter to Rudolph Oldenbourg, Leipzig, 22 February 1899, Bayerische Staatsbibliothek Ratzeliana Supplement 10.

Ratzel, Friedrich. "Lieber Freund." Letter to Rudolph Oldenbourg, S. Florent, 7 April 1899, Bayerische Staatsbibliothek Ratzeliana Supplement 10.

Ratzel, Friedrich. "Lieber Herr Doktor!" Letter to Kurt Hassert, Leipzig, 2 August 1898, Archiv für Geographie, Leibniz Institut für Länderkunde, Nachlass Friedrich Ratzel 163/657.

Ratzel, F. "Lieber Kollege!' Letter to Kurt Hassert, Leipzig, 6 January 1902, Archiv für Geographie, Leibniz Institut für Länderkunde, Nachlass Friedrich Ratzel 163/657.

Ratzel, Friedrich. "Lieber Kollege!" Letter to K. Hassert, Leipzig, 8 January 1900, Archiv für Geographie, Leibniz Institut für Länderkunde, Nachlass Friedrich Ratzel 163/657.

Ratzel, Friedrich. "Lieber Kollege!" Letter to K. Hassert, Leipzig, 8 January 1903, Archiv für Geographie, Leibniz Institut für Länderkunde, Nachlass Friedrich Ratzel 163/657.

Ratzel, Friedrich. "Lieber Kollege!" Letter to Kurt Hassert, Leipzig, 31 July 1899, Archiv für Geographie, Leibniz Institut für Länderkunde, Nachlass Friedrich Ratzel 163/657.

Ratzel, Friedrich. "Mein verehrter Lehrer und lieber Freund!" Letter to Ernst Haeckel, Leipzig, 14 February 1904. Transliterated by Peter Zigman. *Würzburger medizinhistorische Mitteilungen* 28: 408.

Ratzel, Friedrich. "Notitzen zu zwei Reden (eine davon an Dr. Peters gerichtet)." Bayerische Staatsbibliothek, Ratzeliana Supplement 4.

Ratzel, Friedrich. "Reste von Vorlesungsmanuskript: Allgemeine politische Geographie und politische Ethnographie." 1890–1894, Leibniz Institut für Länderkunde, Nachlass Friedrich Ratzel 155/317.

Ratzel, Friedrich. "Sehr geehrter Herr!" Letter to Carl Peters, Ammerland am Würmsee, 28 August 1890, Bundesarchiv, Nachlass Carl Peters N 2223/32.

Ratzel, Marie "Sehr geehrter Herr Professor!" Letter from Marie Ratzel to unknown, Ammerland, 26 January 1905, Archiv für Geographie, Leibniz Institut für Länderkunde, Nachlass Friedrich Ratzel 163/657.

Ratzel, M. "Sehr geehrter Herr Professor!" Letter from Marie Ratzel to unknown, Leipzig, 18 November 1901, Archiv für Geographie, Leibniz Institut für Länderkunde, Nachlass Friedrich Ratzel 163/657.

Ratzel, Marie "Sehr geehrter Herr Professor!" Letter to unknown, Ammerland, 19 June 1906, Archiv für Geographie, Leibniz Institut für Länderkunde, Nachlass Friedrich Ratzel 163/657.

Ratzel, Friedrich. *Tagebuch über Englandreise im Aug/Sep 1881, Abschrift angefertigt von Günter Buttmann*, Bayerische Staatsbibliothek, Ratzeliana Supplement 17a VI 9a.

Ratzel, Friedrich. "Verehrte Freundin!" Letter to Marie Wingens, Boston, 2 November 1873, Bayerische Staatsbibliothek, Ratzeliana, Supplement 10.

Ratzel, Friedrich. "Verehrte Freundin!" Letter to Marie Wingens, Charleston, 2 February 1874, Bayerische Staatsbibliothek, Ratzeliana Supplement 10.

Ratzel, Friedrich. "Verehrte Freundin!" Letter to Marie Wingens, New York, 14 August 1873, Bayerische Staatsbibliothek, Ratzeliana, Supplement 10.

Ratzel, Friedrich. "Verehrte Freundin!" Letter to Marie Wingens, St. Louis, 29 May 1874, Bayerische Staatsbibliothek, Ratzeliana Supplement 10.

Ratzel, Friedrich. "Verehrte Freundin!" Letter to Marie Wingens, Yosemite, 21 August 1874, Bayerische Staatsbibliothek, Ratzeliana Supplement 10.

Ratzel, Friedrich. "Verehrter Freund!" Letter to Ernst Haeckel, Berlin, 20 July 1873. Transliterated by Peter Zigman. *Würzburger medizinhistorische Mitteilungen* 28: 401–402.

Ratzel, Friedrich. "Verehrter Freund u. Kollege." Letter to Karl Lamprecht, Leipzig, 10 July 1893, Universitäts- und Landesbibliothek Bonn, Nachlass Lamprecht, S 2713: Korr. 44.

Ratzel, Friedrich. "Verehrter Herr College!" Letter to Ernst Hasse, Leipzig, 6 August 1889, Archiv für Geographie, Leibniz Institut für Länderkunde, Nachlass Ernst Hasse 437/23.

Ratzel, Friedrich. "Verehrter Herr College!" Letter to Ernst Hasse, Leipzig, 15 December 1890, Archiv für Geographie, Leibniz Institut für Länderkunde, Nachlass Ernst Hasse 437/24.

Ratzel, Friedrich. "Verehrter Herr Kollege!" Letter to Ernst Hasse, Leipzig, 18 October 1888, Archiv für Geographie, Leibniz Institut für Länderkunde, Nachlass Ernst Hasse 437/5.

Ratzel, Friedrich. "Verehrter Herr Professor." Letter to Ernst Haeckel, Petersdorf, 26 May 1871. Transliterated by Peter Zigman. *Würzburger medizinhistorische Mitteilungen* 28: 395–397.

Ratzel, Friedrich. "Verehrter Herr Kollege!" Letter to Wilhelm Wundt, 16 August 1902, Ammerland, Universitätsarchiv Leipzig, Nachlass Wilhelm Wundt; NA Wundt/III/1301–1400/1375/339–342.

Regel, Friedrich. "Ratzel, Fr., Dr., Völkerkunde. Erster Band. Die Naturvölker Afrikas. Mit 494 Abbildungen im Text, 10 Aquarelltafeln und 2 Karten. Leipzig, Bibliographisches Institut, 1885." Bayerische Staatsbibliothek, Ratzeliana II Besprechungen von Ratzelschen Werken.

Semple, Ellen Churchill. "An Abridgement of the Partition of Asiatic Turkey" 30 April 1918, Records of the American Commission to Negotiate Peace, Record Group 256, Inquiry document 500, Records of the Inquiry 1914–19, National Archives.

Semple, Ellen Churchill. "Dear Dr Keltie." Letter to John Scott Keltie, Piccadilly, 30 October 1912, Correspondence Block 1911–1920, Royal Geographical Society.

Semple, Ellen Churchill. "Dear Professor Ratzel." Letter, Louisville, 9 December 1902, Bayerische Staatsbibliothek, Ratzeliana I Supplement 13.

Semple, Ellen Churchill. "Geehrter Herr Professor." Letter to Friedrich Ratzel, n.p., 2 January 1893, Archiv für Geographie, Leibniz Institut für Länderkunde, Nachlass Friedrich Ratzel 163/686.

Semple, Ellen Churchill. "Geehrter Herr Professor." Letter to Friedrich Ratzel, n.p., 30 January 1893, Archiv für Geographie, Leibniz Institut für Länderkunde, Nachlass Friedrich Ratzel 163/686.

Wagner, Moritz. "Lieber Freund." Letter to Friedrich Ratzel, Munich, 1 April 1877, Bayerische Staatsbibliothek, Ratzeliana Supplement 13.

Wagner, Moritz. "Lieber Freund!" Letter to Friedrich Ratzel, Munich, 4 June 1882, Bayerische Staatsbibliothek, Ratzeliana Supplement 13.

Wagner, Moritz. "Lieber Freund!" Letter to Friedrich Ratzel, n.p., 1 July 1885, Bayerische Staatsbibliothek, Ratzeliana Supplement 13.

Wagner, Moritz. "Lieber Freund!" Letter to Friedrich Ratzel, n.p., 21 January 1882, Bayerische Staatsbibliothek, Ratzeliana Supplement 13.

Wagner, Moritz. "Lieber Freund." Letter to Friedrich Ratzel, n.p., 24 December 1884, Bayerische Staatsbibliothek, Ratzeliana Supplement 13.

Wagner, Moritz. "Mein herzlieber theurer Freund!" Letter to Friedrich Ratzel, Ammerland, 5 September 1881, Bayerische Staatsbibliothek, Ratzeliana Supplement 13.

Ziegler, Heinrich Ernst. "Geehrtester Herr Professor!" Letter to Friedrich Ratzel, 10 January 1896, Bayerische Staatsbibliothek, Ratzeliana I, 1–5.

OTHER WORKS CITED

Abrahamsson, Christian. "Book Review Forum: Vibrant Matter: A Political Ecology of Things." *Dialogues in Human Geography* 1, no. 3 (2011): 399–402.

Abrahamsson, Christian. "On the Genealogy of Lebensraum." *Geographica Helvetica* 68 (2013): 37–44.

Achelis, Thomas. "Friedrich Ratzel." *Geographischer Anzeiger* 5 (1904): 243–47.

Achelis, Thomas. "Friedrich Ratzel." *Nord und Süd* 86 (1898): 263–76.

Adams, Ellen E. "Ellen Churchill Semple and American Geography in an Era of Imperialism." PhD dissertation, College of William and Mary, 2011.

Agamben, Giorgio. *Homo Sacer: Sovereign Power and Bare Life*. Stanford, CA: Stanford University Press, 1998.

Agnew, John. "By Words Alone Shall We Know: Is the History of Ideas Enough to Understand the World to Which Our Concepts Refer?" *Dialogues in Human Geography* 4, no. 3 (2014): 311–319.

Agnew, John. "Of Canons and Fanons." *Dialogues in Human Geography* 2, no. 3 (2012): 321–23.

Anderson, Mark. "Ruth Benedict, Boasian Anthropology, and the Problem of the Colour Line." *History and Anthropology* 25, no. 3 (2014): 395–414.

Andrews, Howard F. "The Durkheimians and Human Geography: Some Contextual Problems in the Sociology of Knowledge." *Transactions of the Institute of British Geographers* 9, no. 3 (1984): 315–36.

Antonsich, Marco, Vladimir Kolossov, and Maria Paola Pagnini. *Europe Between Political Geography and Geopolitics: On the Centenary of Ratzel's Politische Geographie*. Rome: Società Geografica Italiana, 2001.

Arbeitskreis Geographie und Faschismus der Bundesfachschaftentagung Geographie. "Beiträge deutscher Geographieprofessoren zum Faschismus." *Geographische Sumpfschau* 4 (1996): 1–124.

Arendt, Hannah. *The Origins of Totalitarianism*. San Diego: Harvest, 1951 (1979).

Ashutosh, Ishan. "Mapping Race and Environment: Geography's Entanglements with Aryanism." *Journal of Historical Geography* 61 (2018): 15–23.

Ashworth, Lucian. "Mapping a New World: Geography and the Interwar Study of International Relations." *International Studies Quarterly* 57, no. 1 (2013): 138–49.

Ashworth, Lucian. "A New World? Isaiah Bowman and the Americanization of Ratzel's Political Geography." Unpublished manuscript, 2018.

Ashworth, Lucian. "Realism and the Spirit of 1919: Halford Mackinder, Geopolitics and the Reality of the League of Nations." *European Journal of International Relations* 17, no. 2 (2010): 279–301.

Atkinson, David. "Geopolitical Imaginations in Modern Italy." In *Geopolitical Traditions: A Century of Geopolitical Thought*, edited by Klaus Dodds and David Atkinson, 93–117. London: Routledge, 2000.

Balakrishnan, Gopal. *The Enemy: An Intellectual Portrait of Carl Schmitt*. London: Verso, 2000.

Banse, Ewald. *Große Forschungsreisende*. Munich: Lehmann, 1933.

Barkin, Kenneth. "W. E. B. Du Bois' Love Affair with Imperial Germany." *German Studies Review* 28, no. 2 (2005): 285–302.

Barnes, Trevor, and Claudio Minca. "Nazi Spatial Theory: The Dark Geographies of Carl Schmitt and Walter Christaller." *Annals of the Association of American Geographers* 103, no. 3 (2013): 669–87.

Barry, Andrew. "Geography and Other Disciplines: Genealogy, Anamnesis and the Canon." *Journal of Historical Geography* 49 (2015): 85–93.

Barua, Maan. "Ratzel, Bio-geography and the More-than-Human." *Journal of Historical Geography* 61 (2018): 102–8.

Bassin, Mark. "Between Realism and the 'New Right': Geopolitics in Germany in the 1990s." *Transactions of the Institute of British Geographers* 28, no. 3 (2003): 350–66.

Bassin, Mark. "Geographical Determinism in Fin-de-siècle Marxism: Georgii Plekhanov and the Environmental Basis of Russian History." *Annals of the Association of American Geographers* 82, no. 1 (1992): 3–22.

Bassin, Mark. "Imperialism and the Nation State in Friedrich Ratzel's Political Geography." *Progress in Human Geography* 11, no. 4 (1987): 473–95.

Bassin, Mark. "Nature, Geopolitics and Marxism: Ecological Contestations in Weimar Germany." *Transactions of the Institute of British Geographers* 21, no. 2 (1996): 315–41.

Bassin, Mark. "Race *contra* Space: The Conflict between German *Geopolitik* and National Socialism." *Political Geography* 6, no. 2 (1987): 115–34.

Bassin, Mark. "Studying Ourselves: History and Philosophy of Geography." *Progress in Human Geography* 24, no. 3 (2000): 475–87.

Bassin, Mark. "Turner, Solov'ev, and the "Frontier Hypothesis": The Nationalist Signification of Open Spaces." *Journal of Modern History* 65, no. 3 (1993): 473–511.

Bauman, Zygmunt. *Modernity and the Holocaust*. Cambridge: Polity, 1989.

Bebel, August. *Woman under Socialism*. New York: New York Labor News Company, 1879 (1904).

Beck, Hanno. "Moritz Wagner als Geograph." *Erdkunde* 7, no. 2 (1953): 125–28.

Beiner, Ronald 2018. "Stephen Bannon's World: Dangerous Minds in Dangerous Times." *The Conversation*, 25 July 2018. https://theconversation.com/stephen-bannons-world-dangerous-minds-in-dangerous-times-100373.

Bekenntnis der Professoren an den deutschen Universitäten und Hochschulen zu Adolf Hitler und dem nationalsozialistischen Staat. Dresden: Nationalsozialistischer Lehrerbund Deutschland/Sachsen, 1933.

Benedict, Ruth. "Obituary: Franz Boas." *Science* 97, no. 25 (1943): 60–62.

Bennett, Jane. *Vibrant Matter: A Political Ecology of Things*. Durham, NC: Duke University Press, 2010.

Bennett, Jane. "Author Response." *Dialogues in Human Geography* 1, no. 3 (2011): 404–6.

Bergson, Henri. *Creative Evolution*. Lanham, MD: University Press of America, 1911 (1983).
"Bericht des Referenten Dr. Kapp." In *Im Bismarckschen Reich 1871–1890*, edited by Hans Fenske, 258–59. Darmstadt: Wissenschaftliche Buchgesellschaft, 1978 (1880).
Berman, Russell A. "Du Bois and Wagner: Race, Nation, and Culture between the United States and Germany." *German Quarterly* 70, no. 2 (1997): 123–35.
Bernasconi, Robert. "Will the Real Kant Please Stand Up." *Radical Philosophy* 117 (2003): 13–23.
Bernhardi, Friedrich von. *Germany and the Next War*. London: Arnold, 1912 (1914).
Bettelheim, A. "Friedrich Ratzel und der deutsche Nekrolog." *Münchner Allgemeine Zeitung* 197 (1904): 399.
Bhambry, Tul'si (Tuesday). "Translator's Introduction." *Journal of Historical Geography* 61 (2018): 59–60.
Bismarck, Otto von. "Rede vor dem Reichstag, 26. Juni 1884." In *Otto von Bismarck, Werke in Auswahl: Reichsgestaltung und Europäische Friedenswahrung, 1883–1890*, edited by Gustav Adolf Rein et al., 162–80. Darmstadt: Wissenschaftliche Buchgesellschaft, 1884 (2001).
Black, Jeremy. *Geopolitics and the Quest for Dominance*. Bloomington: Indiana University Press, 2016.
Blackett, Richard. "Lincoln and Colonization." *OAH Magazine of History* 21, no. 4 (2007): 19–22.
Blight, David W. *Race and Reunion: The Civil War in American Memory*. Cambridge, MA: Harvard University Press, 2001.
Boas, Franz. "The History of Anthropology." *Science* 20, no. 512 (1904): 513–24.
Boas, Franz. "Methods of Research." In *General Anthropology*, edited by Franz Boas, 666–86. Boston: DC Health and Company, 1938.
Bonds, Anne, and Joshua Inwood. "Beyond White Privilege: Geographies of White Supremacy and Settler Colonialism." *Progress in Human Geography* 40, no. 6 (2016): 715–33.
Bowersox, Jeff. *Raising Germans in the Age of Empire: Youth and Colonial Culture*. Oxford: Oxford University Press, 2013.
Bowman, Isaiah. "Geography vs. Geopolitics." *Geographical Review* 32, no. 4 (1942): 646–58.
Bowman, Isaiah. *The New World: Problems in Political Geography*. New York: World Book Company, 1921 (1928).
Bowman, Isaiah. "The Pioneer Fringe." *Foreign Affairs* 6, no. 1 (1927): 49–66.
Brands, H. W. *The Man Who Saved the union: Ulysses S. Grant in War and Peace*. New York: Doubleday, 2012.
Brantlinger, Patrick. *Dark Vanishings: Discourse on the Extinction of Primitive Races, 1800–1930*. Ithaca, NY: Cornell University Press, 2003.
Brantlinger, Patrick. *Taming Cannibals: Race and the Victorians*. Ithaca, NY: Cornell University Press, 2011.
Braun, Bruce. "Environmental Issues: Inventive Life." *Progress in Human Geography* 32, no. 5 (2008): 667–79.
Bunzl, Matti. "Franz Boas and the Humboldtian Tradition: From *Volksgeist* und *Nation-*

alcharacter to an Anthropological Concept of Culture." In *Volksgeist as Method and Ethic: Essays on Boasian Ethnography and the German Anthropological Tradition*, edited by George W. Stocking Jr., 17–78. Madison: University of Wisconsin Press, 1996.

Bushong, Allen D. "Ellen Churchill Semple, 1863–1932." In *Geographers: Biobibliographical Studies 8*, T. W. Freeman, 87–94. London: Mansell, 1984.

Buttmann, Günther. *Friedrich Ratzel: Leben und Werk eines deutschen Geographen*. Stuttgart: Wissenschaftliche Verlagsgesellschaft MBH, 1977.

Campbell, J. A., and D. N. Livingstone, D. N. "Neo-Lamarckism and the Development of Geography in the United States and Great Britain." *Transactions of the Institute of British Geographers* 8, no. 3 (1983): 267–94.

Canguilhem, Georges. "The Living and its Milieu." *Grey Room* 3 (2001): 6–31.

Castellanos, M. Bianet. "Introduction: Settler Colonialism in Latin America." *American Quarterly* 69, no. 4 (2017): 777–81.

Césaire, Aimé. *Discourse on Colonialism*. New York: Monthly Review Press, 1955 (1972).

Chamberlain, Houston Stewart. *Die Grundlagen des neunzehnten Jahrhunderts*. Munich: Bruckmann, 1899 (1912).

Chiantera–Stutte, Patricia. "Ratzel's Stone Guest: The Art of Politics in the Work of Friedrich Ratzel." *Journal of Historical Geography* 61 (2018): 91–96.

Chiantera-Stutte, Patricia. "Space, *Großraum* and *Mitteleuropa* in Some Debates of the Early Twentieth Century." *European Journal of Social Theory* 11, no. 2 (2008): 185–201.

Chickering, Roger. *Karl Lamprecht: A German Academic Life (1856–1915)*. New Jersey: Humanities Press, 1993.

Chickering, Roger. *We Men Who Feel Most German: A Cultural Study of the Pan-German League, 1886–1914*. Boston: Allen & Unwin, 1984.

Christaller, Walter. *Central Places in Southern Germany*. New Jersey: Prentice-Hall, 1933 (1966).

Clark, Christopher. *Kaiser Wilhelm II: A Life in Power*. London: Penguin, 2009.

Clark, Nigel. "Geopolitics of the Threshold." *Political Geography* 37 (2013): 48–50.

Clausewitz, Carl von. *On War*. Princeton, NJ: Princeton University Press, 1832 (1989).

Cohen, Saul B. *Geography and Politics in a World Divided, 2nd edition*. New York: Oxford University Press, 1963 (1973).

Connolly, William. "The 'New Materialism' and the Fragility of Things." *Millennium: Journal of International Studies* 41, no.3 (2013): 399–412.

Conrad, Sebastian. *German Colonialism: A Short History*. Cambridge: Cambridge University Press, 2012.

Contee, Clarence G. "Du Bois, the NAACP, and the Pan-African Congress of 1919." *Journal of Negro History* 57, no. 1 (1972): 13–28.

Conway, Philip. "The Historical Ontology of Environment: From the Unity of Nature to the Birth of Geopolitics." PhD dissertation, University of Aberystwyth, 2019.

Coole, Diana. "Agentic Capacities and Capacious Historical Materialism: Thinking with New Materialisms in the Political Sciences." *Millennium: Journal of International Studies* 41, no. 3 (2013): 451–69.

Coole, Diana, and Samantha Frost. "Introducing New Materialisms." In *New Materialisms: Ontology, Agency, and Politics*, edited by Diana Coole and Samantha Frost, 1–43. Durham, NC: Duke University Press, 2010.

Cox, Michael. *E. H. Carr: A Critical Appraisal*. Basingstoke: Palgrave, 2000.

Daitz, Werner. *Lebensraum und gerechte Weltordnung: Grundlagen einer Anti-Atlantikcharta*. Amsterdam: De Amsterdamsche Keurkamer, 1943.

Dalby, Simon. *Creating the Second Cold War: The Discourse of Politics*. London: Bloomsbury, 1990 (2016).

Darré, Walther. *Das Bauerntum als Lebensquell des nordischen Rasse*. Munich: J. F. Lehmann, 1928 (1942).

Darwin, Charles. *The Descent of Man: Selection in Relation to Sex, Volume I*. Cambridge: Cambridge University Press, 1871 (2009).

Darwin, Charles. *On the Origin of Species by Means of Natural Selection: Or, the Preservation of Favoured Races in the Struggle for Life*. London: Murray, 1859.

Darwin, Charles. *On the Origin of Species by Means of Natural Selection: Or, the Preservation of Favoured Races in the Struggle for Life, Sixth Edition, with Additions and Corrections to 1872*. London: Murray, 1859 (1876).

Darwin, Charles. *The Voyage of the Beagle*. London: Penguin, 1839 (1989).

Darwin, Charles, and Alfred Wallace. "On the Tendency of Species to Form Varieties; and on the Perpetuation of Varieties and Species by Natural Means of Selection." *Journal of the Proceeding of the Linnean Society of London: Zoology* 3, no. 9 (1858): 45–62.

Dawkins, Richard. *The Greatest Show on Earth: The Evidence of Evolution*. New York: Free Press, 2009.

Dawkins, Richard. *The Selfish Gene*. Oxford: Oxford University Press, 1976 (2006).

de Gobineau, Arthur. *The Inequality of Human Races*. London: William Heinemann. 1852 (1915).

Deutsche Kolonialgesellschaft. *Die Deutsche Kolonialgesellschaft: 1882–1907*. Berlin: Dietrich Heimer, 1908.

"Deutschland, wach' auf!" *Deutsche Kolonialzeitung* 15 (1890): 184.

Diamond, Jared. *Guns, Germs and Steel: The Fates of Human Societies*. London: W. W. Norton, 1997.

Dickinson, Edward Ross. "Biopolitics, Fascism, Democracy: Some Reflections on Our Discourse about 'Modernity.'" *Central European History* 37, no. 1 (2004): 1–48.

Die Redaktion. "Friedrich Ratzel." *Archiv für Rassen- und Gesellschaftsbiologie* 1 (1904): 736.

Dijkink, Gertjan. "Ratzel's *Politische Geographie* and Nineteenth-Century German Discourse." In *Europe between Political Geography and Geopolitics: On the Centenary of Ratzel's Politische Geographie*, edited by Marco Antonsich, Vladimir Kolossov, and Maria Paola Pagnini, 115–28. Rome: Società Geografica Italiana, 2001.

Dillon, Michael, and Andrew W. Neal. *Foucault on Politics, Security and War*. Basingstoke: Palgrave, 2008.

Dillon, Michael, and Luis Lobo-Guerrero. "Biopolitics of Security in the 21st Century: An Introduction." *Review of International Studies* 34, no. 2 (2008): 265–92.

Dittmar, Peter. "Die ewige Angst vor dem Chinesen." *Die Welt*, 26 January 2020. https://

www.welt.de/geschichte/article205323439/Gelbe-Gefahr-Die-ewige-Angst-vor-dem-Chinesen.html.
Dittmer, Jason. "Geopolitical Assemblages and Complexity." *Progress in Human Geography* 38, no. 3 (2014): 385–401.
Dittmer, Jason, and Nicholas Gray. "Popular Geopolitics 2.0: Towards New Methodologies of the Everyday." *Geography Compass* 4, no. 11 (2010): 1664–77.
Dixon, Deborah. "The Way of the Flesh: Life, Geopolitics and the Weight of the Future." *Gender, Place & Culture* 21, no. 2 (2014): 136–51.
Dodds, Klaus. "The 1982 Falklands War and a Critical Geopolitical Eye: Steve Bell and the If . . . Cartoons." *Political Geography* 15, no. 6–7 (1996): 571–92.
Dodds, Klaus, and David Atkinson. *Geopolitical Traditions: A Century of Geopolitical Thought*. London: Routledge, 2000.
Drechsler, Horst. *Südwestafrika unter deutscher Kolonialherrschaft: Der Kampf der Herero und Nama gegen den deutschen Imperialismus (1884–1915)*. Berlin: Akademie, 1966.
Driver, Felix. *Geography Militant: Cultures of Exploration and Empire*. Oxford: Blackwell, 2001.
Driver, Felix. "Hidden Histories Made Visible? Reflections on a Geographical Exhibition." *Transactions of the Institute of British Geographers* 38, no. 3 (2013): 420–35.
Driver, Felix, and Elizabeth Baigent. "Biography and the History of Geography: A Response to Ron Johnston." *Progress in Human Geography* 31, no. 1 (2007): 101–6.
Drolet, Jean-François, and Michael C. Williams. "The Radical Right, Realism, and the Politics of Conservatism in Postwar International Thought." *Review of International Studies* 47, no. 3 (2021): 273–93.
Du Bois, W. E. B. *Black Reconstruction in America 1860–1880*. New York: The Free Press, 1935 (1998).
Du Bois, W. E. B. "The Conservation of Races." *American Negro Academy Occasional Papers* 2 (1897): 1–15.
Du Bois, W. E. B. *The Negro*. New York: Henry Holt and Company, 1915.
Du Bois, W. E. B. *The Negro American Family*. Atlanta: Atlanta University Press, 1908.
Du Bois, W. E. B. "The Study of the Negro Problems." *Annals of the American Academy of Political and Social Sciences* 11 (1898): 1–23.
Dugin, Aleksandr. *Last War of the World-Island: The Geopolitics of Contemporary Russia*. London: Arktos, 2015.
Dugin, Aleksandr. *Osnovy geopolitiki: Geopoliticheskoe budushchee Rossii*. Moscow: Arktogeia, 1997.
Durkheim, Émile. "Review of Ratzel's Anthropogeographie." In *Emile Durkheim: Selected Writings*, edited by Anthony Giddens, 87–88. Cambridge: Cambridge University Press, 1899 (1972).
Eckert, Max. "Friedrich Ratzel als politischer Lehrer." *Geographische Zeitschrift* 28, no. 7/8 (1922): 253–61.
Eco, Umberto. 1995. "Ur-facism." *New York Review of Books*, 22 June 1995. http://www.nybooks.com/articles/1995/06/22/ur-fascism/.
E. G. R. "Obituary: Friedrich Ratzel." *The Geographical Journal* 24, no. 4 (1904): 485–87.

Elden, Stuart. "Leibniz and Geography: Geologist, Paleontologist, Biologist, Historian, Political Theorist and Geopolitician." *Geographica Helvetica* 68 (2013): 81–93.

Elden, Stuart. "Necessary, But Not Sufficient: Geography, Territory and the History of Ideas." *Dialogues in Human Geography* 4, no. 3 (2014): 320–23.

Elden, Stuart. "Reading Schmitt Geopolitically." *Radical Philosophy* 161 (2010): 18–26.

Elden, Stuart. "Reassessing Kant's Geography." *Journal of Historical Geography* 35, no. 1 (2009): 3–25.

Elden, Stuart. "Secure the Volume: Vertical Geopolitics and the Depth of Power." *Political Geography* 34 (2013): 35–51.

Eley, Geoff. *Reshaping the German Right: Radical Nationalism and Political Change under Bismarck*. New Haven, CT: Yale University Press, 1980.

Enloe, Cynthia. "Wounds: Militarised Nursing, Feminist Curiosity, and Unending War." *International Relations* 33, no. 3 (2019): 393–412.

Esposito, Roberto. *Bios: Biopolitics and Philosophy*. Minneapolis: University of Minnesota Press, 2008.

Evans, Richard J. "In Search of German Social Darwinism: The History and Historiography of a Concept." In *Medicine and Modernity: Public Health and Medical Care in Nineteenth- and Twentieth-Century Germany*, edited by Manfred Berg and Geoffrey Cocks, 55–89. Cambridge: Cambridge University Press, 1997.

Fabian, Johannes. *Time and the Other: How Anthropology Makes its Object*. New York: Colombia University Press, 1983.

Fabri, Friedrich. *Bedarf Deutschland der Kolonien? Eine politisch-ökonomische Betrachtung*. Gotha: Perthes, 1879.

Fahlbusch, Michael. "Volk ohne Raum—Raum ohne Volk. Der lange Schatten der Deutsch-Völkischen in der Weimarer Republik." In *Demokratiegeschichte als Zäsurgeschichte: Diskurse der frühen Weimarer Republik*, edited by Heidrun Kämper, Peter Haslinger, and Thomas Raithel, 253–84. Berlin: Akademie, 2014.

Fairbairn, Brett. "Economic and Social Developments." In *Imperial Germany, 1871–1918*, edited by James Retallack, 61–82. Oxford: Oxford University Press, 2008.

Fanon, Frantz. *The Wretched of the Earth*. London: Penguin, 1963 (2001).

Featherstone, Mark. "Ruin Value." *Journal for Cultural Research* 9, no. 3 (2005): 301–20.

Fechner, Gustav Theodor. *Einige Ideen zur Schöpfungs- und Entwicklungsgeschichte*. Leipzig: Breitkopf und Härtel, 1873.

Fechner, Gustav Theodor, ed. *The Little Book of Life after Death*. Boston: Little Brown and Company, 1836 (1904).

Ferguson, Niall. *Kissinger: 1923–1968: The Idealist*. London: Penguin, 2015.

Ferretti, Federico. "Rediscovering Other Geographical Traditions." *Geography Compass* 13, no. 3 (2019): 1–15.

Fischer, Fritz. *Germany's Aims in the First World War*. New York: W. W. Norton and Company, 1967.

Flohr, Ernst Friedrich. "Versuch einer Klärung des Begriffs Lebensraum." *Geographische Zeitschrift* 48, 11/12 (1942): 393–404.

Flynn, Curran. "Political Geography and Morgenthau's Early American Works." *Cambridge Review of International Affairs* 29, no. 4 (2016): 1582–1602.

Foner, Philip S. "Alexander von Humboldt on Slavery in America." *Science & Society* 47, no. 3 (1983): 33–34.

Foucault, Michel. *The History of Sexuality Part I: The Will to Knowledge*. London: Penguin, 1978.

Foucault, Michel. *Society Must Be Defended: Lectures at the Collège de France, 1975–1976*. New York: Picador, 1976 (2003).

Freedman, Lawrence. *Strategy: A History*. Oxford: Oxford University Press, 2013.

Frymann, Daniel. *Wenn ich der Kaiser wär': Politische Wahrheiten und Notwendigkeiten*. Leipzig: Dieterich, 1912.

Gasman, Daniel. *Haeckel's Monism and the Birth of Fascist Ideology*. New York: Peter Lang, 1998.

Gavrilova, Sonya. 2019. "The Present Taxidermied: Soviet 'Common Unsaids' in Russian *Krayevedcheskiy* Museums." DPhil diss., University of Oxford.

Geiser, S. W. "Dr. Ernst Kapp, Early Geographer in Texas." *Field and Laboratory* 14, no. 1 (1946): 16–31.

Gerland, Georg. *Ueber das Aussterben der Naturvölker*. Leipzig: Fleischer, 1868.

Giaccaria, Paolo, and Claudio Minca. "Life in Space, Space in Life: Nazi Topographies, Geographical Imaginations, and *Lebensraum*." *Holocaust Studies: A Journal of Culture and History* 22, no. 2–3 (2016): 151–71.

Glick, Leonard B. "Types Distinct from Our Own: Franz Boas on Jewish Identity and Assimilation." *American Anthropologist* 84, no. 3 (1982): 545–65.

Goethe, Johann Wolfgang von. *Wahlverwandtschaften, 1. Band*. Tübingen: J. G. Cottaische Buchhandlung, 1809.

Gooding-Williams, Robert. "Outlaw, Appiah, and Du Bois's 'The Conservation of Races.'" In *W. E. B. Du Bois on Race and Culture*, edited by Bernard W. Bell, Emily Grosholz, and James B. Stewart, 39–56. London: Routledge, 1996.

Grayling, A. C. *War: An Enquiry*. New Haven, CT: Yale University Press, 2017.

Greenhough, Beth. "Vitalist Geographies." In *Taking Place: Non-representational Theories and Geography*, edited by Ben Anderson and Paul Harrison, 37–54. Farnham, Ashgate, 2010.

Gregory, Derek. "Trauma Geographies: Broken Bodies and Lethal Landscapes." The 2018 Antipode RGS-IBG lecture, 29 August 2018, Cardiff.

Gregory, Derek. "Vanishing Points." In *Violent Geographies: Fear, Terror and Political Violence*, edited by Derek Gregory and Allan Pred, 205–23. London: Routledge, 2007.

Griffin, Roger. *Fascism*. Cambridge: Polity, 2018.

Grill, Bartholomäus. *Wir Herrenmenschen: Unser rassistisches Erbe: Eine Reise in die deutsche Kolonialgeschichte*. Munich: Siedler, 2019.

Grill, Johnpeter Horst, and Robert L. Jenkins. "The Nazis and the American South in the 1930s: A Mirror Image?" *Journal of Southern History* 58, no. 4 (1992): 667–94.

Grimm, Hans. *Volk ohne Raum*. Munich: Albert Langen, 1926.

Grimmer-Solem, Erik. *Learning Empire: Globalization and the German Quest for World Status, 1875–1919*. Cambridge: Cambridge University Press, 2019.

Guettel, Jens–Uwe. *German Expansionism, Imperial Liberalism and the United States, 1776–1945*. Cambridge: Cambridge University Press, 2012.

Guettel, Jens-Uwe. "South-West Africa: German Colonialism, Indians, and American Westward Expansion." *Modern Intellectual History* 7 no. 3 (2010): 523–52.

Gumplowicz, Ludwig. *Der Rassenkampf: Sociologische Untersuchungen*. Innsbruck: Verlag der Wagner'schen Univerităts-Buchhandlung, 1883.

Gumplowicz, Ludwig. *Geschichte der Staatstheorien*. Innsbruck: Verlag der Wagner'schen Univerităts-Buchhandlung, 1905.

Günzel, Stephan. *Geophilosophie: Nietzsches philosophische Geographie*. Berlin: Akademie, 2001.

Günzel, Stephan. *Raum: Eine kulturwissenschaftliche Einführung*. Bielefeld: Transcript, 2017.

Guratzsch, Dankwart. "Der Selbstmord der Städte." *Die Welt*, 2 October 2019. https://www.welt.de/kultur/plus201239634/Totale-Verkehrsblockade-Der-Selbstmord-der-Staedte.html.

Guzzini, Stefano. *The Return of Geopolitics in Europe?* Cambridge: Cambridge University Press, 2012.

Haaß, Wolfgang. *Vom Dorf in die Welt: Gedanken zu dem großen Geographen Friedrich Ratzel*. Kronau: Gemeinde Angelbachtal, 2013.

Haeckel, Ernes. *The History of Creation Vol. 1*. London: Henry S. King, 1876.

Haeckel, Ernst. *Natürliche Schöpfungsgeschichte*. Berlin: Georg Reimer, 1868.

Hajdú, Zoltán. "The Influence of Friedrich Ratzel on Hungarian Geography." In *Europe between Political Geography and Geopolitics: On the Centenary of Ratzel's Politische Geographie*, edited by Marco Antonsich, Vladimir Kolossov, and Maria Paola Pagnini, 129–43. Rome: Società Geografica Italiana, 2001.

Halas, Matus. "Searching for the Perfect Footnote: Friedrich Ratzel and the Others at the Roots of Lebensraum." *Geopolitics* 19, no. 1 (2014): 1–18.

Hampe, Erich. *Im Spannungsfeld der Luftmächte: Eine Einführung in Luftgefahr und Bevölkerungsschutz*. Cologne: Maximilian, 1956.

Hare, J. Laurence, and Fabian Link. "The Idea of *Volk* and the Origins of *Völkisch* Research, 1800–1930s." *Journal of the History of Ideas* 80, no. 4 (2019): 575–96.

Harrington, Anne. *Reenchanted Science: Holism in German Culture from Wilhelm II to Hitler*. Princeton, NJ: Princeton University Press, 1996.

Harris, Kimberly Ann. "W. E. B. Du Bois's 'Conservation of Races': A Metaphilosophical Text." *Metaphilosophy* 50, no. 5 (2019): 670–87.

Hartman, Saidiya. *Scenes of subjection: Terror, Slavery, and Self-Making in Nineteenth-Century America*. New York: Oxford University Press, 1997.

Hartmann, Georg. "Die Mischrassen in unseren Kolonien, besonders in Südwestafrika." In *Verhandlungen des Deutschen Kolonialkongresses 1905*, n.e., 906–32. Berlin: Georg Reimer, 1910.

Hartmann, Georg. *Einfluss des Treibeises auf die Bodengestalt der Polargebiete: Inaugural-Dissertation zur Erlangung der Doktorwürde*. Leipzig: Verein für Erdkunde, 1891.

Hartshorne, Richard. "Recent Developments in Political Geography I." *American Political Science Review* 29, no. 5 (1935): 785–804.

Hassert, Kurt. "Friedrich Ratzel. Sein Leben und Wirken." *Geographische Zeitschrift* 11, no. 6 (1905): 305–25.

Hassert, Kurt. "Friedrich Ratzel. Sein Leben und Wirken." *Geographische Zeitschrift* 11, no. 7 (1905): 361–80.

Haushofer, Karl. *Der Nationalsozialistische Gedanke in der Welt*. Munich: Georg Callwey, 1933.

Haushofer, Karl. "Die Einheit der Monsoonländer." *Zeitschrift für Geopolitik* 1 (1924): 20–27.

Haushofer, Karl. "Einleitung: Friedrich Ratzel als raum—und volkspolitischer Gestalter." In *Erdenmacht und Völkerschicksal, herausgegeben und eingeleitet von Generalmajor a. D. Prof. Dr. Karl Haushofer*, Friedrich Ratzel, i–xxvii. Stuttgart: Alfred Kröner, 1941.

Haushofer, Karl. "Friedrich Ratzel im raumpolitischen Umbruch seiner und unserer Zeit." *Geographische Zeitschrift* 50, no. 3/4 (1944): 81–86.

Haushofer, Karl. "Friedrich Ratzel im raumpolitischen Umbruch unserer Zeit." *Geographische Zeitschrift* 50, no. 3/4 (1944): 81–86.

Haushofer, Karl. *Geopolitik der Pan-Ideen*. Berlin: Zentral, 1931.

Haushofer, Karl. "Geopolitische Einflüsse bei den Verkörperungsversuchen von nationalem Sozialismus und sozialer Aristokratie." *Zeitschrift für Geopolitik* 1 (1924): 127–34.

Haushofer, Karl. "Karl Haushofer an Max-Hildebert Böhm, 8. Februar 1931." In *Karl Haushofer: Leben und Werk, Band II*, edited by Hans-Adolf Jacobsen, 103–4. Boppard am Rhein: Boldt, 1931 (1979).

Haushofer, Karl. "Politische Erdkunde und Geopolitik." In *Freie Wege vergleichender Erdkunde*, edited by Erich von Drygalski, 86–103. Munich: R. Oldenbourg, 1925.

Haushofer, Karl. *Ratzel: Erdenmacht und Völkerschicksal, herausgegeben und eingeleitet von Generalmajor a. D. Prof. Dr. Karl Haushofer*. Stuttgart: Alfred Kröner, 1941.

Haushofer, Karl. *Wehrgeopolitik: Geographische Grundlagen einer Wehrkunde*. Berlin: Junker und Dünnhaupt, 1941.

Haushofer, Karl. *Weltpolitik von heute*. Berlin: Zeitgeschichte, 1934.

Hayes, Bascom Barry. *Bismarck and Mitteleuropa*. London: Associated University Presses, 1994.

Heffernan, Michael. "Mapping Schmitt." In *Spatiality, Sovereignty and Carl Schmitt: Geographies of the Nomos*, edited by Stephen Legg, 234–43. London: Routledge, 2011.

Heffernan, Michael. *The Meaning of Europe: Geography and Geopolitics*. London: Arnold, 2000.

Heffernan, Michael. "Professor Penck's Bluff: Geography, Espionage and Hysteria in World War I." *Scottish Geographical Journal* 116, no. 4 (2008): 267–82.

Heinrich, Michael. *Karl Marx and the Birth of Modern Society*. New York: Monthly Review Press, 2019.

Hell, Julia. *The Conquest of Ruins: The Third Reich and the Fall of Rome*. Chicago: University of Chicago Press, 2019.

Hell, Julia, and George Steinmetz. "Ruinopolis: Post-imperial Theory and Learning from Las Vegas." *International Journal of Urban and Regional Research* 38, no. 3 (2014): 1047–68.

Helmolt, Hans. *Kleine Schriften von Friedrich Ratzel, 1. Band*. Munich: R. Oldenbourg, 1906.

Helmolt, Hans. *Kleine Schriften von Friedrich Ratzel, 2. Band*. Munich: R. Oldenbourg, 1906.

Helmolt, Hans, ed. *Zu Friedrich Ratzels Gedächtnis. Geplant als Festschrift zum 60. Geburtstag, nun als Grabspende dargebracht von Fachgenossen und Schülern, Freunden und Verehrern*. Leipzig: Seele, 1904.

Hennig, Richard. *Einführung in die Geopolitik*. Leipzig: B. G. Teubner, 1935.

Henniges, Norman. "'Naturgesetze der Kultur': Die Wiener Geographen und die Ursprünge der 'Volks- und Kulturbodentheorie.'" *ACME: An International Journal for Critical Geographers* 14, no. 4 (2015): 1309–51.

Hepple, Leslie W. "Dudley Stamp and the *Zeitschrift für Geopolitik*." *Geopolitics* 13, no. 2 (2009): 386–95.

Hepple, Leslie W. "Metaphor, Geopolitical Discourse and the Military in South America." In *Writing Worlds: Discourse, Text and the Metaphor in the Representations of Landscape*, edited by Trevor J. Barnes and James S. Duncan, 136–54. London: Routledge, 1992.

Hepple. Leslie W. "South American Heartland: The Charcas, Latin American Geopolitics and Global Strategies." *The Geographical Journal* 170, no. 4 (2004): 359–67.

Herbertson, A. J. "Review: Dr. Reich's Handbook of Geography." *Geographical Journal* 34 (1909): 203–6.

Hertzberg, Heinrich. "Friedrich Ratzel's politische Geographie (Fortsetzung)." *Geographische Zeitschrift* 4, no. 8 (1898): 455–65.

Herwig, Holger. *The Demon of Geopolitics: How Karl Haushofer 'Educated' Hitler and Hess*. Lanham, MD: Rowman and Littlefield, 2016.

Herwig, Holger. "Geopolitik: Haushofer, Hitler and Lebensraum." *Journal of Strategic Studies* 22, no. 2–3 (1999): 218–41.

Herzl, Theodor. *Der Judenstaat: Staatsschrift*. n.p.: Gutenberg-DE, 1896 (2005).

Hesse, Fritz. "Das Gesetz der wachsenden Räume." *Zeitschrift für Geopolitik* 1 (1924): 1–4.

Hettner, A. "Bücherbesprechungen: Friedrich Ratzel. Glücksinseln und Träume; Kleine Schriften." *Geographische Zeitschrift* 13 (1907): 526–28.

Hettner, Alfred. *Der Friede und die deutsche Zukunft*. Stuttgart: Deutsche Verlags-Anstalt, 1917.

Hettner, Alfred. "Der Friede und die politische Geographie." *Geographische Zeitschrift* 25, no. 8/9 (1919): 233–35.

Hettner, Alfred. "Die Geographie des Menschen." *Geographische Zeitschrift* 13, no. 8 (1907): 401–25.

Heuser, Beatrice. *The Evolution of Strategy: Thinking War from Antiquity to the Present*. Cambridge: Cambridge University Press, 2010.

Heyden, Günther. *Kritik der deutschen Geopolitik: Wesen und soziale Funktion einer reaktion- ären soziologischen Studie*. Berlin: Dietz, 1958.

Hippler, Bruno. *Hitlers Lehrmeister: Karl Haushofer als Vater der NS Ideologie*. St. Ottilien: EOS, 1996.

Hitchens, Christopher. *The Trial of Henry Kissinger*. London: Verso, 2001.

Hitler, Adolf. *Hitler's Second Book: The Unpublished Sequel to Mein Kampf.* New York: Enigma, 1928 (2006).

Hitler, Adolf. *Mein Kampf: Zwei Bände in einem Band.* Munich: Eher, 1925/1927 (1936).

Hobson, John M. *The Eurocentric Conception of World Politics: Western International Theory, 1760–2010.* Cambridge: Cambridge University Press, 2012.

Hobson, Rolf. *Imperialism at Sea: Naval Strategic Thought, the ideology of sea power, and the Tirpitz plan, 1875–1914.* Boston and Leiden: Brill, 2002.

Hodge, F. W. "The History of Mankind, by Professor Friedrich Ratzel, Translated from the Second German Edition by A. J. Butler, M.A." *American Anthropologist* A9, no. 11 (1896): 381–82.

Holdar, Sven. "The Ideal State and the Power of Geography: The Life-Work of Rudolf Kjellén." *Political Geography* 11, no. 3 (1921): 307–23.

Hönsch, Ingrid. "Friedrich Ratzel's Collected Papers in the Archive for Geography in Leipzig." In *Europe between Political Geography and Geopolitics: On the Centenary of Ratzel's Politische Geographie*, edited by Marco Antonsich, Vladimir Kolossov, and Maria Paola Pagnini, 93–100. Rome: Società Geografica Italiana, 2001.

Howard, Michael. *The Franco-Prussian War.* London: Routledge, 1961 (2000).

Hull, Isabel. *Absolute Destruction: Military Culture and the Practices of War in Imperial Germany.* Ithaca, NY: Cornell University Press, 2005.

Hunter, James M. *Perspective on Ratzel's Political Geography.* Lanham, MD: University Press of America, 1983.

Huxley, Thomas H. *Handbuch der Anatomie der Wirbeltiere.* Breslau: Kern, 1873.

Huyssen, Andreas. "Fortifying the Heart—Totally Ernst Jünger's Armoured Texts." *German Critique* 59 (1993): 3–23.

Jacobsen, Hans-Adolf. *Karl Haushofer: Leben und Werk.* Boppard: Boldt, 1979.

James, Joy. "All Power to the People! Hannah Arendt's Theory of Communicative Power in a Racialized Democracy." In *Race and Racism in Continental Philosophy*, edited by Robert Bernasconi and Sybol Cook, 249–67. Bloomington: Indiana University Press, 2003.

James, William. "Introduction." In *The Little Book of Life after Death*, edited by Gustav Theodor Fechner, vii–xix. Boston: Little Brown and Company, 1904.

Jelly-Schapiro, Joshua. "What Are borders For?" *New Yorker*, 27 November 2019. https://www.newyorker.com/books/under-review/what-are-borders-for/amp.

Johnston, Tom. "Environmental Determinism." In *AAG International Encyclopaedia of Geography*, edited by Douglas Richardson et al., 1–3. Oxford: Wiley-Blackwell, 2017.

Jünger, Ernst. *Storm of Steel.* London: Penguin, 1920 (2004).

Jureit, Ulrike. *Das Ordnen von Räumen: Territorium und Lebensraum im 19. und 20. Jahrhundert.* Hamburg: Hamburger Edition, 2012.

Jureit, Ulrike. "Mastering Space: Laws of Movement and the Grip on the Soil." *Journal of Historical Geography* 61 (2018): 81–85.

Jütersonke, Oliver. *Morgenthau, Law and Realism.* Cambridge: Cambridge University Press, 2010.

Kaemmel, Otto. "Friedrich Ratzel." *Die Grenzboten* 63 (1904): 420–21.

Kahn, Herman. *On Thermonuclear War*. New Brunswick, NJ: Transaction, 1960 (2007).

Kaldor, Mary. *New and Old Wars: Organised Violence in a Global Era*. Cambridge: Polity, 2012.

Kaplan, Robert D. *The Revenge of Geography: What the Map Tells Us about Coming Conflicts and the Battle against Fate*. New York: Random House, 2012.

Kapp, Friedrich. *Geschichte der Sklaverei in den Vereinigten Staaten von Amerika*. Hamburg: Otto Meißner, 1854 (1861).

Kearns, Gerry. *Geopolitics and Empire: The Legacy of Halford Mackinder*. Oxford: Oxford University Press, 2009.

Kehr, Eckart. *Schlachtflottenbau und Parteipolitik 1894–1901*. Berlin: Emil Ebering, 1930.

Keighren, Innes M. *Bringing Geography to Book: Ellen Semple and the Reception of Geographical Knowledge*. London: I. B. Tauris, 2010.

Keighren, Innes M. "History and Philosophy of Geography III: The Haunted, the Reviled, and the Plural." *Progress in Human Geography* 44, no. 1 (2020): 160–67.

Keighren, Innes M., Christian Abrahamsson, and Veronica della Dora. "On Canonical Geographies." *Dialogues in Human Geography* 2, no. 3 (2012): 296–312.

Keith, Leeanna. *The Colfax Massacre: The Untold Story of Black Power, White Terror and the Death of Reconstruction*. Oxford: Oxford University Press, 2008.

Kelly, Alfred. *The Descent of Darwin: The Popularization of Darwinism in Germany, 1860–1914*. Chapel Hill: University of North Carolina Press, 1981.

Kelly, Patrick J. *Tirpitz and the Imperial German Navy*. Bloomington: Indiana University Press, 2011.

Kendi, Ibram X. *Stamped from the Beginning: The Definitive History of Racist Ideas in America*. London: Bodley Head, 2016.

Kerslake, Christian. *Deleuze and the Unconscious*. London: Continuum, 2007.

Kittel, Rudolf. "Zu Friedrich Ratzels Gedächtnis." *Die Grenzboten* 63 (1904): 516–24.

Kjellén, Rudolf. *Der Staat als Lebensform*. Leipzig: S. Hirzel, 1917.

Kjellén, Rudolf. *Die Ideen von 1914: Eine weltgeschichtliche Perspektive*. Leipzig: S. Hirzel, 1915.

Kjellén, Rudolf. *Grundriss zu einem System der Politik*. Leipzig: S. Hirzel, 1920.

Klinke, Ian. *Cryptic Concrete: A Subterranean Journey into Cold War Germany*. Oxford: Wiley-Blackwell, 2018.

Klinke, Ian. "Friedrich Ratzel, Lebensraum and the Death Motif." *Journal of Historical Geography* 61 (2018): 97–101.

Klinke, Ian. "Geopolitics and the Political Right: Lessons from Germany." *International Affairs* 94, no. 3 (2018): 495–514.

Klinke, Ian. "On the History of a Subterranean Geopolitics." *Geoforum* 127 (2021): 356–63.

Klinke, Ian. "Vitalist Temptations: Life, Earth and the Nature of War." *Political Geography* 72 (2019): 1–9.

Klinke, Ian, and Mark Bassin. "*Lebensraum* and its Discontents." *Journal of Historical Geography* 61 (2018): 53–58.

Knieper, Franz. *Geopolitik für die Unterrichtspraxis*. Bochum: Verlags–und Lehrmittel-Anstalt Ferdinand Kamp, 1937.

Knieper, Franz. *Politische Geographie (Geopolitik) für die Unterrichtspraxis*. Ahlen: Sommer, 1957.

Koelsch, William A. "Franz Boas, Geographer, and the Problem of Disciplinary Identity." *Journal of the History of the Behavioural Sciences* 40, no. 1 (2004): 1–22.

Kost, Klaus. "Anti-semitism in German Geography 1900–1945." *GeoJournal* 46 (1998): 285–91.

Kramer, Fritz L. "A Note on Carl Ritter, 1779–1859." *Geographical Review* 49, no. 3 (1959): 406–9.

Kruck, Alfred. *Geschichte des Alldeutschen Verbandes 1890–1939*. Wiesbaden: Franz Steiner, 1954.

Krug-Genthe, Martha. "Tributes to Friedrich Ratzel." *Bulletin of the American Geographical Society* 36, no. 9 (1904): 550–53.

Kundrus, Birthe. "German Colonialism: Some Reflections on Reassessments, Specificities, and Constellations." In *German Colonialism: Race, the Holocaust, and Postwar Germany*, edited by Volker Langbehn and Mohammad Salama, 29–47. New York: Colombia University Press, 2011.

Kuss, Susanne. *German Colonial Wars and the Context of Military Violence*. Cambridge, MA: Harvard University Press, 2017.

Kutschera, Ulrich, Georgy S. Levit, and Uwe Hossfeld. "Ernst Haeckel (1834–1919): The German Darwin and his Impact on Modern Biology." *Theory in Biosciences* 138 (2019): 1–7.

Lamprecht, Karl. "Friedrich Ratzel." *Berichte Über die Verhandlungen der Königlich Sächsischen Gesellschaft der Wissenschaften zu Leipzig*, Sitzung vom 10. Dezember 1904: 259–69.

Lange, Karl. "Der Terminus 'Lebensraum' in Hitlers 'Mein Kampf.'" *Vierteljahrshefte für Zeitgeschichte* 13, no. 4 (1965): 426–37.

Lautensach, Hermann. "Die Mittelmeere als geopolitische Kraftfelder." *Zeitschrift für Geopolitik* 1 (1924): 36–47.

Lawrence, Ken. "The Ku Klux Klan and Fascism." In *The U.S. Antifascism Reader*, edited by Bill V. Mullen and Christopher Vials, 289–99. London: Verso, 1982 (2020).

Legg, Stephen. *Spatiality, Sovereignty and Carl Schmitt: Geographies of the Nomos*. London: Routledge, 2011.

Lemke, Thomas. "An Alternative Model of Politics? Prospects and Problems of Jane Bennett's Vital Materialism." *Theory, Culture and Society* 35, no. 6 (2018): 31–54.

Lemke, Thomas. *Biopolitics: An Advanced Introduction*. New York: New York University Press, 2011.

Lerman, Katherine A. "Bismarckian Germany." In *Imperial Germany, 1871–1918*, edited by James Retallack, 18–39. Oxford: Oxford University Press, 2008.

Lerp, Dörte. "Farmers to the Frontier: Settler Colonialism in the Eastern Prussian Provinces and German Southwest Africa." *Journal of Imperial and Commonwealth History* 41, no. 4 (2013): 567–83.

Leser, Paul. "Zur Geschichte des Wortes Kulturkreis." *Anthropos* 58, no. 1/2 (1963): 1–36.

Leshem, Noam. "'Over Our Dead Bodies': Placing Necropolitical Activism." *Political Geography* 45 (2015): 34–44.

Lewis, Daniel Levering. *W. E. B. Du Bois: A Biography of Race 1868–1919*. New York: Henry Holt and Company, 1993.

Lewis-Stempel, John. "The Women Made to Boil Heads of Their Own People in Germany's First Holocaust." *Sunday Express*, 19 January 2014. https://www.express.co.uk/news/world-war-1/454733/The-women-made-to-boil-heads-of-their-own-people-in-Germany-s-first-Holocaust.

Liebert, Eduard von. "Die politische, militärische und volkswirtschaftliche Bedeutung einer starken Seemacht." In *Verhandlungen des Deutschen Kolonialkongresses 1905*, n.e., 928–39. Berlin: Georg Reimer, 1906.

Livingstone, David N. *Adam's Ancestors: Race, Religion and the Politics of Human Origins*. Baltimore: Johns Hopkins University Press, 2008.

Livingstone, David N. "Climate's Moral Economy: Science, Race and Place in Post-Darwinian British and American Geography." In *Geography and Empire*, edited by Anne Godlewska and Neil Smith, 132–54. Oxford: Blackwell, 1994.

Livingstone, David N. "Environmental Determinism." In *The Sage Handbook of Geographical Knowledge*, edited by John Agnew and David N. Livingstone, 368–80. London: Sage, 2011.

Livingstone, David. N. *The Geographical Tradition: Episodes in the History of a Contested Enterprise*. Oxford: Blackwell, 1992.

Livingstone, David. N. "*The Geographical Tradition* and the Challenges of Geography Geographised." *Transactions of the Institute of British Geographers* 44, no. 3 (2019): 458–62.

Livingstone, David. N. "The Geography of Darwinism." *Interdisciplinary Science Reviews* 31, no. 1 (2006): 32–41.

Löfken, Alexander. "Überleben in Schutzraumbauten." *Ziviler Luftschutz* 24, no. 5 (1960): 162–68.

Löfken, Alexander. "Warum 'Baulicher Luftschutz'?" *Ziviler Luftschutz* 17, no. 2 (1954): 33–34.

Lorenz, Konrad. "Biologische Fragestellung in der Tierpsychologie." *Zeitschrift für Tierpsychologie* 1, no. 1 (1937): 24–32.

Lorenz, Konrad. "Ein neuer Wissenschaftszweig—die Kulturethologie." In *Otto Koenig 70 Jahre: Kulturwsisenschaftliche Beiträge zur Verhaltensforschung*, edited by Gesellschaft der Freunde der Forschungsgemeinschaft Wilhelminenberg, 7–9. Vienna: Ueberreuter, 1984.

Lüdecke, Cornelia. "Ein genußreiches Zusammenleben und–arbeiten' Friedrich Ratzels Zeit in München (1875–1886)." *Berichte zur Wissenschaftsgeschichte* 25, no. 1 (2002): 25–39.

Lüdecke, Cornelia. *Militärische und zivile Geographie: Von Carl Ritters Geographie in Berlin bis zu Karl Haushofers Geopolitik in München*. Norderstedt: Die Bibliothek der Erde, 2019.

Lurie, Edward. "Louis Agassiz and the Races of Man." *History of Science Society* 45, no. 3 (1954): 227–42.

Mackinder, Halford. *Britain and the British Seas*. London: William Heinemann, 1902.

Mackinder, Halford. *Democratic Ideals and Reality: A Study in the Politics of Reconstruction*. London: Faber and Faber, 1919 (1942, reprinted 2009).

Mackinder, Halford. *The First Ascent to Mount Kenya*. Athens: Ohio University Press, 1991.

Mackinder, Halford. "The Geographical Pivot of History." *The Geographical Journal* 23, no. 4 (1904): 421–44.

Mackinder, Halford. "A Journey to the Summit of Mount Kenya, British East Africa." *The Geographical Journal* 15, no. 5 (1900): 453–76.

Mackinder, Halford. "Modern Geography, German and English." *The Geographical Journal* 6, no. 4 (1895): 367–79.

Mackinder, Halford. "The Round World and the Winning of Peace." *Foreign Affairs* 21, no. 4 (1943): 595–605.

Maddrell, Avril. "To Read or Not to Read? The Politics of Overlooking Gender in the Geographical Canon." *Journal of Historical Geography* 49 (2015): 31–38.

Maddrell, Avril, and James D. Sidaway, eds. *Deathscapes: Spaces for Death, Dying, Mourning and Remembrance*. Farnham: Ashgate, 2010.

Madley, Benjamin. "From Africa to Auschwitz: How German South West Africa Incubated Ideas and Methods Adopted and Developed by the Nazis in Eastern Europe." *European History Quarterly* 35, no. 3 (2005): 429–64.

Madley, Benjamin. "Understanding Genocide in California under United States Rule, 1846–1873." *Western Historical Quarterly* 47, no. 4 (2017): 449–61.

Mahan, Alfred Thayer. *The Influence of Sea Power upon History 1660–1805*. London: Bison, 1890 (1980).

Malthus, Thomas Robert. *An Essay on the Principle of Population*. Oxford: Oxford University Press, 1798 (2004).

Maltzahn, Curt von. "Friedrich Ratzel. Ein Gedenkwort." *Marine-Rundschau* 2 (1905): 217–20.

Mamadouh, Virginie. "Geographers and War, Geographers and Peace." In *The Geography of War and Peace: From Death Camps to Diplomats*, edited by Colin Flint, 26–60. Oxford: Oxford University Press, 2005.

Marsh, Alex. *John Kasper and Ezra Pound: Saving the Republic*. London: Bloomsbury, 2015.

Marshall, Tim. *Prisoners of Geography: Ten Maps That Tell You Everything You Need to Know about Global Politics*. London: Elliot and Thompson, 2015.

Maull, Otto. "Friedrich Ratzel zum Gedächtnis." *Zeitschrift für Geopolitik* 5 (1929): 628.

Maull, Otto. *Politische Geographie*. Berlin: Safari, 1956.

May, Jon, and Nigel Thrift. "Introduction." In *Timespace: Geographies of Temporality*, edited by Jon May and Nigel Thrift, 1–46. London: Routledge, 2001.

May, Todd. *Death*. London: Routledge, 2009.

Mayr, E. "Isolation as an Evolutionary Factor." *Proceedings of the American Philosophical Society* 103, no. 2 (1943): 221–30.

Mazower, Mark. "Foucault, Agamben: Theory and the Nazis." *Boundary 2* 35, no. 1 (2008): 23–34.

Mazower, Mark. *Hitler's Empire: Nazi Rule in Occupied Europe*. London: Penguin, 2008.

Mbembe, Achille. *Necropolitics*. Durham, NC: Duke University Press, 2019.

Mbembe, Achille. "Necropolitics." *Public Culture* 15, no. 1 (2003): 11–40.
Mehring, Reinhard. *Carl Schmitt: A Biography*. London: Polity, 2014.
Mendieta, Eduardo. "War as the School of Space: The Space of War and the War of Space." *Ethics, Place and Environment* 9, no. 2 (2006): 207–29.
Metz, Friedrich. "Friedrich Ratzel, Begründer der politischen Geographie." *Mein Heimatland* 25 (1938): 314–18.
Meyers, Jeffrey. "Henry Morton Stanley and Emin Pasha: New historical sources for *Heart of Darkness*." *Conradiana* 48, no. 1 (2016): 77–82.
Michel, Boris. "Anti-semitism in Early Twentieth Century German Geography. From a 'Spaceless' People to the Root of the 'Ills' of Urbanization." *Political Geography* 65 (2018): 1–7.
Michel, Boris. "Making Mount Kilimanjaro German: Nation Building and Heroic Masculinity in the Colonial Geographies of Hans Meyer." *Transactions of the Institute of British Geographers* 44, no. 3 (2019): 493–508.
Michelini, Francesca, and Kristian Köchy. *Jakob von Uexküll and Philosophy: Life, Environments, Anthropology*. Abingdon: Routledge, 2019.
Mildenberger, Florian. "Lebensraum oder Umwelt? Friedrich Ratzel (1844–1904) und Jakob von Uexküll (1864–1944)." In *Jahrbuch für europäische Wissenschaftskultur* 5, edited by Olaf Breidbach and Stefano Poggi, 249–283. Stuttgart: Steiner, 2009.
Mills, Charles W. *Black Rights/White Wrongs: The Critique of Racial Liberalism*. Oxford: Oxford University Press, 2017.
Minca, Claudio. "Geographies of the Camp." *Political Geography* 49 (2015): 74–83.
Minca, Claudio. "Giorgio Agamben and the New Biopolitical Nomos." *Geografiska Annaler: Series B* 88, no. 4 (2006): 387–403.
Minca, Claudio, and Rory Rowan. *On Schmitt and Space*. London: Routledge, 2015.
Minca, Claudio, and Nick Vaughan-Williams. "Carl Schmitt and the Concept of the Border." *Geopolitics* 17, no. 4 (2012): 756–72.
"Möge Gott helfen." *Der Spiegel*, 17 March 1980. https://www.spiegel.de/spiegel/print/d-14319788.html.
Monbiot, George. "Colonised and Coloniser, Empire's Poison Infects Us All." *The Guardian*, 8 October 2012. https://amp.theguardian.com/commentisfree/2012/oct/08/empire-torture-kenya-catastrophe-europe.
Moore, James. "Darwin's Progress and the Problem of Slavery." *Progress in Human Geography* 34, no. 5 (2010): 555–83.
Morgenthau, Hans. *Politics among Nations: The Struggle for Power and Peace*. New York: Knopf, 1948 (1993).
Moyd, Michelle R. *Violent Intermediaries: African Soldiers, Conquest, and Everyday Colonialism in German East Africa*. Athens: Ohio University Press, 2014.
Müller, Gerhard H. *Friedrich Ratzel (1844–1904), Naturwissenschaftler, Geograph, Gelehrter*. Stuttgart: GNT, 1996.
Müller, Gerhard H. "Zur Geschichte des Begriffs 'Anthropogeographie.'" *Geographische Zeitschrift* 80, no. 3 (1992): 184–90.
Murphy, David T. *Heroic Earth: Geopolitical Thought in Weimar Germany 1918–1933*. Kent, OH: Kent State University Press, 1997.
Murphy, David T. "Hitler's Geostrategist? The Myth of Karl Haushofer and the 'Institut für Geopolitik.'" *The Historian* 76, no. 1 (2014): 1–25.

Murphy, David T. "'Retroactive Effects': Ratzel's Spatial Dynamics and the Expansionist Imperative in Interwar Germany." *Journal of Historical Geography* 61 (2018): 86–90.

Murphy, David T. "Space, Race and Geopolitical Necessity: Geopolitical Rhetoric in German Colonial Revanchism, 1919–1933." In *Geography and Empire*, edited by Anne Godlewska and Neil Smith, 173–87. Oxford: Blackwell, 1994.

Nationaldemokratische Partei Deutschlands. "Portfolio: Identität." Accessed 20 September 2019. Available at https://npd.de/portfolio/identitaet/.

Natter, Wolfgang. "Friedrich Ratzel's Spatial Turn." In *B/ordering space*, edited by Henk van Houtum, Oliver Kramsch, and Wolfgang Zierhofer, 171–87. Aldershot: Ashgate, 2005.

Naumann, Friedrich. *Mitteleuropa*. Berlin: Georg Reimer, 1915.

Neocleous, Mark. "Long Live Death! Fascism, Resurrection, Immortality." *Journal of Political Ideologies* 10, no. 1 (2005): 31–49.

Neumann, Sigmund. "Fashions in Space." *Foreign Affairs* 21, no. 2 (1943): 276–88.

Ó Tuathail, Gearóid. *Critical Geopolitics: The Politics of Writing Global Space*. Minneapolis: University of Minnesota Press, 1996.

Ó Tuathail, Gearóid, and John Agnew. "Geopolitics and Discourse: Practical Geopolitical Reasoning in American Foreign Policy." *Political Geography* 11, no. 2 (1992): 190–204.

Oberhummer, Eugen. "Anhang. Die politische Geographie vor Ratzel und ihre jüngste Entwicklung." In Friedrich Ratzel, *Politische Geographie*, 597–618. Munich: R. Oldenbourg, 1923.

Obst, Erich. "Die geopolitischen Leitlinien des europäischen Russlands." *Zeitschrift für Geopolitik* 1 (1924): 5–19.

Oldfield, Jon D., and Denis J. J. Shaw. "A Russian Geographical Canon? The Contested Canon of Russian and Soviet Geography, 1884–1953." *Journal of Historical Geography* 49 (2015): 75–84.

Olusoga, David, and Casper W. Erichsen. *The Kaiser's Holocaust: Germany's Forgotten Genocide*. London: Faber and Faber, 2010.

Osterhammel, Jürgen. *Die Verwandlung der Welt: Eine Geschichte des 19. Jahrhunderts*. Munich: C. H. Beck, 2009.

Owens, Patricia. "Racism in the Theory Canon: Hannah Arendt and 'the one great crime in which America was never involved.'" *Millennium: Journal of International Studies* 45, no. 3 (2017): 403–24.

Owens, Patricia, and Katharina Rietzler. *Women's International Thought: A New History*. Cambridge: Cambridge University Press, 2021.

Ozkan, Behlül. "Turkey, Davutoglu and the Idea of Pan-Islamism." *Survival* 56, no. 4 (2014): 119–40.

Parker, Geoffrey. "Ratzel, the French School and the Birth of Alternative Geopolitics." *Political Geography* 19, no. 8 (2000): 957–69.

Paxton, Robert O. *The Anatomy of Fascism*. London: Alfred A. Knopf, 2004.

Pearson, Karl. *The Chances of Death and Other Studies in Evolution*. London: Arnold, 1897.

Peet, Richard. "The Social Origins of Environmental Determinism." *Annals of the Association of American Geographers* 75, no. 3 (1985): 309–33.

Penck, Albrecht. *Morphologie der Erdoberfläche: Erster Teil*. Stuttgart: Engelhorn, 1894.

Penck, Albrecht. *Über politische Grenzen: Rede zum Antritt des Rektorates der Königlichen Friedrich Wilhelms-Universität in Berlin*. Berlin: Norddeutsche Buchdruckerei und Verlagsanstalt, 1917.

Perera, Suvendrini. "Claiming Truganini: Australian National Narratives in the Year of Indigenous Peoples." *Cultural Studies* 10, no. 3 (1996): 393–412.

Perras, Arne. *Carl Peters and German Colonialism 1856–1918: A Political Biography*. Oxford: Oxford University Press, 2004.

Peschel, Oscar. "Ursprung und Verschiedenheit der Menschenrassen." *Das Ausland* 33 (1860): 391–94.

Peschel, Oscar. *Völkerkunde*. Leipzig: Duncker und Humblot, 1874.

Pfeffer, Karl Heinz, and Kurt Vowinckel. "Was will Geopolitik heute?" *Zeitschrift für Geopolitik* 24 (1954): 193–200.

Pipes, Richard. "Militarism and the Soviet State." *Daedalus* 109, no. 4 (1980): 1–12.

Plöckinger, Othmar. *Geschichte eines Buches: Adolf Hitlers Mein Kampf: 1922–1945*. Munich: Oldenbourg, 2011.

Povinelli, Elizabeth A. *Geontologies: A Requiem to Late Liberalism*. Durham, NC: Duke University Press, 2016.

Powell, Richard C. "Echoes of the New Geography? History and Philosophy of Geography I." *Progress in Human Geography* 36, no. 4 (2009): 518–26.

Powell, Richard C. "Notes on a Geographical Canon? Measures, Models and Scholarly Enterprise." *Journal of Historical Geography* 49 (2015): 2–8.

Powell, Richard C. "The Study of Geography? Franz Boas and his Canonical Returns." *Journal of Historical Geography* 49 (2015): 21–30.

Rash, Felicity. *The Discourse Strategies of Imperialist Writing: The German Colonial Idea and Africa, 1848–1945*. London: Routledge, 2017.

Reclus, Élisée. "Recent Books on the United States." *The Geographical Journal* 6, no. 5 (1895): 448–53.

Reich, Emil. *Handbook of Geography: Descriptive & Mathematical*. London: Duckworth, 1908.

Reich, Emil. *Selected Documents Illustrating Mediæval and Modern History*. London: P. S. King and Son, 1905.

Reynolds, Andrew. "Ernst Haeckel and the Theory of the Cell State: Remarks on the History of a Bio-political Metaphor." *History of Science* 46, no. 2 (2008): 123–52.

Richards, Robert J. "Ernst Haeckel's Alleged Anti-semitism and Contributions to Nazi Biology." *Biological Theory* 2, no. 1 (2007): 97–103.

Richards, Robert J. *The Tragic Sense of Life: Ernst Haeckel and the Struggle over Evolutionary Thought*. Chicago: Chicago University Press, 2008.

Ritter, Carl. *Allgemeine Erdkunde: Vorlesungen an der Universität zu Berlin gehalten*. Berlin: Reimer, 1862.

Ritter, Carl. *Ueber das historische Element in der geographischen Wissenschaft*. Berlin: Königliche Akademie der Wissenschaften, 1834.

Romanillos, José Luis. "Geography, Death, and Finitude." *Environment and Planning A* 43, no. 11 (2011): 2533–53.

Rose, Gillian. "Tradition and Paternity: Same Difference?" *Transactions of the Institute of British Geographers* 20, no. 4 (1995): 414–16.
Rose, Mitch. "The Question of Culture in Cultural Geography: Latent Legacies and Potential Futures." *Progress in Human Geography* 45, no. 5 (2021): 951–71.
Rosenberg, Alfred. *Der Mythus des 20. Jahrhunderts*. Munich: Hoheneichen, 1930.
Rosenboim, Or. "Geopolitics and Nationalism: Interpreting Friedrich Ratzel in Italian, 1898–1916." In *Radicals and Reactionaries in Twentieth-Century International Thought*, edited by Ian Hall, 17–41. Basingstoke: Palgrave, 2015.
Ruddick, Sue. "Rethinking the Subject, Reimagining Worlds." *Dialogues in Human Geography* 7, no. 2 (2017): 119–39.
Ruge, Friedrich. "Die vergessene See." *Zeitschrift für Geopolitik* 26 (1955): 355–60.
Ruge, Friedrich. *In vier Marinen. Lebenserinnerungen als Beitrag zur Zeitgeschichte*. Munich: Bernard & Graefe, 1979.
Ruge, Friedrich. *Seemacht und Sicherheit: Eine Schicksalsfrage für alle Deutschen*. Tübingen: Fritz Schlichtenmeyer, 1955.
Rumley, Dennis, Minghi Julian, and Frank M. Grimm. "The Content of Ratzel's Politische Geographie." *The Professional Geographer* 25, no. 3 (1973): 271–77.
Sandner, Gerhard, and Mechthild Rössler. "Geography and Empire in Germany, 1871–1945." In *Geography and Empire*, edited by Anne Godlewska and Neil Smith, 115–27. Oxford: Blackwell, 1994.
Santini, Carlotta. "At the Origins of Modern Geography. The Oecumene: An Anthropogeographical Pattern." *History of European Ideas* 43, no. 6 (2017): 560–69.
Santini, Carlotta. "One Sea, One Humanity: Modelling the Man-Sea Relationship in Friedrich Ratzel's Anthropogeographical Project." *Journal of Interdisciplinary History of Ideas* 6, no. 12 (2017): 1–24.
Sauer, Carl O. "The Formative Years of Ratzel in the United States." *Annals of the Association of American Geographers* 61 (1971): 245–54.
Sax, Boria, and Peter H. Klopfer. "Jakob von Uexküll and the Anticipation of Sociobiology." *Semiotica* 134, no. 1 (2001): 767–78.
Scheffauer, Herman George. "Der Ku Klux Klan." *Zeitschrift für Geopolitik* 1 (1924): 700–704.
Scherzer, Karl von. "Biographische Skizze." In *Die Entstehung der Arten durch räumliche Sonderung: Gesammelte Aufsätze von Moriz Wagner*, edited by Moritz Wagner, 9–32. Basel: Benno Schwabe, 1899.
Scheuerman, William E. *Hans Morgenthau: Realism and Beyond*. Cambridge: Polity, 2009.
Schilling von Cannstadt, Wilhelm. *Das Großherzoglich Badische 5. Infranterie-Regiment, jetzt Königlich Preußische 5. Badische Infranterie-Regiment Nr.113 im Feldzuge 1870–71*. Berlin: Ernst Siegfried Mittler und Sohn, 1876.
Schlögel, Karl. *Im Raume lesen wir die Zeit: über Zivilisationsgeschichte und Geopolitik*. Munich: Hanser, 2003.
Schmitt, Carl. "Dialogue on New Space." In *Dialogues on Space and Power*, edited by Andreas Kalyvas and Federico Finchelstein. Cambridge: Polity, 2015 (1958).
Schmitt, Carl. "Die Bibliothek Carl Schmitt." Accessed 10 February 2023. Available at https://www.carl-schmitt.de/wp-content/uploads/CSG_Bibliothek_H2_2021.pdf.

Schmitt, Carl. "The *Großraum* Order of International Law with a Ban on Intervention for Spatially Foreign Powers: A Contribution to the Concept of *Reich* in International Law (1939–1941)." In *Carl Schmitt: Writings on War*, edited by Timothy Nunan, 75–124. Cambridge: Polity, 1941 (2011).

Schmitt, Carl. *Land and Sea*. Washington, DC: Plutarch, 1942 (1997).

Schmitt, Carl. *The Nomos of the Earth in the International Law of the Jus Publicum Europaeum*. New York: Telos, 1950 (2006).

Schmitt, Carl. *Völkerrechtliche Großraumordnung mit Interventionsverbot für raumfremde Mächte*. Berlin: Duncker and Humblot, 1941 (1991).

Schmitt, Carl. "Zum Gedächtnis von Sege Maiwald." *Zeitschrift für Geopolitik* 23 (1952): 447–48.

Schmitthenner, Heinrich. *Lebensräume im Kampf der Kulturen*. Heidelberg: Quelle and Meyer, 1938 (1951).

Schmitthenner, Heinrich. *Lebensräume im Kampf der Kulturen*. Leipzig: Quelle and Meyer, 1938.

Schmitthenner, Heinrich. "Zum Begriff Lebensraum." *Geographische Zeitschrift* 48, no. 11/12 (1942): 405–17.

Schöller, Peter. "Wege und Irrwege der politischen Geographie und Geopolitik." *Erdkunde* 11, no. 1 (1957): 1–20.

Schrepfer, Hans. "Was heißt Lebensraum? Eine notwendige begriffliche Klärung." *Geographische Zeitschrift* 48, no. 11–12 (1942): 417–24.

Schulte-Althoff, Franz-Josef. *Studien zur politischen Wissenschaftsgeschichte der deutschen Geographie*. Paderborn: Schöningh, 1971.

Schultz, Hans-Dietrich. "Albrecht Penck: Vorbereiter und Wegbereiter der NS-Lebensraumpolitik?" *E & G Quaternary Science Journal* 66, no. 2 (2018): 115–29.

Schultz, Hans-Dietrich. "'Eines schickt sich nicht für alle': Rassen, Völker und Nationen im länderkundlichen Denken Friedrich Ratzels." In *Prisma und Spiegel: Ethnologie zwischen postkolonialer Kritik und Deutung der eigenen Gesellschaft*, edited by Dorothea E. Schulz and Jochen Seebode, 115–32. Hamburg: Argument, 2010.

Schultz, Hans-Dietrich. "Fantasies of *Mitte*: *Mittellage* and *Mitteleuropa* in German Geographical Discussion in the 19th and 20th century." *Political Geography Quarterly* 8, no. 4 (1989): 315–39.

Schultz, Hans-Dietrich. "Friedrich Ratzel: Bellizistischer Raumtheoretiker mit Naturgefühl oder Vorläufer der NS Lebensraumpolitik?" In *Auf der Suche nach Vielfalt: Ethnologie und Geographie in Leipzig*, edited bv Claus Deimel, Sebastian Lentz, and Bernhard Streck, 125–42. Leipzig: Leibniz-Institut für Länderkunde, 2009.

Schultz, Hans–Dietrich. "Friedrich Ratzel: (k)ein Rassist?" *Geographische Hochschulmanuskripte* 2 (2006): 1–46.

Schultz, Hans–Dietrich. "Friedrich Ratzel—Pionier moderner Sozialgeographie? Ein Missverständnis." *Kultursoziologie* 23, no. 3 (2014): 54–66.

Schultz, Hans-Dietrich. "'Hätte doch die Erde mehr Raum!' Friedrich Ratzel und sein (politisch-) geographisches Weltbild." *Mitteilungen der Geographischen Gesellschaft in München* 89 (2007): 3–45.

Schultz, Hans-Dietrich. "Herder und Ratzel: Zwei Extreme, ein Paradigma?" *Erdkunde* 52, no. 2 (1998): 127–43.

Schultz, Hans-Dietrich. "Pax geographica: Räumliche Konzepte für Krieg und Frieden in der geographischen Tradition." *Geographische Zeitschrift* 75, no. 1 (1987): 1–22.
Schultz, Hans–Dietrich. "'Rasse' und Raum von Ritter zu Ratzel." In *'Rasse' und Raum: Topologien zwischen Kolonial-, Geo- und Biopolitik: Geschichte, Kunst, Erinnerung*, edited by Claudia Bruns, 221–49. Wiesbaden: Reichert, 2017.
Schweinfurth, Georg August, and Friedrich Ratzel. *Emin-Pascha. Eine Sammlung von Reisebriefen und Berichten Dr. Emin Paschas aus den ehemals ägyptischen Äquatorialprovinzen und deren Grenzländern.* Leipzig: F. A. Brockhaus, 1888.
Seegel, Steven. *Map Men: Transnational Lives and Deaths of Geographers in the Making of East Central Europe*. Chicago: University of Chicago Press, 2018.
Seegraef, Werner. "Politische Geographie in Schulbüchern für den Erdkundeunterricht seit 1945—ein Überblick." *Internationale Schulbuchforschung* 5, no. 3 (1983): 285–300.
Semple, Ellen Churchill. *American History and its Geographic Conditions*. Boston: Houghton, Mifflin and Company, 1903.
Semple, Ellen Churchill. *Influences of Geographic Environment on the Basis of Ratzel's System of Anthropo-geography*. New York: Holt, 1911.
Semple, Ellen Churchill. "The North-shore Villages of the Lower St. Lawrence." In *Zu Friedrich Ratzels Gedächtnis. Geplant als Festschrift zum 60. Geburtstag, nun als Grabspende dargebracht von Fachgenossen und Schülern, Freunden und Verehrern*, edited by Hans Helmolt, 349–60. Leipzig: Seele, 1904.
Semple, Ellen Churchill. "Slavery: A Study in Sociology." MA thesis, Vassar College, 1891.
Semple, Ellen Churchill. "Tributes to Friedrich Ratzel." *Bulletin of the American Geographical Society* 36, no. 9 (1904): 553.
Semple, Ellen Churchill. "Two Works on Political Geography." *Geographical Review* 14, no. 4 (1924): 665–667.
Sharp, Joanne P. "Publishing American Identity: Popular Geopolitics, Myth and the Reader's Digest." *Political Geography* 12, no. 6 (1993): 491–503.
Sidaway, James D. "The (Re)making of the Western 'Geographical Tradition': Some Missing Links." *Area* 29, no. 1 (1997): 72–80.
Simms, Brendan. *Hitler: Only the World Was Enough*. London: Penguin, 2019.
Simons, Oliver. "Carl Schmitt's Spatial Rhetoric." In *The Oxford Handbook of Carl Schmitt*, edited by Jens Meierhenrich and Oliver Simons, 776–802. Oxford: Oxford University Press, 2016.
Smith, J. Russell. "Geography in Germany. II. The University." *Journal of Geography* 1, no. 10 (1902): 448–57.
Smith, Neil. *American Empire: Roosevelt's Geographer and the Prelude to Globalization*. Oakland: University of California Press, 2004.
Smith, Woodruff D. "Friedrich Ratzel and the Origins of Lebensraum." *German Studies Review* 3, no. 1 (1980): 51–68.
Smith, Woodruff D. *The Ideological Origins of Nazi Imperialism*. Oxford: Oxford University Press, 1986.
Smith, Woodruff D. *Politics and the Sciences of Culture in Germany, 1840–1920*. Oxford: Oxford University Press, 1991.

Smith, Woodruff D. "Review of Jürgen Zimmerer: Von Windhuk nach Auschwitz? Beiträge zum Verhältnis von Kolonialismus und Holocaust." *German History* 30, no. 2 (2012): 312–14.

Smith, Woodruff D. "'Weltpolitik' und 'Lebensraum,'" In *Das Kaiserreich transnational: Deutschland in der Welt 1871–1914*, edited by Sebastian Conrad and Jürgen Osterhammel, 29–48. Göttingen: Vandenhoeck & Ruprecht. 2004.

Snyder, Timothy. *Black Earth: The Holocaust as History and Warning*. London: Vintage, 2015.

Somavilla, Fabian. "Ohne Impfpass kein Grenzübertritt: Wie Corona Grenzen militarisiert." *Der Standard*, 3 May 2020. https://www.derstandard.at/story/2000117194363/ohne-impfpass-kein-grenzuebertritt-wie-corona-grenzen-mililtarisiert?amplified=True.

Spahn, Martin. *Die Großmächte: Richtlinien ihrer Geschichte, Maßstäbe ihres Wesens*. Berlin: Ullstein, 1918.

Spang, Christian. *Karl Haushofer und die OAG: Deutsch–japanische Netzwerke in der ersten Hälfte des 20. Jahrhunderts*. Munich: Iudicium, 2018.

Spang, Christian. *Karl Haushofer und Japan: Die Rezeption seiner geopolitischen Theorien in der deutschen und japanischen Politik*. Munich: Iudicium, 2013.

Specter, Matthew. *The Atlantic Realists: Empire and International Political Thought between Germany and the United States*. Stanford, CA: Stanford University Press, 2021.

Specter, Matthew. "Grossraum and Geopolitics: Resituating Schmitt in an Atlantic Context." *History and Theory* 56, no. 3 (2017): 398–406.

Speer, Albert. *Inside the Third Reich*. London: Phoenix, 1970.

Speitkamp, Winfried. *Deutsche Kolonialgeschichte*. Stuttgart: Reclam, 2008.

Spencer, Herbert. *The Principles of Sociology, Vol. 1*. London: Williams and Norgate, 1877.

Spencer, Herbert. *Social Statics, or the Conditions Essential to Human Happiness*. London: Williams and Norgate, 1851 (1868).

Sprengel, Rainer. *Kritik der Geopolitik. Ein deutscher Diskurs, 1914–1944*. Berlin: Akademie, 1996.

Squire, Vicki. "Reshaping Critical Geopolitics? The Materialist Challenge." *Review of International Studies* 41, no. 1 (2015): 139–59.

Stehlin, Stewart. "Introduction." In *Sketches of Urban and Cultural Life in North America*, edited by Stuart Stehlin, xiii–xxiv. New Brunswick, NJ: Rutgers University Press, 1988.

Steigmann-Gall, Richard. "Star-Spangled Fascism: American Interwar Political Extremism in Comparative Perspective." *Social History* 42, no. 1 (2017): 94–119.

Stein, Ernst. "Die Zeit auf dem Buckel." *Die Zeit*, 25 November 1966, https://www.zeit.de/1966/48/die-zeit-auf-dem-buckel/komplettansicht.

Steinmetz, George. "Geopolitics." In *The Wiley-Blackwell Encyclopedia of Globalization*, edited by George Ritzer, 1–23. Hoboken: Wiley-Blackwell, 2012.

Steinmetzler, Johannes. *Die Anthropogeographie Friedrich Ratzels und ihre ideengeschichtlichen Wurzeln*. Bonn: Im Selbstverlag des geographischen Instituts, 1956.

Stieler, Karl. *Durch Krieg zum Frieden: Stimmungsbilder aus den Jahren 1870–1871*. Stuttgart: Bonz und Comp., 1895.

Stoakes, Geoffrey. *Hitler and the Quest for World Domination: Nazi Ideology and Foreign Policy in the 1920s*. Lemington Spa: Berg, 1986.

Stoddart, D. R. "Darwin's Impact on Geography." *Annals of the Association of American Geographers* 56, no. 4 (1966): 683–98.

Stogiannos, Alexandros. *The Genesis of Geopolitics and Friedrich Ratzel: Dismissing the Myth of the Ratzelian Geodeterminism*. Cham: Springer, 2019.

Stokes, Curtis. "Tocqueville and the Problem of Racial Inequality." *Journal of Negro History* 75, no. 1/2 (1990): 1–15.

Stoler, Ann Laura. "Rethinking Colonial Categories: European Communities and the Boundaries of Rule." *Comparative Studies in Society and History* 31, no. 1 (1989): 134–61.

Strauch, Christoph. "Von Brest–Litowsk zum Generalplan Ost?" *Frankfurter Allgemeine Zeitung*, 13 March 2018. https://www.faz.net/aktuell/politik/der-erste-weltkrieg/erster-weltkrieg-von-brest-litowsk-zum-generalplan-ost-15471514.html.

Streck, Bernhard. "Diffusionism and Geopolitics in the Work of Friedrich Ratzel." In *Europe between Political Geography and Geopolitics: On the Centenary of Ratzel's Politische Geographie*, edited by Marco Antonsich, Vladimir Kolossov, and Maria Paola Pagnini, 51–66. Rome: Società Geografica Italiana, 2001.

Streck, Bernhard. "Gutachten zu Wolfgang Gedeon: 'Christlich-europäische Leitkultur' (3 Bd.) unter besonderer Berücksichtigung des Antisemitismusfrage." 2016. Accessed 7 October 2020. http://www.wolfgang-gedeon.de/wp-content/uploads/2016/11/Gutachten_Prof_Streck_zu_Gedeon.pdf.

Sulloway, Frank J. "Geographic Isolation in Darwin's Thinking: The Vicissitudes of a Crucial Idea." *Studies in History of Biology* 3 (1979): 23–65.

Sundberg, Juanita. "Decolonizing Posthumanist Geographies." *Cultural Geographies* 21, no. 1 (2014): 33–47.

Svirsky, Marcelo, and Simone Bignall. *Agamben and Colonialism*. Edinburgh: University of Edinburgh Press, 2012.

Takeuchi, Keiichi. "Japanese Geopolitics in the 1930s and 40s." In *Geopolitical Traditions: A Century of Geopolitical Thought*, edited by Klaus Dodds and David Atkinson, 72–92. London: Routledge, 2000.

Termer, Franz. "Die natürlichen Grundlagen amerikanischer Staatsentwicklung." *Zeitschrift für Geopolitik* 1 (1924): 28–35.

Termer, Franz. "Einführung." In *Aus Mexico: Reiseskizzen aus den Jahren 1874 und 1875*, Friedrich Ratzel, v–xii. Stuttgart: Brockhaus, 1969.

Thorpe, Benjamin J. "The Time and Space of Richard Coudenhove-Kalergi's Pan-Europe, 1923–1939." PhD dissertation, University of Nottingham, 2018.

Töppel, Roman. "'Volk und Rasse': Hitlers Quellen auf der Spur." *Vierteljahrshefte für Zeitgeschichte* 64, no. 1 (2016): 1–35.

Traverso, Enzo. "Warping history." *Jacobin*, 8 October 2016. https://www.jacobinmag.com/2016/08/timothy-snyder-black-earth-bloodlands-holocaust-soviets-nazis/.

Troll, Carl. "Geographic Science in Germany during the Period 1933–1945: A Critique and Justification." *Annals of the Association of American Geographers* 39, no. 2 (1949): 99–137.

Turnbull, Thomas M. "From Incommensurability to Ubiquity: An Energy History of Geographic Thought." *Journal of Historical Geography* 73 (2021): 11–23.

Uexküll, Jakob von. *Nie geschaute Welten*. München: Paul List, 1936 (1957).

Uexküll, Jakob von. *Staatsbiologie: Anatomie—Physiologie—Pathologie des Staates*. Berlin: Gebrüder Paetel, 1920.

Uhl, Karsten. "The Idea of Lebensraum and the Spatial Order of Power at German Factories, 1900–1945." *European Review of History* 20, no. 2 (2013): 287–307.

Ulmen, G. L. "Translator's Introduction." In *The Nomos of the Earth in the International Law of the Jus Publicum Europaeum*, Carl Schmitt, 9–34. New York: Telos, 2006.

Unagst, Matthew. "Men of Science and Action: The Celebrity of Explorers and German National Identity, 1870–1895." *Central European History* 50, no. 3 (2017): 305–27.

Usher, Mark. "Territory *Incognita*." *Progress in Human Geography* 44, no. 6 (2020): 1019–46.

van der Wusten, Herman, and Gertjan Dijkink. "German, British and French Geopolitics: The Enduring Differences." *Geopolitics* 7, no. 3 (2002): 19–38.

Verne, Julia. "The Neglected 'Gift' of Ratzel for/from the Indian Ocean: Thoughts on mobilities, Materialities and Relational Spaces." *Geographica Helvetica* 72 (2017): 85–92.

Vidal de la Blache, Paul. "Nécrologie: Friedrich Ratzel." *Annales de Géographie* 13, no. 72 (1904): 466–67.

von Humboldt, Alexander. *Cosmos: A Sketch of a Physical Description of the Universe, Vol. 1*. Baltimore: The Johns Hopkins University Press, 1845 (1997).

Vorzimmer, Peter J. *Charles Darwin: The Years of Controversy. The Origin of Species and its Critics 1859–1882*. Philadelphia: Temple University Press, 1970.

Vowinckel, Kurt. "Ein Seher vom Oberrhein. Fr. Ratzel, der Schöpfer der deutschen Geopolitik." *Straßburger Neueste Nachrichten*, 13 April 1944.

Vowinckel, Kurt. "Zum Begriff Lebensraum." *Zeitschrift für Geopolitik* 16 (1939): 638–39.

Wagner, Hermann. "Bericht über die Entwicklung der Methodik und des Studiums der Erdkunde (1889–91)." *Geographisches Jahrbuch* 14 (1891): 371–462.

Wagner, Hermann. "F. Ratzels Anthropogeographie II. Oder die geographische Verbreitung des Menschen." *Zeitschrift der Gesellschaft für Erdkunde zu Berlin* 16 (1891): 465–78.

Wagner, Moritz. *The Darwinian Theory and the Law of the Migration of Organisms*. London: Edward Stanford, 1873.

Wagner, Moritz. *Darwin'sche Theorie und das Migrationsgesetz der Organismen*. Munich: Akademische Buchdruckerei von F. Straub, 1868.

Wagner, Moritz. *Die Entstehung der Arten durch räumliche Sonderung: Gesammelte Aufsätze von Moriz Wagner*. Basel: Benno Schwabe, 1889.

Wagner, Moritz. "Die Kulturzüchtung des Menschen gegenüber der Naturzüchtung im Tierreich." In *Die Entstehung der Arten durch räumliche Sonderung: Gesammelte Aufsätze von Moriz Wagner*, Moritz Wagner, 519–39. Basel: Benno Schwabe, 1886 (1889).

Walsh, Edmund A. "The Mystery of Haushofer." *Life*, 16 September 1946.

Wanklyn, Harriet. *Friedrich Ratzel: A Biographical Memoir and Bibliography*. Cambridge: Cambridge University Press, 1961.

Wardenga, Ute. "Friedrich Ratzel zum 100. Todestag am 9. August 2004." In *Jubiläen 2004. Personen-Ereignisse*, edited by Rektor der Universität Leipzig, 47–51. Leipzig: Universität Leipzig, 2004.

Wardenga, Ute. "Writing the History of Geography: What We Have Learnt—and Where to Go Next." *Geographica Helvetica* 68 (2013): 27–35.

Weidling, Paul. "Ernst Haeckel, Darwinismus and the Secularization of Nature." In *History, Humanity and Evolution: Essays for John C. Greene*, edited by John R. Moore, 311–53. Cambridge: Cambridge University Press, 1989.

Weikart, Hans. "Review: Haushofer and the Pacific." *Foreign Affairs* 20 (1942): 732–42.

Weikart, Richard. *From Darwin to Hitler: Evolutionary Ethics, Eugenics, and Racism in Germany*. Basingstoke: Palgrave, 2004.

Weissman, Charlotte. "The Origins of Species: The Debate between August Weismann and Moritz Wagner." *Journal of the History of Biology* 43, no. 4 (2010): 727–66.

Weißmann, Karlheinz. *Rubikon: Deutschland vor der Entscheidung*. Berlin: JF Edition, 2016.

Weißmann, Karlheinz. *Rückruf in die Geschichte. Die deutsche Herausforderung: Alte Gefahren—neue Chancen*. Berlin: Ullstein, 1992.

Werber, Niels. *Die Geopolitik der Literatur: Eine Vermessung der medialen Weltraumordnung*. Munich: Hanser, 2007.

Werber, Niels. *Geopolitik zur Einführung*. Hamburg: Junus, 2014.

Werner, Fritz. "Die Grenzboten: Aus der Geschichte einer achtzigjährigen Zeitschrift nationaler Bedeutung." *Die Grenzboten* 81 (1922): 448–52.

Werner, Lothar. *Der Alldeutsche Verband, 1890–1918*. Berlin: E. Ebering, 1935.

Weule, Karl. "Friedrich Ratzel." *Deutsche Erde* 3 (1904): 150–51.

Weule, Karl. "Friedrich Ratzel." *Mitteilungen des Vereins für Erdkunde zu Leipzig* (1904): 1–29.

Whatmore, Sarah. "Materialist Returns: Practicing Cultural Geography in and for a More-Than-Human World." *Cultural Geographies* 13, no. 4 (2006): 600–609.

Whitman, James Q. *Hitler's American Model: The United States and the Making of Nazi Race Law*. Princeton, NJ: Princeton University Press, 2017.

Williams, Michael. "Hitler's Holocaust Blueprint." *The Daily Mail*, 23 September 2010. https://www.dailymail.co.uk/news/article-1314399/amp/Hitlers-Holocaust-blueprint-Africa-concentration-camps-used-advance-racial-theories.html.

Williams, Michael C. *Realism Reconsidered: The Legacy of Hans Morgenthau in International Relations*. Oxford: Oxford University Press, 2007.

Withers, Charles W. J. "Geography's Evolving Traditions and Textual Critique." *Dialogues in Human Geography* 2, no. 3 (2012): 317–20.

Wittfogel, Karl A. "Geopolitics, Geographical Materialism and Marxism." *Antipode* 17, no. 1 (1929 [1985]): 21–53.

Wolfe, Patrick. "Settler Colonialism and the Elimination of the Native." *Journal of Genocide Research* 8, no. 4 (2006): 387–409.

Woodward, William R. "Fechner's Panpsychism: A Scientific Solution to the Mind-Body Problem." *Journal of the History of the Behavioural Sciences* 8 (1972): 367–86.

Wulf, Andrea. *The Invention of Nature: The Adventures of Alexander von Humboldt, the Lost Hero of Science*. London: Hodder and Stoughton, 2015.

Yoshida, Haruyo. "Streitfall 'Kulturkreise': Zur Aktualität der Positionen von Spengler, Frobenius und Bloch." *Neue Beiträge zur Germanistik* 3, no. 3 (2004): 58–72.

Yusoff, Kathryn. *A Billion Black Anthropocenes or None*. Minneapolis: University of Minnesota Press, 2018.

Zigman, Peter M. "'. . . und meine Anschauungen sind natürlich selbstständiger geworden'—Friedrich Ratzels Briefe an Ernst Haeckel, seinen Mentor und akademischen Kollegen." *Würzburger medizinhistorische Mitteilungen* 28 (2009): 375–410.

Zimmerer, Jürgen. "The Birth of the *Ostland* Out of the Spirit of Colonialism: A Postcolonial Perspective on Nazi Policy of Conquest and Extermination." *Patterns of Prejudice* 39, no. 2 (2005): 197–219.

Zimmerer, Jürgen. "In the Service of Empire: Geographers at Berlin's University between Colonial Studies and Ostforschung (*Eastern research*)." In *Hitler's Geographies: The Spatialities of the Third Reich*, edited by Paolo Giaccaria and Claudio Minca, 67–92. Chicago and London: Chicago University Press, 2016.

Index

"FR" indicates Friedrich Ratzel. Titles of books are italicized; titles of articles appear within quotation marks; books and articles are by FR unless otherwise indicated. Page numbers in italic indicate figures; references to notes are indicated by "n" (e.g., "230n57" indicates note 57 on page 230).

Printed and bound by CPI Group (UK) Ltd, Croydon, CR0 4YY

16/09/2026

14963488-0001